Problem Books in Mathematics

Series Editors:
Peter Winkler
Department of Mathematics
Dartmouth College
Hanover, NH 03755
USA

For further volumes:
http://www.springer.com/series/714

Ovidiu Furdui

Limits, Series, and Fractional Part Integrals

Problems in Mathematical Analysis

 Springer

Ovidiu Furdui
Department of Mathematics
Technical University of Cluj-Napoca
Cluj-Napoca, Romania

ISSN 0941-3502
ISBN 978-1-4899-9243-7   ISBN 978-1-4614-6762-5 (eBook)
DOI 10.1007/978-1-4614-6762-5
Springer New York Heidelberg Dordrecht London

Mathematics Subject Classification (2010): 11B73, 11M06, 33B15, 26A06, 26A42, 40A10, 40B05, 65B10
© Springer Science+Business Media New York 2013
Softcover re-print of the Hardcover 1st edition 2013

Printed on acid-free paper

Springer is part of Springer Science+Business Media (www.springer.com)

Don't put peace and understanding beyond the truth.
—St. John Chrysostom (347–407)

Read to get wise and teach others when it will be needed.
—St. Basil the Great (329–378)

Where things are simple, there are a hundred angels; but where things are complicated, there is not a single one.
—St. Ambrose of Optina (1812–1891)

To my parents, Traian and Maria

A room with no books is like a body without a soul.
—Elder Vichentie Mălău (1887–1945)

Preface

There is no subject so old that something new cannot be said about it.

Fyodor Dostoevsky (1821–1881)

This book is the fruit of my work in the last decade teaching, researching, and solving problems. This volume offers an unusual collection of problems specializing in three topics of mathematical analysis: limits, series, and fractional part integrals. The book is divided into three chapters, each dealing with a specific topic, and two appendices. The first chapter of the book collects non-standard problems on limits of special sequences and integrals. Why limits? First, because in analysis, most things reduce to the calculation of a limit; and second, because limits are the most fundamental basic problems of analysis. Why non-standard limits? Because the standard problems on limits are known, if not very well known, and have been recorded in other books, they might not be so attractive and interesting anymore. The problems vary in difficulty and specialize in different aspects of calculus: from the study of the asymptotic behavior of a sequence to the evaluation of a limit involving a special function, an integral or a finite sum.

The second chapter of the book introduces the reader to a collection of problems that are rarely seen: the evaluation of exotic integrals involving a fractional part term, called fractional part integrals. The problems of this chapter were motivated by the interesting formula $\int_0^1 \{1/x\}\,\mathrm{d}x = 1 - \gamma$, which connects an exotic integral to the Euler–Mascheroni constant. One may wonder: are there any other similar formulae? What happens when the integrand function is changed from $\{1/x\}$ to $\{1/x\}^2$? Is the integral $\int_0^1 \{1/x\}^2\,\mathrm{d}x$ calculable in terms of exotic constants or this integral formula is a singular case? Is it possible to extend this equality from the one-dimensional case to the multiple case? The reader will find the answer to these questions by going through the problems of this chapter. The novelty of the problems stands in the fact that, comparatively to the classical integrals that can be calculated by the well-known techniques, integration by parts or substitution, many of these integrals invite the reader to use a host of mathematical techniques that involve elegant connections between integrals, infinite series, exotic constants, and special functions. This chapter has a special section called "Quickies" which contains problems that have an unexpected succint solution. The quickies are solved by using symmetry combined with tricks involving properties of the fractional part function.

The last chapter of the book offers the reader a bouquet of problems with a flavor towards the computational aspects of infinite series and special products, many of these problems being new in the literature. These series, linear or quadratic, single or multiple, involve combinations of exotic terms, special functions, and harmonic numbers and challenge the reader to explore the ability to evaluate an infinite sum, to discover new connections between a series and an integral, to evaluate a sum by using the modern tools of analysis, and to investigate further. In general, the classes of series that can be calculated exactly are widely known and such problems appear in many standard books that have topics involving infinite sums, so by this chapter we offer the reader a collection of interesting and unconventional problems for solutions.

Each chapter contains a section of difficult problems, motivated by other problems in the book, which are collected in a special section entitled "Open problems" and few of them are listed in the order in which the problems appear in the book. These problems may be considered as research problems or projects for students with a strong background in calculus and for the readers who enjoy mathematical research and discovery in mathematics. The intention of having the open problems recorded in the book is to stimulate creativity and the discovery of original methods for proving known results and establishing new ones.

There are two appendices which contain topics of analysis and special function theory that appear throughout the book. In the first part of Appendix A we review the special constants involved in the computations of series and products, the second part of Appendix A contains a bouquet of special functions, from Euler's Gamma function to the celebrated Riemann zeta function, and in the last part of Appendix A we collect some lemmas and theorems from integration theory concerning the calculation of limits of integrals. Appendix B is entirely devoted to the Stolz–Cesàro lemma, a classical tool in analysis, which has applications to the calculations of limits of sequences involving sums.

This volume contains a collection of challenging problems; many of them are new and original. I do not claim originality of all the problems included in the book and I am aware that some may be either known or very old. Other problems, by this volume, are revived and brought into light. Most of the problems are statements to be proved and others are challenges: *calculate, find*. Each chapter contains a very short section, consisting of hints. The hints help the reader to point to the heart of the problem. Detailed solutions are given for nearly all of the problems and for the remaining problems references are provided. I would like to hear about other solutions as well as comments, remarks, and generalizations on the existing ones.

I have not attempted to document the source of every problem. This would be a difficult task: on the one hand, many of the problems of this volume have been discovered by the author over the last decade, some of them have been published in various journals with a problem column, and others will see the light of publication for the first time. Also, there are problems whose history is either lost, with the passing of time, or the author was not aware of it. I have tried to avoid collecting too many problems that are well known or published elsewhere, in order to keep a high level of originality. On the other hand, other problems of this book arose in a natural way: either as generalizations or motivated by known results that have long been forgotten; see the nice alternating series, due to Hardy, recorded in the first part of Problem **3.35**. For such problems, when known, the source of the problem is mentioned, either as a remark, included in the solution, or as a small footnote which contains a brief comment on it.

As I mentioned previously, this book specializes on three selected topics of mathematical analysis: limits, series, and special integrals. I have not collected problems on all topics of analysis because of many problem books, both at the elementary and at the advanced level, that cover such topics. Instead, I tried to offer the reader problems that don't overlap with the existing ones in the literature and others that have received little or no coverage in other texts. Whether I succeeded or not in accomplishing this task is left to the reader to decide, I accept the criticism.

The level of the problems is appropriate for Putnam exams and for problem sections of journals like The American Mathematical Monthly and other journals that have problem sections addressed to undergraduate students. The problems require thorough familiarity with sequences, limits, Riemann integrals, and infinite series and no advanced topics of analysis are required. Anyone with strong knowledge in calculus should be ready for almost everything to be found here.

This book is mainly addressed to undergraduate students with a strong background in analysis, acquired through an honors calculus class, who prepare for prize exams like the Putnam exam and other high-level mathematical contests. Mathematicians and students interested in problem solving will find this collection of topics appealing. This volume is a must-have for instructors who are involved in math contests as well as for individuals who wish to enrich and test their knowledge by solving problems in analysis. It could also be used by anyone for independent study courses. This book can be used by students in mathematics, physics, and engineering and by anyone who wants to explore selected topics of mathematical analysis.

I also address this work to the first and the second year graduate students who want to learn more about the application of certain techniques, to do calculations which happen to have interesting results. Pure and applied mathematicians, who confront certain difficult computations in their research, might find this book attractive.

This volume is accessible to anyone who knows calculus well and to the reader interested in solving challenging problems at the monthly level. However, it is not expected that the book will be an easy reading for students who don't have at their fingertips some classical results of analysis.

I would like to express my great appreciation to Alina Sîntămărian, who read substantial portions of the manuscript and provided many helpful comments and spotted numerous misprints.

I also thank my parents for all of their support during the preparation of this manuscript. Without their effort this book would not have been written.

I am grateful to anonymous referees for their comments and suggestions that led to the improvement of the presentation of the final version of this volume.

Thank you all!

In conclusion, I say to the readers who may use this book: good luck in problem solving since there is no better way to approach mathematics.

Câmpia Turzii, Romania   Ovidiu Furdui
December 2012

Contents

Notations

$\mathbb{N}$	The set of natural numbers ($\mathbb{N} = \{1,2,3,\ldots\}$)
$\mathbb{Z}$	The set of integers ($\mathbb{Z} = \{\ldots,-2,-1,0,1,2,\ldots\}$)
$\mathbb{R}$	The set of real numbers
$\mathbb{R}^*$	The set of nonzero real numbers ($\mathbb{R}^* = \mathbb{R} \setminus \{0\}$)
$\overline{\mathbb{R}}$	The completed real line ($\overline{\mathbb{R}} = \mathbb{R} \cup \{-\infty,\infty\}$)
$\mathbb{C}$	The set of complex numbers
$\lfloor a \rfloor$	The integer part (the floor) of a, that is the greatest integer not exceeding a
$\{a\}$	The fractional part of the real number a, defined by $\{a\} = a - \lfloor a \rfloor$
$\|f\|_\infty$	The supremum norm of f defined by $\|f\|_\infty = \sup_{x \in [a,b]} |f(x)|$
Landau's notations	$f(x) = o(g(x))$, as $x \to x_0$, if $f(x)/g(x) \to 0$, as $x \to x_0$
	$f(x) = O(g(x))$, as $x \to x_0$, if $f(x)/g(x)$ is bounded in a neighborhood of x_0
	$f \sim g$, as $x \to x_0$, if $\lim_{x \to x_0} f(x)/g(x)$ exists and is finite
$\Re(z)$	The real part of the complex number z
$n!$	n factorial, equal to $1 \cdot 2 \cdot 3 \cdots (n-1) \cdot n \quad (n \in \mathbb{N})$
$(2n)!!$	$2 \cdot 4 \cdots (2n-2) \cdot 2n \quad (n \in \mathbb{N})$
$(2n+1)!!$	$1 \cdot 3 \cdots (2n-1) \cdot (2n+1) \quad (n \in \mathbb{N} \cup \{0\})$
$\binom{n}{k}$	The binomial coefficient indexed by n and k is the coefficient of the x^k term in the polynomial expansion of the binomial power $(1+x)^n$
A	The Glaisher–Kinkelin constant $A = \lim_{n \to \infty} (1^1 2^2 \cdots n^n)/(n^{n^2/2+n/2+1/12} e^{-n^2/4})$ $= 1.28242\,71291\,00622\,63687\ldots$
e	The Natural Logarithmic Base (Napier's Constant) $\lim_{n \to \infty} (1 + 1/n)^n = 2.71828\,18284\ldots$
γ	The Euler–Mascheroni constant $\gamma = \lim_{n \to \infty} (1 + 1/2 + \cdots + 1/n - \ln n) = 0.57721\,56649\,01532\,86060\ldots$
γ_n	The Stieltjes constants $\gamma_n = \lim_{m \to \infty} \left(\sum_{k=1}^{m} (\ln k)^n/k - (\ln m)^{n+1}/(n+1) \right)$
G	Catalan's constant $G = \sum_{n=0}^{\infty} (-1)^n/(2n+1)^2 = 0.91596\,55941\,772\ldots$

B_k	The kth Bell number,
	$B_k = e^{-1} \sum_{n=1}^{\infty} n^k/n!$, for $k \geq 1$ and $B_0 = 1$
δ_n	The logarithmic constants
	$\delta_n = \lim_{m\to\infty} \left(\sum_{k=1}^{m} \ln^n k - \int_1^m \ln^n x\, dx - \frac{1}{2} \ln^n m \right) = (-1)^n (\zeta^{(n)}(0) + n!)$
$s(n,k)$	The Stirling numbers of the first kind are defined by the generating function
	$z(z-1)(z-2)\cdots(z-n+1) = \sum_{k=0}^{n} s(n,k) z^k$
H_n	The nth harmonic number
	$H_n = 1 + 1/2 + \cdots + 1/n$
Γ	The Gamma function (Euler's Gamma function)
	$\Gamma(z) = \int_0^{\infty} x^{z-1} e^{-x} dx, \quad \Re(z) > 0$
B	The Beta function (Euler's Beta function)
	$B(a,b) = \int_0^1 x^{a-1}(1-x)^{b-1} dx, \quad \Re(a) > 0, \quad \Re(b) > 0$
ψ	The Digamma function (the Psi function)
	$\psi(z) = \Gamma'(z)/\Gamma(z), \quad z \neq 0, -1, -2, \ldots$
ζ	The Riemann zeta function
	$\zeta(z) = \sum_{n=1}^{\infty} 1/n^z = 1 + 1/2^z + \cdots + 1/n^z + \cdots, \quad \Re(z) > 1$
Ci	The Cosine integral function
	$\mathrm{Ci}(x) = -\int_x^{\infty} \frac{\cos t}{t} dt, \quad x > 0$
Li_n	The Polylogarithm function
	$\mathrm{Li}_n(z) = \sum_{k=1}^{\infty} z^k/k^n = \int_0^z \frac{\mathrm{Li}_{n-1}(t)}{t} dt, \quad |z| \leq 1$ and $n \neq 1, 2$
$\overline{\lim} x_n$	The limit superior of $(x_n)_{n\in\mathbb{N}}$ defined by $\overline{\lim} x_n = \inf_n \sup_{k\geq n} x_k$
$\underline{\lim} x_n$	The limit inferior of $(x_n)_{n\in\mathbb{N}}$ defined by $\underline{\lim} x_n = \sup_n \inf_{k\geq n} x_k$
$\sum_{n_1,\ldots,n_k=1}^{\infty} a_{n_1,\ldots,n_k}$	the iterated sum $\sum_{n_1=1}^{\infty} \cdots \sum_{n_k=1}^{\infty} a_{n_1,\ldots,n_k}$
$C(\Omega)$	The space of continuous functions on Ω
$C^n(\Omega)$	Functions with n continuous derivatives on Ω
$L^1(\Omega)$	The space of Lebesgue integrable functions on Ω

Chapter 1
Special Limits

> The last thing one knows when writing a book is what to put
> first.
>
> Blaise Pascal (1623–1662)

1.1 Miscellaneous Limits

> One says that a quantity is the limit of another quantity if the
> second approaches the first closer than any given quantity,
> however small...
>
> Jean le Rond d'Alembert (1717–1783)

1.1. Let α be a positive number. Find the value of

$$\lim_{n \to \infty} \sum_{k=1}^{n} \frac{1}{n + k^{\alpha}}.$$

1.2. Calculate

$$\lim_{n \to \infty} \frac{1}{n} \left(\frac{n}{\frac{1}{2} + \frac{2}{3} + \cdots + \frac{n}{n+1}} \right)^n.$$

1.3. Find the value of

$$\lim_{n \to \infty} \frac{1}{n} \sum_{i=1}^{n} \sum_{j=1}^{n} \frac{i+j}{i^2 + j^2}.$$

O. Furdui, *Limits, Series, and Fractional Part Integrals: Problems in Mathematical Analysis*, Problem Books in Mathematics, DOI 10.1007/978-1-4614-6762-5_1, © Springer Science+Business Media New York 2013

1.4. Find the value of

$$\lim_{n \to \infty} \frac{1}{n} \sum_{k=1}^{n} \left\{ \frac{n}{k} \right\}^2,$$

where $\{a\}$ denotes the fractional part of a.

1.5. The behavior of a remarkable geometric mean. Calculate

$$\lim_{n \to \infty} \frac{\sqrt[n]{\binom{n}{1} \binom{n}{2} \cdots \binom{n}{n}}}{e^{n/2} n^{-1/2}}.$$

1.6. Let $p > 1/2$ be a real number. Calculate

$$\lim_{n \to \infty} \left(\sum_{k=1}^{n} \frac{1}{\binom{n}{k}^p} \right)^{n^p}.$$

1.7. Find

$$\lim_{n \to \infty} \left(\left(\sum_{k=n+1}^{2n} (2 \sqrt[2k]{2k} - \sqrt[k]{k}) \right) - n \right).$$

1.8. Let α be a positive real number. Calculate

$$\lim_{n \to \infty} n^\alpha \left(\frac{\sqrt[n+1]{(n+1)!}}{n+1} - \frac{\sqrt[n]{n!}}{n} \right).$$

1.9. Let $a \in (0, \infty)$. Calculate

$$\lim_{n \to \infty} \left(e^{\frac{1}{a} + \frac{1}{a+1} + \cdots + \frac{1}{a+n}} - e^{\frac{1}{a} + \frac{1}{a+1} + \cdots + \frac{1}{a+n-1}} \right).$$

1.10. For positive integers n, let $g_n = 1 + 1/2 + \cdots + 1/n - \ln n$. Prove that

$$\lim_{n \to \infty} \left(\frac{g_n^\gamma}{\gamma g_n} \right)^{2n} = \frac{e}{\gamma}.$$

1.11. Find

$$\lim_{n \to \infty} n \prod_{m=1}^{n} \left(1 - \frac{1}{m} + \frac{5}{4m^2} \right).$$

1.12. Let α be a positive real number. Calculate

$$\lim_{n \to \infty} \sum_{k=1}^{n} \left(\frac{k}{n} \right)^{\alpha k}.$$

1.13. A Wolstenholme limit. Let s be any positive real number. Prove that

$$\lim_{m\to\infty}\sum_{k=1}^{m}\left(\frac{k}{m}\right)^{sm}=\frac{e^s}{e^s-1}.$$

1.14. Prove that

$$\lim_{n\to\infty}\left(\sqrt[n+1]{\Gamma\left(\frac{1}{1}\right)\Gamma\left(\frac{1}{2}\right)\cdots\Gamma\left(\frac{1}{n+1}\right)}-\sqrt[n]{\Gamma\left(\frac{1}{1}\right)\Gamma\left(\frac{1}{2}\right)\cdots\Gamma\left(\frac{1}{n}\right)}\right)=\frac{1}{e},$$

where Γ denotes the Gamma function.

1.15. Let α and β be positive numbers. Calculate

$$\lim_{n\to\infty}\prod_{k=1}^{n}\left(1+\frac{k^\alpha}{n^\beta}\right).$$

1.16. Let α be a real number and let $p>1$. Find the value of

$$\lim_{n\to\infty}\prod_{k=1}^{n}\frac{n^p+(\alpha-1)k^{p-1}}{n^p-k^{p-1}}.$$

1.17. Let k and α be positive real numbers. Find the value of

$$\lim_{n\to\infty}n^k\left(1-\prod_{j=n}^{\infty}\frac{j^{k+1}-\alpha}{j^{k+1}+\alpha}\right).$$

1.18. Let $k>1$ and $p\geq 2$ be integers and let $(x_n)_{n\in\mathbb{N}}$ be a sequence of positive numbers such that $\lim_{n\to\infty}x_n/\sqrt[p]{n}=L\in(0,\infty)$. Calculate

$$\lim_{n\to\infty}\frac{x_n+x_{n+1}+\cdots+x_{kn}}{nx_n}.$$

1.19. If $(x_n)_{n\geq 1}$ is a sequence which converges to zero and $\lim_{n\to\infty}n^k(x_n-x_{n+1})=l\in[-\infty,\infty]$, with $k>1$, then $\lim_{n\to\infty}n^{k-1}x_n=l/(k-1)$.

1.20. Let a and b be two nonnegative real numbers. Calculate

$$\lim_{n\to\infty}\sum_{k=1}^{n}\frac{1}{n+k+b+\sqrt{n^2+kn+a}}.$$

1.21. Let a, b, and c be positive numbers. Find the value of

$$\lim_{n\to\infty} \sum_{k=1}^{n} \frac{k+a}{\sqrt{n^2+kn+b}\cdot\sqrt{n^2+kn+c}}.$$

1.22. Let a, b, and c be nonnegative real numbers. Find the value of

$$\lim_{n\to\infty} \sum_{k=1}^{n} \frac{\sqrt{n^2+kn+a}}{\sqrt{n^2+kn+b}\cdot\sqrt{n^2+kn+c}}.$$

1.23. Let a, b, and c be nonnegative real numbers. Find the value of

$$\lim_{n\to\infty} \frac{1}{n^3} \sum_{i=1}^{n}\sum_{j=1}^{n} \frac{ij}{\sqrt{i^2+j^2+ai+bj+c}}.$$

1.24. Let $x \in \mathbb{R}$. Prove that if $\lim_{n\to\infty}\cos(nx) = 1$, then x is a multiple of 2π.

1.25. Let a, b, and α be nonnegative real numbers and let β be a positive number. Find the limit

$$\lim_{n\to\infty} \sum_{k=1}^{n} \frac{(n^2+kn+a)^\alpha}{(n^2+kn+b)^\beta}.$$

1.26. An iterated exponential limit. Let $n \geq 1$ be a natural number and let

$$f_n(x) = x^{x^{\cdot^{\cdot^{\cdot^{x}}}}},$$

where the number of x's in the definition of f_n is n. For example,

$$f_1(x) = x, \quad f_2(x) = x^x, \quad f_3(x) = x^{x^x},\ldots.$$

Calculate the limit

$$\lim_{x\to1} \frac{f_n(x) - f_{n-1}(x)}{(1-x)^n}.$$

1.27. Iterated exponentials.

(a) Let $(x_k)_{k\geq1}$ be the sequence defined as follows:

$$\frac{1^{3^{\frac{1}{3}}}}{3}, \ \frac{1^{3^{5^{\frac{1}{5}}}}}{5}, \ \frac{1^{7^{7^{7^{\frac{1}{7}}}}}}{7}, \ \ldots.$$

(continued)

(continued)

In other words the sequence $(x_k)_{k\geq 1}$ is defined as "the fraction $1/(2k+1)$ appears as an exponent in the definition of x_k exactly $2k+1$ times." Prove that the following inequalities hold

$$\frac{1}{2k+1} < x_k < \frac{1}{\ln(2k+1)}, \quad k \geq 1.$$

(b) Let $(y_k)_{k\geq 1}$ be the sequence defined by

$$y_k = \frac{1^{\frac{1}{2k}^{\cdots^{\frac{1}{2k}}}}}{2k},$$

where the fraction $1/(2k)$ appears exactly $2k$ times in the definition of y_k. For example,

$$y_1 = \frac{1^{\frac{1}{2}}}{2}, \quad y_2 = \frac{1^{\frac{1}{4}^{\frac{1}{4}}}}{4}, \quad y_3 = \frac{1^{\frac{1}{6}^{\frac{1}{6}^{\frac{1}{6}}}}}{6}.$$

Prove that, for $k \geq 10$, the following inequalities hold

$$1 - \frac{1}{\ln(2k)} < y_k < 1 - \frac{1}{2k}.$$

Open problem. Find the asymptotic expansion, or the first few terms, of the sequences $(x_k)_{k\geq 1}$ and $(y_k)_{k\geq 1}$.

(c) **Does the limit exist?** Let $f : (0,1) \to \mathbb{R}$ be the function defined by $f(x) = x^{x^{\cdot^{\cdot^{x}}}}$, where the number of $x's$ in the definition of $f(x)$ is $\lfloor \frac{1}{x} \rfloor$. For example, $f\left(\frac{1}{3}\right) = \frac{1}{3}^{\frac{1}{3}^{\frac{1}{3}}}$. Find $\lim_{x\to 0^+} f(x)$ or prove that it does not exist.

1.28. A double Euler–Mascheroni sequence. Let

$$x_n = \sum_{i=1}^{n} \sum_{j=1}^{n} \frac{1}{i+j}.$$

Prove that

(a) $x_n = 1 + (2n+1)H_{2n+1} - (2n+2)H_{n+1}$, where H_n denotes the nth harmonic number.

(b) $\lim_{n\to\infty}(x_n - (2n+1)\ln 2 + \ln n) = -1/2 - \gamma$.

(c) $\lim_{n\to\infty} x_n/n = 2\ln 2$.

1.29. The limit of a multiple sum. Let $k \geq 2$ be an integer and let m be a nonnegative integer. Prove that

$$\lim_{n \to \infty} \frac{1}{n^{k-1}} \sum_{i_1=1}^{n} \cdots \sum_{i_k=1}^{n} \frac{1}{i_1 + i_2 + \cdots + i_k + m} = \frac{k}{(k-1)!} \sum_{j=2}^{k} (-1)^{k-j} j^{k-2} \binom{k-1}{j-1} \ln j.$$

1.30. Let $a > 1$ be a real number and let k be a positive number. Prove that

$$\lim_{x \to \infty} \frac{x^k}{\ln x} \sum_{n=0}^{\infty} \frac{1}{a^n + x^k} = \frac{k}{\ln a}$$

and

$$\lim_{x \to \infty} \ln \ln x \left(\frac{x^k}{\ln x} \sum_{n=0}^{\infty} \frac{1}{a^n + x^k} - \frac{k}{\ln a} \right) = 0.$$

The Riemann zeta function, ζ, is a function of a complex variable defined, for $\Re(z) > 1$, by $\zeta(z) = \sum_{n=1}^{\infty} 1/n^z$.

1.31. (a) Let a be a positive real number. Calculate

$$\lim_{n \to \infty} a^n \left(n - \zeta(2) - \zeta(3) - \cdots - \zeta(n) \right).$$

(b) Determine

$$\lim_{n \to \infty} \frac{n - \zeta(2) - \zeta(3) - \cdots - \zeta(n)}{n - 1 - \zeta(2) - \zeta(3) - \cdots - \zeta(n-1)}.$$

1.32. (a) Let $f : [0,1] \to \mathbb{R}$ be a continuously differentiable function and let

$$x_n = f\left(\frac{1}{n}\right) + f\left(\frac{2}{n}\right) + \cdots + f\left(\frac{n-1}{n}\right).$$

Calculate $\lim_{n \to \infty} (x_{n+1} - x_n)$.

(b) **Open problem.** What is the limit when f is only continuous?

1.33. Find

$$\lim_{n \to \infty} \left(\sum_{k=1}^{n} \arctan \frac{1}{k} - \ln n \right).$$

1.2 Limits of Integrals

> *The infinite! No other question has ever moved so profoundly
> the spirit of man.*
>
> David Hilbert (1862–1943)

1.34. Find the value of

$$\lim_{n\to\infty} \int_0^{\pi/2} \sqrt[n]{\sin^n x + \cos^n x}\,dx.$$

1.35. Let $\alpha \geq 2$ be an even integer and let $x \in [0,1]$. Find the value of

$$\lim_{n\to\infty} n^{\frac{1}{\alpha}} \int_0^1 (1-(t-x)^\alpha)^n\,dt.$$

1.36. Let a and b be real numbers and let c be a positive number. Find the value of

$$\lim_{n\to\infty} \int_a^b \frac{dx}{c + \sin^2 x \cdot \sin^2(x+1)\cdots \sin^2(x+n)}.$$

1.37. Let k be a positive integer. Find the value of

$$L = \lim_{n\to\infty} \int_0^1 \frac{\ln(1+x^{n+k})}{\ln(1+x^n)}\,dx \quad \text{and} \quad \lim_{n\to\infty} n\left(\int_0^1 \frac{\ln(1+x^{n+k})}{\ln(1+x^n)}\,dx - L\right).$$

1.38. A Frullani limit.

(a) Let $f : [0,1] \to \mathbb{R}$ be a continuous function and let $0 < a < b \leq 1$ be real numbers. Prove that

$$\int_0^1 \frac{f(ax)-f(bx)}{x}\,dx = f(0)\ln\frac{b}{a} - \int_a^b \frac{f(x)}{x}\,dx.$$

(b) Let $f : [0,1] \to (-1,1]$ be a continuous function. Find the limit

$$\lim_{n\to\infty} \int_0^1 \frac{f^n(ax)-f^n(bx)}{x}\,dx.$$

1.39. (a) Let k be a positive integer. Prove that

$$\lim_{n\to\infty} n^2 \int_0^1 \frac{x^n}{1+x^k+x^{2k}+\cdots+x^{nk}}\,dx = k\sum_{m=0}^{\infty} \frac{1}{(1+km)^2}.$$

(b) More generally, let k be a positive integer and let $f : [0,1] \to \mathbb{R}$ be a continuous function. Prove that

$$\lim_{n\to\infty} n^2 \int_0^1 \frac{x^n f(x)}{1+x^k+x^{2k}+\cdots+x^{nk}}\, dx = k \cdot f(1) \cdot \sum_{m=0}^{\infty} \frac{1}{(1+km)^2}.$$

1.40. Let $f : [0, \pi/2] \to \mathbb{R}$ be a continuous function. Find the value of

$$\lim_{n\to\infty} \frac{1}{n} \int_0^{\frac{\pi}{2}} \frac{\sin^2 nx}{\sin^2 x} f(x)\, dx.$$

1.41. Let $f : [0,1] \to \mathbb{R}$ be an integrable function which is continuous at 1 and let $k \geq 1$. Find the value of

$$\lim_{n\to\infty} \frac{1}{n^k} \int_0^1 \left(x + 2^k x^2 + 3^k x^3 + \cdots + n^k x^n \right) f(x)\, dx.$$

1.42. Let $f : [0,1] \to \mathbb{R}$ be an integrable function which is continuous at 1. Find the value of

$$\lim_{n\to\infty} n \int_0^1 \left(\sum_{k=n}^{\infty} \frac{x^k}{k} \right) f(x)\, dx.$$

1.43. Let $f : [0,1] \to \mathbb{R}$ be an integrable function which is continuous at 1. Calculate

$$\lim_{n\to\infty} n \int_0^1 \left(\sum_{k=n}^{\infty} \frac{x^k}{k} \right)^2 f(x)\, dx.$$

1.44. Let $a > 0$ and $b > 1$ be real numbers and let $f : [0,1] \to \mathbb{R}$ be a continuous function. Calculate

$$\lim_{n\to\infty} n^{\frac{a}{b}} \int_0^1 \frac{f(x)}{1+n^a x^b}\, dx.$$

1.45. Let $g : [0,1] \to [0,\infty)$ be an integrable function such that $\int_0^1 g(x)\, dx = 1$ and let $f : [0,1] \to (0,\infty)$ be a continuous function. Find the value of

$$\lim_{n\to\infty} \left(\int_0^1 g(x) \sqrt[n]{f(x^n)}\, dx \right)^n.$$

1.46. (a) Let $f : [0,1] \to \mathbb{R}$ be a continuous function. Find the value of

$$\lim_{n\to\infty} \sqrt{n} \int_0^{\pi/2} \sin^n x f(\cos x)\, dx.$$

(b) Let $g : [0, \pi/2] \to \mathbb{R}$ be a continuous function. Find the value of

$$\lim_{n\to\infty} \sqrt{n} \int_0^{\pi/2} g(x) \sin^n x\, dx.$$

1.47. A limit of Pólya and Szegö. Let $a < b$ and let $f : [a,b] \to [0,\infty)$ be a continuous function, with $f \neq 0$. Calculate

$$\lim_{n\to\infty} \frac{\int_a^b f^{n+1}(x)dx}{\int_a^b f^n(x)dx}.$$

1.48. The limit of n times a difference of two nth roots. Let $f : [a,b] \to [0,\infty)$ be a continuous function. Calculate

$$\lim_{n\to\infty} n \left(\sqrt[n]{\int_a^b f^{n+1}(x)dx} - \sqrt[n]{\int_a^b f^n(x)dx} \right).$$

1.49. If $h, g : [0,\infty) \to (0,\infty)$ are integrable functions such that for all nonnegative integers n,

$$\int_0^\infty h(x)x^n \, dx < \infty \quad \text{and} \quad \int_0^\infty g(x)x^n \, dx < \infty,$$

then

$$\lim_{n\to\infty} \frac{\int_0^\infty h(x)x^n \, dx}{\int_0^\infty g(x)x^n \, dx} = \lim_{x\to\infty} \frac{h(x)}{g(x)},$$

provided that the right-hand side limit exists.

1.50. Let $f : [0,\infty) \to (0,\infty)$ be positive and monotone decreasing such that for all nonnegative integers n,

$$\int_0^\infty x^n f(x) \, dx < \infty,$$

and let $F(x) = \int_x^\infty f(t)dt$. Then

$$\lim_{n\to\infty} \frac{n \int_0^\infty x^n f(x) \, dx}{\int_0^\infty x^{n+1} f(x) \, dx} = \lim_{x\to\infty} \frac{f(x)}{F(x)},$$

provided that the right-hand side limit exists.

1.51. When is the limit equal to the supremum norm of f?

(a) Let $f : [0,1] \to [0,\infty)$ be a continuous function that attains its maximum either at 0 or at 1. Prove that

$$\lim_{n\to\infty} \sqrt[n]{\int_0^1 f(x)f(x^2)\cdots f(x^n)dx} = ||f||_\infty. \tag{1.1}$$

Conjecture. Prove that if f is a function that does not attain its maximum at 0 or 1, then strict inequality holds in (1.1).

(b) Let $g : [0,1] \to [0,\infty)$ be a continuous function which attains its maximum at 1. Prove that

$$\lim_{n\to\infty} \sqrt[n]{\int_0^1 g(x)g(\sqrt{x})\cdots g(\sqrt[n]{x})dx} = ||g||_\infty.$$

Open problem. Does the result hold if g is a function which does not attain its maximum at 1?

1.52. Evaluating an integral limit. Let $f : [a,b] \to [0,1]$ be an increasing function which is differentiable at b with $f(b) = 1$ and $f'(b) \neq 0$, and let g be a bounded function on $[a,b]$ with $\lim_{x\to b-} g(x) = L$. Find the value of

$$\lim_{n\to\infty} n \int_a^b f^n(x)g(x)dx,$$

assuming that the integral exists for each positive integer n.

1.53. Let $f : [0,1] \to [0,1]$ be an increasing function which is differentiable at 1 with $f(1) = 1$ and $f'(1) \neq 0$, and let g be a bounded function on $[0,1]$. Find the value of

$$\lim_{n\to\infty} n \int_0^1 f^n(x)g(x^n)dx,$$

assuming that the integral exists for each positive integer n.

1.54. Let $k \geq 2$ be an integer and let $f : [0,\infty) \to \mathbb{R}$ be a bounded function which is continuous at 0. If x is a positive real number, find the value of

$$\lim_{n\to\infty} \sqrt[k]{n} \int_0^x \frac{f(t)}{(1+t^k)^n}dt.$$

1.55. Let $\alpha > 0$ be a real number and let $\beta \geq 2$ be an even integer. Let $x \in \mathbb{R}$ be a fixed real number and let $f : \mathbb{R} \to \mathbb{R}$ be a bounded function on $\mathbb{R}$ such that $\lim_{t\to x} f(t) = L$. Prove that

$$\lim_{n\to\infty} n^{\alpha/\beta} \int_{-\infty}^\infty \frac{f(t)}{1+n^\alpha(t-x)^\beta}dt = \frac{2\pi L}{\beta \sin \pi/\beta}.$$

1.56. Let $h : \mathbb{R} \to \mathbb{R}$ be bounded and continuous, let x be a fixed real number, and let $k \geq 2$ be an even integer. Find the limit

$$\lim_{n\to\infty} \sqrt[k]{n} \int_{-\infty}^\infty h(x+t)e^{-nt^k}dt.$$

1.57. Let $k \geq 1$ be an integer. Calculate

$$\lim_{n \to \infty} \int_0^{\sqrt[k]{n}} \left(1 - \frac{x^k}{n}\right)^n dx.$$

1.58. Let $k \geq 0$ be an integer. Calculate

$$\lim_{n \to \infty} n^{k+1} \int_0^1 \left(\frac{1-x}{1+x}\right)^n x^k dx.$$

1.59. Let $a > 1$ be a real number and let $f : [1, \infty) \to \mathbb{R}$ be a continuous function such that, for some $\alpha > 0$, $\lim_{x \to \infty} x^\alpha f(x)$ exists and is finite. Calculate

$$l = \lim_{n \to \infty} n \int_1^a f(x^n) dx \quad \text{and} \quad \lim_{n \to \infty} n \left(n \int_1^a f(x^n) dx - l\right).$$

1.60. The limit is a log integral. Let $f : [0, 1] \to (0, \infty)$ be a function such that $(\ln f(x))/x$ is integrable over $[0, 1]$ and let $g : [0, 1] \to \mathbb{R}$ be an integrable function which is continuous at 1. Prove that

$$\lim_{n \to \infty} n^2 \left(\int_0^1 \sqrt[n]{f(x^n)} g(x) dx - \int_0^1 g(x) dx\right) = g(1) \int_0^1 \frac{\ln f(x)}{x} dx.$$

1.61. (a) Let p and k be nonnegative real numbers. Prove that

$$\lim_{n \to \infty} n^2 \left(\int_0^1 \sqrt[n]{(1 + x^{n+k})(1 + x^{n+p})} dx - 1\right) = \zeta(2).$$

(b) More generally, if $f : [0, 1] \to (0, \infty)$ is a function such that $(\ln f(x))/x$ is integrable over $[0, 1]$, then

$$\lim_{n \to \infty} n^2 \left(\int_0^1 \sqrt[n]{f(x^{n+k}) f(x^{n+p})} dx - 1\right) = 2 \int_0^1 \frac{\ln f(x)}{x} dx.$$

1.62. Let a and b be positive real numbers. Prove that

$$\lim_{n \to \infty} n^2 \left(\int_0^1 \sqrt[n]{\frac{1 + ax^n}{1 + bx^n}} dx - 1\right) = -\int_a^b \frac{\ln(1+x)}{x} dx.$$

1.63. (a) Let $f : [0, 1] \to \mathbb{R}$ be an integrable function such that $\lim_{x \to 0^+} f(x)/x$ exists and is finite, and let $g : [0, 1] \to \mathbb{R}$ be an integrable function which is continuous at 1. Prove that

$$\lim_{n \to \infty} n \int_0^1 f(x^n) g(x) dx = g(1) \int_0^1 \frac{f(x)}{x} dx.$$

(b) Let f and g be two integrable functions which are continuous at 1, with $g(1) \neq 0$. Calculate

$$\lim_{n \to \infty} \frac{\int_0^1 x^n f(x) dx}{\int_0^1 x^n g(x) dx}.$$

1.64. Let $f : [0,1] \to \mathbb{R}$ be continuous and let $g : [0,1] \to \mathbb{R}$ be a continuously differentiable function. Prove that

$$L = \lim_{n \to \infty} n \int_0^1 x^n f(x^n) g(x) dx = g(1) \int_0^1 f(x) dx$$

and

$$\lim_{n \to \infty} n \left(n \int_0^1 x^n f(x^n) g(x) dx - L \right) = -(g(1) + g'(1)) \int_0^1 \frac{\int_0^x f(t) dt}{x} dx.$$

1.65. Two forms of Abel's lemma for integrals.

(a) Prove that

$$\lim_{\sigma \to 0^+} \sigma \int_0^\infty e^{-\sigma t} f(t) dt = \lim_{x \to \infty} f(x),$$

where f is a locally integrable[1] function for which both the integral and the limit are finite.

(b) Let α be a positive real number and let $g : (0, \infty) \to \mathbb{R}$ be a locally integrable function such that $\int_0^\infty g(t) dt < \infty$. Prove that

$$\lim_{p \to 0^+} \int_0^\infty g(t) e^{-pt^\alpha} dt = \int_0^\infty g(t) dt.$$

1.66. Let $f : [0,1] \to \mathbb{R}$ be an integrable function and let $\alpha \in (0,1)$. Prove that

$$\lim_{n \to \infty} \int_0^1 (1 - x^n)^n f(x) dx = \int_0^1 f(x) dx$$

and

$$\lim_{n \to \infty} n^\alpha \left(\int_0^1 (1 - x^n)^n f(x) dx - \int_0^1 f(x) dx \right) = 0.$$

1.67. Let $f, g : [0,1] \to \mathbb{R}$ be two continuous functions. Prove that

$$\lim_{n \to \infty} \int_0^1 f(x^n) g(x) dx = f(0) \int_0^1 g(x) dx.$$

---

[1]A measurable function $\varphi(t)$ on $\mathbb{R}$ is *locally integrable*, $\varphi \in L^1_{loc}$, if $|\varphi|$ is integrable over any compact set (see [56, p. 215]).

1.68. Let $f : [0,1]^2 \to \mathbb{R}$ be an integrable function over $[0,1]^2$. Find the value of

$$\lim_{n\to\infty} \int_0^1 \int_0^1 (x-y)^n f(x,y) dx dy.$$

1.3 Non-standard Limits

> *I will be sufficiently rewarded if when telling it to others you will not claim the discovery as your own, but will say it was mine.*
>
> Thales (circa 600 BC)

1.69. [2] **A Lalescu limit.** Let $m \geq 0$ be an integer and let f be a polynomial of degree m whose coefficient in the leading term is positive. Then

$$\lim_{x\to\infty} \left((f(x+2)\Gamma(x+2))^{\frac{1}{1+x}} - (f(x+1)\Gamma(x+1))^{\frac{1}{x}} \right) = \frac{1}{e}$$

and

$$\lim_{x\to\infty} x \left((f(x+2)\Gamma(x+2))^{\frac{1}{1+x}} - (f(x+1)\Gamma(x+1))^{\frac{1}{x}} - \frac{1}{e} \right) = \frac{1}{e}\left(m + \frac{1}{2}\right),$$

where Γ denotes the Gamma function.

1.70. (a) Let f and g be two continuous functions on $[0,1]$. Prove that

$$\lim_{n\to\infty} \int_0^1 f\left(\left\{\frac{n}{x}\right\}\right) g(x) dx = \int_0^1 f(x) dx \int_0^1 g(x) dx,$$

where $\{a\} = a - \lfloor a \rfloor$ denotes the fractional part of a.
(b) **A special case.** Let $f : [0,1] \to \mathbb{R}$ be a continuous function. Prove that

$$\lim_{n\to\infty} n \left(\int_0^1 f\left(\left\{\frac{n}{x}\right\}\right) dx - \int_0^1 f(x) dx \right) = \frac{1}{2} \int_0^1 f(x) dx - \int_0^1 x f(x) dx.$$

---

[2]The problem is motivated by the following limit of Traian Lalescu (see [82])

$$\lim_{n\to\infty} \left(\sqrt[n+1]{(n+1)!} - \sqrt[n]{n!} \right) = 1/e.$$

1.71. A Fejer limit. Let $f \in L^1[0,1]$ and let $g : [0,1] \to \mathbb{R}$ be a continuous function. Prove that

$$\lim_{n \to \infty} \int_0^1 f(x)g(\{nx\})dx = \int_0^1 f(x)dx \int_0^1 g(x)dx,$$

where $\{a\} = a - \lfloor a \rfloor$ denotes the fractional part of a.

1.72. Let $f,g : [0,1] \times [0,1] \to \mathbb{R}$ be two continuous functions. Prove that

$$\lim_{n \to \infty} \int_0^1 \int_0^1 f(\{nx\},\{ny\})g(x,y)dxdy = \int_0^1 \int_0^1 f(x,y)dxdy \cdot \int_0^1 \int_0^1 g(x,y)dxdy,$$

where $\{a\} = a - \lfloor a \rfloor$ denotes the fractional part of a.

1.73. Let $f : [-1,1] \times [-1,1] \to \mathbb{R}$ be a continuous function. Prove that

$$\lim_{t \to \infty} \int_{-1}^1 f(\sin(xt),\cos(xt)) \, dx = \frac{1}{\pi} \int_0^{2\pi} f(\sin x, \cos x)dx$$

and

$$\lim_{t \to \infty} \int_{-1}^1 \int_{-1}^1 f(\sin(xt),\cos(yt)) \, dxdy = \frac{1}{\pi^2} \int_0^{2\pi} \int_0^{2\pi} f(\sin x, \cos y)dxdy.$$

1.74. The limit of a harmonic mean. Let $f : [0,1] \to \mathbb{R}$ be a continuous function. Find the limit

$$\lim_{n \to \infty} \int_0^1 \cdots \int_0^1 f\left(\frac{n}{\frac{1}{x_1} + \cdots + \frac{1}{x_n}}\right) dx_1 \cdots dx_n.$$

1.75. Let $f : [0,1] \to \mathbb{R}$ be a continuous function. Find the limit

$$\lim_{n \to \infty} \int_0^1 \cdots \int_0^1 f(\sqrt[n]{x_1 x_2 \cdots x_n}) \, dx_1 \cdots dx_n.$$

1.76. The limit of a harmonic and a geometric mean. Let $f,g : [0,1] \to \mathbb{R}$ be two continuous functions. Find the limit

$$\lim_{n \to \infty} \int_0^1 \cdots \int_0^1 f\left(\frac{n}{\frac{1}{x_1} + \cdots + \frac{1}{x_n}}\right) \cdot g(\sqrt[n]{x_1 x_2 \cdots x_n}) \, dx_1 \cdots dx_n.$$

1.77. Let $f : [0,1] \to \mathbb{R}$ be a continuous function. Find the limit

$$\lim_{n\to\infty} \int_0^1 \cdots \int_0^1 f\left(\sqrt[n]{x_1 x_2 \cdots x_n} - \frac{n}{\frac{1}{x_1} + \cdots + \frac{1}{x_n}} \right) dx_1 \cdots dx_n.$$

1.78. Let $f, g : [0,1] \to \mathbb{R}$ be two continuous functions. Find the value of

$$\lim_{n\to\infty} \int_0^1 \cdots \int_0^1 f\left(\frac{n\sqrt[n]{x_1 \cdots x_n}}{\frac{1}{x_1} + \cdots + \frac{1}{x_n}} \right) dx_1 \cdots dx_n.$$

1.4 Open Problems

"Obvious" is the most dangerous word in mathematics.

Eric Temple Bell (1883–1960)

In this section we collect several open problems that are motivated by some of the exercises and problems from the previous sections. The first problem is motivated by Problem **1.51**, with $f(x) = x(1-x)$ being a function that attains its maximum at $1/2$. Problem **1.80**, which gives the behavior of a finite product, is closely related to an exercise from the theory of infinite series (see Problem **3.3**), Problem **1.82** is motivated by Problems **1.70** and **1.71**, and Problem **1.83** generalizes Problems **1.42** and **1.43**.

1.79. Are the limits equal to $1/4$? Calculate

$$\lim_{n\to\infty} \sqrt[n]{\int_0^1 x^{\frac{n(n+1)}{2}}(1-x)(1-x^2)\cdots(1-x^n)\,dx}$$

and

$$\lim_{n\to\infty} \sqrt[n]{\int_0^1 x^{1+\frac{1}{2}+\cdots+\frac{1}{n}}(1-x)(1-\sqrt{x})\cdots(1-\sqrt[n]{x})\,dx}.$$

1.80. (a) Let $n \geq 2$ be a natural number. Prove that

$$\frac{2-\sqrt{e}}{n-1} \leq \prod_{k=2}^{n}(2-\sqrt[k]{e}) < \frac{1}{n}.$$

(b) Prove that $\lim_{n\to\infty} n \prod_{k=2}^{n}(2-\sqrt[k]{e})$ exists.

(c) **Open problem.** Determine, if possible, in terms of well-known constants the limit

$$\lim_{n\to\infty} n \prod_{k=2}^{n} (2 - \sqrt[k]{e}).$$

1.81. A Gamma limit. Prove that

$$\lim_{n\to\infty} \sum_{k=1}^{n} \left(\Gamma\left(\frac{k}{n}\right) \right)^{-k} = \frac{e^{\gamma}}{e^{\gamma} - 1}.$$

1.82. Let $f, g : [0,1] \to \mathbb{R}$ be two continuous functions. Calculate

$$\lim_{n\to\infty} \int_0^1 f(\{nx\}) g\left(\left\{ \frac{n}{x} \right\} \right) dx,$$

where $\{a\} = a - \lfloor a \rfloor$ denotes the fractional part of a.

1.83. Let α, β, and p be positive real numbers and let $f : [0,1] \to \mathbb{R}$ be an integrable function which is continuous at 1. Calculate

$$\lim_{n\to\infty} n^{\alpha} \int_0^1 \left(\sum_{k=n}^{\infty} \frac{x^k}{k^p} \right)^{\beta} f(x) dx.$$

1.5 Hints

A hint is a word, or a paragraph, usually intended to help the reader to find a solution. The hint itself is not necessarily a condensed solution of the problem; it may just point to what I regard as the heart of the matter.

Paul R. Halmos (1916–2006)

1.5.1 Miscellaneous Limits

Constants don't vary—unless they are parameters.

Anon.

1.1. $1/(n+n^{\alpha}) \le 1/(n+k^{\alpha}) \le 1/(n+1)$ for $1 \le k \le n$.

1.2. Observe that $1/2 + 2/3 + \cdots + n/(n+1) = n + 1 - \gamma_{n+1} - \ln(n+1)$, where $\gamma_{n+1} = 1 + 1/2 + \cdots + 1/(n+1) - \ln(n+1)$, and study the behavior of the logarithm of the sequence in question.

1.3. Use Stolz–Cesàro lemma (the ∞/∞ case).

1.4. Observe the limit equals $\int_0^1 \{1/x\}^2 \, dx$.

1.5. Study the behavior of the logarithm of the binomial sequence and use the definition of Glaisher–Kinkelin constant.

1.6. Observe that, for $n \geq 6$, one has

$$1 + \frac{2}{n^p} \leq \sum_{k=1}^{n} \frac{1}{\binom{n}{k}^p} \leq 1 + \frac{2}{n^p} + \frac{2^{p+1}}{n^p(n-1)^p} + \frac{n-5}{\binom{n}{3}^p}.$$

1.8. Let $x_n = \sqrt[n]{n!}/n$. Calculate $\ln x_{n+1} - \ln x_n$ by using Stirling's formula and apply Lagrange's Mean Value Theorem to the difference $\ln x_{n+1} - \ln x_n$.

1.9. Factor out $\exp(1/a + 1/(a+1) + \cdots + 1/(a+n-1))$.

1.11. Note that

$$x_n = \prod_{m=1}^{n} \left(1 - \frac{1}{m} + \frac{5}{4m^2}\right) = \prod_{m=1}^{n} \frac{(2m-1)^2 + 4}{4m^2}$$

and

$$\prod_{k=1}^{2n} \left(1 + \frac{4}{k^2}\right) = \left(\prod_{m=1}^{n} \frac{m^2+1}{m^2}\right) \cdot \frac{4^{2n}(n!)^4}{((2n)!)^2} \cdot x_n.$$

Approximate $4^{2n}(n!)^4/((2n)!)^2$ by Stirling's formula and use the well-known limits $\lim_{n\to\infty} \prod_{k=1}^{2n}(1 + 4/k^2) = (e^{2\pi} - e^{-2\pi})/4\pi$ and $\lim_{n\to\infty} \prod_{k=1}^{n}(1 + 1/k^2) = (e^\pi - e^{-\pi})/2\pi$.

1.14. Prove that

$$\lim_{n\to\infty} \frac{\sqrt[n]{\Gamma\left(\frac{1}{1}\right)\Gamma\left(\frac{1}{2}\right)\cdots\Gamma\left(\frac{1}{n}\right)}}{n} = \frac{1}{e},$$

and then apply Lagrange's Mean Value Theorem to $f(x) = \ln x$ on the interval $\left[\sqrt[n+1]{\Gamma(1/1)\Gamma(1/2)\cdots\Gamma(1/n+1)}/n, \sqrt[n]{\Gamma(1/1)\Gamma(1/2)\cdots\Gamma(1/n)}/n\right]$.

1.15. Let $x_n = \sum_{k=1}^{n} \ln\left(1 + k^\alpha/n^\beta\right)$ and note that

$$\min_{k=1,\ldots,n} \left(\frac{\ln\left(1 + \frac{k^\alpha}{n^\beta}\right)}{\frac{k^\alpha}{n^\beta}}\right) \sum_{k=1}^{n} \frac{k^\alpha}{n^\beta} \leq x_n \leq \max_{k=1,\ldots,n} \left(\frac{\ln\left(1 + \frac{k^\alpha}{n^\beta}\right)}{\frac{k^\alpha}{n^\beta}}\right) \sum_{k=1}^{n} \frac{k^\alpha}{n^\beta}.$$

1.16. Let $x_n = \sum_{k=1}^n \ln\left(1 + \alpha \cdot k^{p-1}/(n^p - k^{p-1})\right)$ and observe that

$$\min_{k=1,\ldots,n}\left(\frac{\ln\left(1 + \frac{\alpha k^{p-1}}{n^p - k^{p-1}}\right)}{\frac{\alpha k^{p-1}}{n^p - k^{p-1}}}\right)\sum_{k=1}^n \frac{\alpha k^{p-1}}{n^p - k^{p-1}} \leq x_n$$

$$\leq \max_{k=1,\ldots,n}\left(\frac{\ln\left(1 + \frac{\alpha k^{p-1}}{n^p - k^{p-1}}\right)}{\frac{\alpha k^{p-1}}{n^p - k^{p-1}}}\right)\sum_{k=1}^n \frac{\alpha k^{p-1}}{n^p - k^{p-1}}.$$

1.17. Apply Lagrange's Mean Value Theorem to $f(x) = e^x$ and note that

$$1 - \prod_{j=n}^\infty \frac{j^{k+1} - \alpha}{j^{k+1} + \alpha} = 1 - e^{\sum_{j=n}^\infty \ln\left(1 - \frac{2\alpha}{j^{k+1}+\alpha}\right)} = -\sum_{j=n}^\infty \ln\left(1 - \frac{2\alpha}{j^{k+1}+\alpha}\right)\cdot e^{\theta_n},$$

where $\theta_n \in \left(\sum_{j=n}^\infty \ln\left(1 - 2\alpha/(j^{k+1}+\alpha)\right), 0\right)$.

1.18. Prove that, for $0 < \varepsilon < L$ and large n, one has

$$\frac{L-\varepsilon}{L+\varepsilon}\cdot\frac{\sqrt[p]{n} + \sqrt[p]{n+1} + \cdots + \sqrt[p]{kn}}{n\sqrt[p]{n}} < \frac{x_n + x_{n+1} + \cdots + x_{kn}}{nx_n}$$

$$< \frac{L+\varepsilon}{L-\varepsilon}\cdot\frac{\sqrt[p]{n} + \sqrt[p]{n+1} + \cdots + \sqrt[p]{kn}}{n\sqrt[p]{n}}.$$

1.19. Apply Stolz–Cesàro lemma (the $0/0$ case).

1.20. Prove that the sequence

$$x_n = \sum_{k=1}^n \frac{1}{n + k + b + \sqrt{n^2 + kn + a}}$$

is bounded above and below by two sums that are the Riemann sums of the function $f(x) = 1/(1+x+\sqrt{1+x})$ on the interval $[0,1]$ associated with the partition $0, 1/n, \ldots, (n-1)/n, 1$ and special systems of intermediary points.

1.21. Note that if $b \leq c$, then

$$\frac{k+a}{n^2 + kn + c} \leq \frac{k+a}{\sqrt{n^2 + kn + b}\cdot\sqrt{n^2 + kn + c}} \leq \frac{k+a}{n^2 + kn + b}$$

and show that if α and β are nonnegative real numbers, one has

$$\lim_{n\to\infty}\sum_{k=1}^n \frac{k+\alpha}{n^2 + kn + \beta} = 1 - \ln 2.$$

1.22. Note that if $b \le c$, then

$$\frac{\sqrt{n^2+kn+a}}{n^2+kn+c} \le \frac{\sqrt{n^2+kn+a}}{\sqrt{n^2+kn+b}\cdot\sqrt{n^2+kn+c}} \le \frac{\sqrt{n^2+kn+a}}{n^2+kn+b}$$

and prove that if α and β are nonnegative real numbers, one has

$$\lim_{n\to\infty}\sum_{k=1}^{n}\frac{\sqrt{n^2+kn+\alpha}}{n^2+kn+\beta}=2\sqrt{2}-2.$$

1.23. It suffices to prove that if $k \ge 0$ is a real number, then

$$\lim_{n\to\infty}\frac{1}{n^3}\sum_{i=1}^{n}\sum_{j=1}^{n}\frac{ij}{\sqrt{(i+k)^2+(j+k)^2}}=\frac{2(\sqrt{2}-1)}{3}.$$

1.24. Use that $\sin^2(nx)+\cos^2(nx)=1$.

1.25. Observe that

$$\sum_{k=1}^{n}\frac{(n^2+kn+a)^\alpha}{(n^2+kn+b)^\beta}=\frac{n^{2\alpha}}{n^{2\beta-1}}\cdot\frac{1}{n}\sum_{k=1}^{n}\frac{\left(1+\frac{k}{n}+\frac{a}{n^2}\right)^\alpha}{\left(1+\frac{k}{n}+\frac{b}{n^2}\right)^\beta},$$

and prove that if $a,b,u \ge 0$ are real numbers and $v > 0$, one has

$$\lim_{n\to\infty}\frac{1}{n}\sum_{k=1}^{n}\frac{\left(1+\frac{k}{n}+\frac{a}{n^2}\right)^u}{\left(1+\frac{k}{n}+\frac{b}{n^2}\right)^v}=\int_{0}^{1}(1+x)^{u-v}dx.$$

1.26. Denote the value of the limit by L_n, use that $f_n(x)=\exp(f_{n-1}(x)\ln x)$, and apply Lagrange's Mean Value Theorem to the exponential function to prove that $L_n=-L_{n-1}$.

1.28. Use that $1/(i+j)=\int_0^1 x^{i+j-1}dx$ and interchange the order of integration and the order of summation.

1.29. Observe the limit equals

$$\int_{0}^{1}\cdots\int_{0}^{1}\frac{dx_1 dx_2\cdots dx_k}{x_1+x_2+\cdots+x_k}.$$

This integral can be calculated by using that $1/a=\int_0^\infty e^{-ax}dx$.

1.30. Show, by applying Lagrange's Mean Value Theorem to the function $F(n)=\int_0^n dt/(a^t+x^k)$ on the interval $[n,n+1]$, that

$$\int_{0}^{\infty}\frac{dt}{a^t+x^k}\le\sum_{n=0}^{\infty}\frac{1}{a^n+x^k}\le\int_{0}^{\infty}\frac{dt}{a^t+x^k}+\frac{1}{1+x^k}.$$

1.32. Apply Lagrange's Mean Value Theorem to the difference $f(k/(n+1)) - f(k/n)$ and use the uniform continuity of f'.

1.33. Add and substract $1/k$ from the finite sum and then use the power series expansion of arctan to prove that

$$\sum_{k=1}^{\infty} (\arctan 1/k - 1/k) = \sum_{n=1}^{\infty} \frac{(-1)^n}{2n+1} \zeta(2n+1).$$

The last series can be calculated in terms of the Gamma function.

1.5.2 Limits of Integrals

> *Mathematics is trivial, but I can't do my work without it.*
>
> Richard Feynman (1918–1988)

1.34. Prove that $\sqrt{2} \le \int_0^{\pi/2} \sqrt[n]{\sin^n x + \cos^n x}\, dx \le \sqrt{2} \cdot \sqrt[n]{2}$.

1.35. Reduce the problem to proving that if $k \in (0,1]$ is a fixed real number, one has

$$\lim_{n \to \infty} n^{\frac{1}{\alpha}} \int_0^k (1 - y^\alpha)^n\, dy = \Gamma(1 + 1/\alpha).$$

1.36. Use Arithmetic Mean–Geometric Mean Inequality to prove that

$$\int_a^b \frac{dx}{c + \sin^2 x \cdot \sin^2(x+1) \cdots \sin^2(x+n)} \ge \int_a^b \frac{dx}{c + \left(\frac{1}{2} - \frac{\sin(n+1)\cos(n+2x)}{2(n+1)\sin 1}\right)^{n+1}}.$$

Apply Fatou's lemma and show that

$$\underline{\lim} \int_a^b \frac{dx}{c + \left(\frac{1}{2} - \frac{\sin(n+1)\cos(n+2x)}{2(n+1)\sin 1}\right)^{n+1}} \ge \frac{b-a}{c}.$$

1.38. Fix $\varepsilon > 0$ and prove that

$$\int_\varepsilon^1 \frac{f(ax) - f(bx)}{x}\, dx = f(c) \ln \frac{b}{a} - \int_a^b \frac{f(t)}{t}\, dt,$$

where $c \in [a\varepsilon, b\varepsilon]$.

1.40. First, show that if n is a nonnegative integer, then

$$\int_0^{\frac{\pi}{2}} \frac{\sin^2 nx}{\sin^2 x}\, dx = \frac{n\pi}{2}.$$

Second, prove that if $k \geq 0$ is an integer, one has

$$\lim_{n \to \infty} \frac{1}{n} \int_0^{\frac{\pi}{2}} \frac{\sin^2 nx}{\sin^2 x} x^k dx = \begin{cases} \frac{\pi}{2} & \text{if } k = 0, \\ 0 & \text{if } k \geq 1. \end{cases}$$

Then, use that any continuous function defined on a compact set can be uniformly approximated by polynomials.

1.41. Use Stolz–Cesàro lemma (the ∞/∞ case) and prove that if f is integrable and continuous at 1, one has $\lim_{n \to \infty} n \int_0^1 x^n f(x)dx = f(1)$.

1.42. Apply Stolz–Cesàro lemma (the $0/0$ case).

1.44. Use the substitution $x = (y/n^a)^{1/b}$ to prove that

$$n^{\frac{a}{b}} \int_0^1 \frac{f(x)}{1 + n^a x^b} dx = \frac{1}{b} \int_0^\infty \frac{y^{1/b-1}}{1+y} f\left(\left(\frac{y}{n^a} \right)^{1/b} \right) \chi_n(y) dy$$

and calculate the limit of the last integral by using Lebesgue Convergence Theorem.

1.45. Use the technique for calculating limits of indeterminate form of type 1^∞.

1.46. First, prove that if $p \geq 0$ is an integer, then

$$\lim_{n \to \infty} \sqrt{n} \int_0^{\pi/2} \sin^n x \cos^p x dx = \begin{cases} \sqrt{\frac{\pi}{2}} & \text{if } p = 0, \\ 0 & \text{if } p \geq 1. \end{cases}$$

Second, use that a continuous function on a compact set can be uniformly approximated by polynomials.

1.47. Observe that $\int_a^b f^{n+1}(x)dx \leq M \int_a^b f^n(x)dx$, where $M = \sup_{x \in [a,b]} f(x)$. To prove the reverse limit inequality apply Hölder's inequality, with $p = (n+1)/n$ and $q = n+1$, to the integral $\int_a^b f^n(x)dx$.

1.48. Express the two nth roots as exponential functions and then apply Lagrange's Mean Value Theorem.

1.52. Use the $\varepsilon - \delta$ definition of the differentiability of f and of the limit of g at b and determine the behavior of the product $f''(x)g(x)$ on $(b - \delta, b)$. Then, study the limits of the two integrals

$$n \int_a^{b-\delta} f''(x)g(x)dx \quad \text{and} \quad n \int_{b-\delta}^b f''(x)g(x)dx.$$

1.53. Use the substitution $x^n = y$ and study the limits of the two integrals

$$\int_0^\varepsilon f^n(\sqrt[n]{y}) y^{\frac{1}{n}-1} g(y)dy \quad \text{and} \quad \int_\varepsilon^1 f^n(\sqrt[n]{y}) y^{\frac{1}{n}-1} g(y)dy,$$

where $\varepsilon > 0$.

1.54. Use the substitution $t = y/\sqrt[k]{n}$ and calculate the limit of the integral

$$\int_0^{x\sqrt[k]{n}} \frac{f(y/\sqrt[k]{n})}{(1+y^k/n)^n}\,dy,$$

by using Lebesgue Convergence Theorem.

1.56. The problem reduces to the calculation of the limit of the integral

$$\int_0^\infty \left(h\left(x + \sqrt[k]{\frac{u}{n}}\right) + h\left(x - \sqrt[k]{\frac{u}{n}}\right) \right) e^{-u} u^{\frac{1}{k}-1}\,du,$$

which can be done by using Lebesgue Convergence Theorem.

1.57. Use the substitution $x^k = y$ and the inequalities $0 \le e^{-x} - (1-x/n)^n \le x^2 e^{-x}/n$, for $0 \le x \le n$.

1.58. Use the substitution $(1-x)/(1+x) = y$ and integrate by parts, k times, the integral

$$\int_0^1 y^n \frac{(1-y)^k}{(1+y)^{k+2}}\,dy.$$

1.59. Prove that l equals $\int_1^\infty f(x)/x\,dx$ by making the substitution $x^n = y$ and by calculating the limits of the two integrals

$$\int_1^\infty \frac{f(y)}{y} y^{\frac{1}{n}}\,dy \quad \text{and} \quad \int_{a^n}^\infty \frac{f(y)}{y} y^{\frac{1}{n}}\,dy.$$

The second limit equals $\int_1^\infty (f(y)\ln y)/y\,dy$ and reduces to the calculation of the limits of the integrals

$$n\int_1^\infty \frac{f(y)}{y}\left(y^{\frac{1}{n}} - 1 \right) dy \quad \text{and} \quad n\int_{a^n}^\infty \frac{f(y)}{y} y^{\frac{1}{n}}\,dy.$$

1.60. Write

$$n^2 \left(\int_0^1 \sqrt[n]{f(x^n)} g(x)\,dx - \int_0^1 g(x)\,dx \right) = \int_0^1 \frac{\ln f(y)}{y} \exp(\theta_n(y)) g\left(y^{\frac{1}{n}} \right) y^{\frac{1}{n}}\,dy,$$

where the equality follows based on Lagrange's Mean Value Theorem applied to the function $x \to \exp(x)$ and $\theta_n(y)$ is between $\frac{1}{n}\ln f(y)$ and $\frac{1}{n}\ln f(0) = 0$. Then, calculate the limit of the last integral by using Lebesgue Convergence Theorem.

1.61. Follow the ideas from the hint of Problem **1.60**.

1.63. Use the $\varepsilon - \delta$ definition of $\lim_{x\to 0^+} f(x)/x$ and calculate the limit of the integral

$$\int_0^1 \frac{f(y)}{y} g(y^{1/n}) y^{1/n}\,dy.$$

1.64. For proving the first limit, make the substitution $x^n = y$ and use the Bounded Convergence Theorem. Write the second limit as

$$n \int_0^1 f(x) \left(x^{1/n} g(x^{1/n}) - g(1) \right) dx,$$

integrate by parts, and use the Bounded Convergence Theorem.

1.65. (a) Use the $\varepsilon - \delta$ definition of $\lim_{x \to \infty} f(x)$ and the fact that $\sigma \int_0^\infty e^{-\sigma t} dt = 1$.
(b) First, observe that if α is a positive real number and $f : (0, \infty) \to \mathbb{R}$ is a locally integrable function such that $\lim_{t \to \infty} f(t) = L$, one has

$$\lim_{p \to 0^+} p \int_0^\infty f(t) \alpha t^{\alpha - 1} e^{-pt^\alpha} dt = L.$$

Second, integrate by parts to get that

$$\int_0^\infty g(t) e^{-pt^\alpha} dt = p\alpha \int_0^\infty \left(\int_0^t g(s) ds \right) e^{-pt^\alpha} t^{\alpha - 1} dt$$

and use the observation to show that the limit equals $\int_0^\infty g(t) dt$.

1.66. First, show that

$$\int_0^1 (1 - x^n)^n dx = (1/n) \cdot B(n + 1, 1/n) \quad \text{and} \quad \lim_{n \to \infty} \int_0^1 (1 - x^n)^n dx = 1.$$

Second, observe that

$$\left| \int_0^1 (1 - x^n)^n f(x) dx - \int_0^1 f(x) dx \right| \le M \int_0^1 (1 - (1 - x^n)^n) dx,$$

where $M = \sup_{x \in [0,1]} |f(x)|$ and prove that $\lim_{n \to \infty} n^\alpha \int_0^1 (1 - (1 - x^n)^n) dx = 0$.

1.67. Use the Bounded Convergence Theorem.

1.68. Prove the limit equals 0 by using Lebesgue Convergence Theorem.

1.5.3 Non-standard Limits

*There is no branch of mathematics, however abstract, which
may not some day be applied to phenomena of the real world.*

Nikolai Lobachevsky (1792–1856)

1.70. (a) First, prove that if k and l are two nonnegative integers, one has

$$\lim_{n \to \infty} \int_0^1 \left\{ \frac{n}{x} \right\}^k x^l dx = \frac{1}{(k+1)(l+1)}.$$

Second, approximate f and g by polynomials.

(b) Prove, by using the substitution $n/x = y$, that

$$n\left(\int_0^1 f\left(\left\{\frac{n}{x}\right\}\right)dx - \int_0^1 f(x)dx\right) = n\int_0^1 f(t)\left(\sum_{k=n}^\infty \frac{n}{(k+t)^2} - 1\right)dt$$

and calculate the limit of the last integral by using Lebesgue Convergence Theorem.

1.71. Follow the steps:

Step 1. Prove that if n is an integer and k and m are two nonnegative integers, then

$$\int_n^{n+1} y^k(\{y\})^m dy = \sum_{p=0}^k \binom{k}{p}\frac{n^p}{k+m-p+1}.$$

Step 2. Use Step 1 to prove that if k and m are two nonnegative integers, one has

$$\lim_{n\to\infty}\int_0^1 x^k(\{nx\})^m dx = \frac{1}{(k+1)(m+1)} = \int_0^1 x^k dx \int_0^1 x^m dx.$$

Step 3. Solve the problem for the case when f is a continuous function. In this case, approximate both f and g by polynomials and use the previous step.

Step 4. Use that any L^1 function is approximated in norm by a continuous function.

1.72. Approximate f and g by polynomials.

1.73. Prove, by using Euler's formulae,

$$\sin(tx) = \frac{e^{itx} - e^{-itx}}{2i} \quad \text{and} \quad \cos(tx) = \frac{e^{itx} + e^{-itx}}{2},$$

that if m and n are two nonnegative integers, one has

$$\lim_{t\to\infty}\int_{-1}^1 \sin^n(tx)\cos^m(tx)dx = \frac{1}{\pi}\int_0^{2\pi}\sin^n x\cos^m x\,dx.$$

Then, approximate f by a polynomial.

1.76. Show that if $k \geq 0$ and $l > 0$ are integers, then

$$\lim_{n\to\infty}\int_0^1\cdots\int_0^1 (\sqrt[n]{x_1 x_2\cdots x_n})^k\left(\frac{n}{\frac{1}{x_1}+\cdots+\frac{1}{x_n}}\right)^l dx_1\cdots dx_n = 0$$

and if $k \geq 0$ is an integer, then

$$\lim_{n\to\infty}\int_0^1\cdots\int_0^1 (\sqrt[n]{x_1 x_2\cdots x_n})^k dx_1\cdots dx_n = \left(\frac{1}{e}\right)^k.$$

Next, approximate f and g by polynomials.

1.6 Solutions

> *You are never sure whether or not a problem is good unless you*
> *actually solve it.*
>
> Mikhail Gromov (Abel Prize, 2009)

This section contains the solutions of the problems from the first chapter.

1.6.1 Miscellaneous Limits

> *When asked about his age: I was x years old in the year x^2.*
>
> Augustus De Morgan (1806–1871)

1.1. The limit equals 1 when $0 < \alpha < 1$, $\ln 2$ when $\alpha = 1$, and 0 when $\alpha > 1$. We have

$$\frac{n}{n+n^\alpha} < \frac{1}{n+1^\alpha} + \frac{1}{n+2^\alpha} + \cdots + \frac{1}{n+n^\alpha} < \frac{n}{n+1}$$

and hence, when $\alpha < 1$, the limit equals 1.
When $\alpha = 1$, we have

$$\lim_{n\to\infty}\left(\frac{1}{n+1} + \frac{1}{n+2} + \cdots + \frac{1}{2n}\right) = \lim_{n\to\infty}\frac{1}{n}\sum_{k=1}^{n}\frac{1}{1+k/n} = \int_0^1 \frac{dx}{1+x} = \ln 2.$$

Let $\alpha > 1$. We have, since $n + k^\alpha \geq 2\sqrt{n}\cdot k^{\alpha/2}$, that

$$0 < \frac{1}{n+1^\alpha} + \frac{1}{n+2^\alpha} + \cdots + \frac{1}{n+n^\alpha} < \frac{1}{2\sqrt{n}}\left(\frac{1}{1^{\alpha/2}} + \frac{1}{2^{\alpha/2}} + \cdots + \frac{1}{n^{\alpha/2}}\right).$$

An application of Stolz–Cesàro lemma (the ∞/∞ case) shows that

$$\lim_{n\to\infty}\frac{\frac{1}{1^{\alpha/2}} + \frac{1}{2^{\alpha/2}} + \cdots + \frac{1}{n^{\alpha/2}}}{\sqrt{n}} = \lim_{n\to\infty}\frac{\frac{1}{(n+1)^{\alpha/2}}}{\sqrt{n+1} - \sqrt{n}} = \lim_{n\to\infty}\frac{\sqrt{n+1} + \sqrt{n}}{(n+1)^{\alpha/2}} = 0.$$

1.2. The limit equals $e^{\gamma-1}$, where γ denotes the Euler–Mascheroni constant.[3] We have

$$\frac{1}{2}+\frac{2}{3}+\cdots+\frac{n}{n+1} = n - \left(\frac{1}{2}+\frac{1}{3}+\cdots+\frac{1}{n+1}\right) = n+1-\gamma_{n+1}-\ln(n+1),$$

where $\gamma_{n+1} = 1+1/2+\cdots+1/(n+1)-\ln(n+1)$. It follows that

$$x_n = \frac{1}{n}\left(\frac{1}{1-\left(\frac{\gamma_{n+1}+\ln(n+1)-1}{n}\right)}\right)^n = \frac{1}{n}\left(\frac{1}{1-a_n}\right)^n,$$

where $a_n = (\gamma_{n+1}+\ln(n+1)-1)/n$. Also, we note that $a_n = O\left(\ln n/n\right)$. Thus,

$$\ln x_n = -n\ln(1-a_n) - \ln n = n\left(a_n+\frac{a_n^2}{2}+\cdots\right) - \ln n$$

$$= n\left(a_n+O\left(\left(\frac{\ln n}{n}\right)^2\right)\right) - \ln n$$

$$= na_n - \ln n + n\cdot O\left(\left(\frac{\ln n}{n}\right)^2\right)$$

$$= \gamma_{n+1} - 1 + \ln\frac{n+1}{n} + n\cdot O\left(\left(\frac{\ln n}{n}\right)^2\right).$$

1.3. The limit equals $\ln 2 + \pi/2$. We apply Stolz–Cesàro lemma (the ∞/∞ case) and we have that

$$\lim_{n\to\infty}\frac{1}{n}\sum_{i=1}^{n}\sum_{j=1}^{n}\frac{i+j}{i^2+j^2} = \lim_{n\to\infty}\left(\sum_{i=1}^{n+1}\sum_{j=1}^{n+1}\frac{i+j}{i^2+j^2} - \sum_{i=1}^{n}\sum_{j=1}^{n}\frac{i+j}{i^2+j^2}\right)$$

$$= 2\left(\frac{(n+1)+1}{(n+1)^2+1^2}+\frac{(n+1)+2}{(n+1)^2+2^2}+\cdots+\frac{(n+1)+n}{(n+1)^2+n^2}\right)+\frac{1}{n+1}$$

$$= \frac{2}{n+1}\left(\frac{1+\frac{1}{n+1}}{1+\left(\frac{1}{n+1}\right)^2}+\frac{1+\frac{2}{n+1}}{1+\left(\frac{2}{n+1}\right)^2}+\cdots+\frac{1+\frac{n}{n+1}}{1+\left(\frac{n}{n+1}\right)^2}\right)+\frac{1}{n+1}$$

$$= 2\int_0^1\frac{1+x}{1+x^2}dx = \pi/2 + \ln 2.$$

1.4. The limit equals $\ln(2\pi) - 1 - \gamma$. In what follows we solve a more general problem. Let $f:[0,1]\to\mathbb{R}$ be a continuous function and we calculate the limit

---

[3]The Euler–Mascheroni constant, γ, is defined by $\gamma = \lim_{n\to\infty}(1+1/2+\cdots+1/n-\ln n)$.

$$\lim_{n\to\infty}\frac{1}{n}\sum_{k=1}^{n}f\left(\left\{\frac{n}{k}\right\}\right)=\int_{0}^{1}f\left(\left\{\frac{1}{x}\right\}\right)dx$$

$$=\int_{1}^{\infty}\frac{f(\{x\})}{x^{2}}dx$$

$$=\int_{0}^{1}f(x)\left(\sum_{n=1}^{\infty}\frac{1}{(n+x)^{2}}\right)dx$$

$$=\int_{0}^{1}f(x)\left(\ln\Gamma(x+1)\right)''dx.$$

Thus, if $f\in C^{2}[0,1]$, one has that

$$\int_{0}^{1}f(x)\left(\ln\Gamma(x+1)\right)''dx=\gamma(f(0)-f(1))+f(1)+\int_{0}^{1}f''(x)\ln\Gamma(x+1)dx.$$

In particular, when $f(x)=x^{2}$, we have that the limit equals $1-\gamma+2\int_{0}^{1}\ln\Gamma(x+1)dx$, and the result follows based on *Raabe's integral*[4] (see [136, pp. 446–447])

$$\int_{0}^{1}\ln\Gamma(x)dx=\ln\sqrt{2\pi}.$$

For the calculation of this limit see also Problem **2.2**.

Remark. This problem, as well as its generalization, is due to Alexandru Lupaş and Mircea Ivan. We proved that if $f\in C[0,1]$, the following limit holds

$$\lim_{n\to\infty}\frac{1}{n}\sum_{k=1}^{n}f\left(\left\{\frac{n}{k}\right\}\right)=\int_{0}^{1}f(x)\psi'(x+1)dx.$$

1.5. The limit equals $e/\sqrt{2\pi}$. We need the definition of the Glaisher–Kinkelin constant, A, defined by the limit

$$A=\lim_{n\to\infty}A_{n}=\lim_{n\to\infty}\frac{1^{1}2^{2}\cdots n^{n}}{n^{n^{2}/2+n/2+1/12}e^{-n^{2}/4}}=1.28242\ldots.$$

---

[4]This improper integral is attributed to the Swiss mathematician and physicist J.L. Raabe (1801–1859). We learned from the work of Ranjan Roy [110, Exercise 2, p. 489] that an interesting proof of this integral, involving Euler's reflection formula $\Gamma(x)\Gamma(1-x)=\pi/\sin(\pi x)$, was published by Stieltjes in 1878 (see [125, pp. 114–118]). Stieltjes has also proved that for $x\geq 0$ one has

$$\int_{0}^{1}\ln\Gamma(x+u)du=x\ln x-x+\frac{1}{2}\ln(2\pi).$$

Let $x_n = \sqrt[n]{\binom{n}{1}\binom{n}{2}\cdots\binom{n}{n}}/e^{n/2}n^{-1/2}$. We have

$$\ln x_n = \frac{1}{n}\sum_{k=1}^{n}\ln\binom{n}{k} - \frac{n}{2} + \frac{\ln n}{2}$$

$$= \frac{1}{n}\left(n\ln n! - \sum_{k=1}^{n}\ln k! - \sum_{k=1}^{n}\ln(n-k)!\right) - \frac{n}{2} + \frac{\ln n}{2}$$

$$= \frac{1}{n}\left((n+1)\ln n! - 2\sum_{k=1}^{n}\ln k!\right) - \frac{n}{2} + \frac{\ln n}{2}.$$

We note that $\sum_{k=1}^{n}\ln k! = \sum_{k=1}^{n}(n+1-k)\ln k = (n+1)\ln n! - \sum_{k=1}^{n}k\ln k$, and hence,

$$\ln x_n = \frac{2}{n}\sum_{k=1}^{n}k\ln k - \frac{n+1}{n}\ln n! - \frac{n}{2} + \frac{1}{2}\ln n.$$

Since $\ln A_n = \sum_{k=1}^{n}k\ln k - \left(n^2/2 + n/2 + 1/12\right)\ln n + n^2/4$, it follows that

$$\ln x_n = \frac{2}{n}\ln A_n + \left(n + \frac{3}{2} + \frac{1}{6n}\right)\ln n - n - \frac{n+1}{n}\ln n!.$$

An application of Stirling's formula, $\ln n! = \ln\sqrt{2\pi} + \frac{1}{2}\ln n + n\ln n - n + O(1/n)$, implies that

$$\ln x_n = \frac{2}{n}\ln A_n - \frac{1}{3n}\ln n + 1 - \frac{n+1}{n}\ln\sqrt{2\pi} - \left(1 + \frac{1}{n}\right)O\left(\frac{1}{n}\right).$$

Thus, $\lim_{n\to\infty}\ln x_n = 1 - \ln\sqrt{2\pi}$, and the result follows.

Remark. This problem, which deals with the asymptotic behavior of the geometric mean of the binomial coefficients, is a stronger version of a famous problem that goes back to Pólya and Szegö [104, Problem 51, p. 45] which states that if G_n denotes the geometric mean of the binomial coefficients $\binom{n}{1}$, $\binom{n}{2}$, ..., $\binom{n}{n}$, then $\lim_{n\to\infty}\sqrt[n]{G_n} = \sqrt{e}$. Our problem implies that for large n, one has

$$G_n = \sqrt[n]{\binom{n}{1}\binom{n}{2}\cdots\binom{n}{n}} = O\left(e^{n/2}n^{-1/2}\right).$$

1.6. The limit equals e^2. Let $a_n = \sum_{k=1}^{n}1/\binom{n}{k}^p$. Then for $n \geq 3$, one has

$$a_n \geq \frac{1}{\binom{n}{n}^p} + \frac{1}{\binom{n}{1}^p} + \frac{1}{\binom{n}{n-1}^p} = 1 + \frac{2}{n^p}.$$

Also, for $n \geq 6$ we have

$$a_n = \frac{1}{\binom{n}{n}^p} + \left(\frac{1}{\binom{n}{1}^p} + \frac{1}{\binom{n}{n-1}^p}\right) + \left(\frac{1}{\binom{n}{2}^p} + \frac{1}{\binom{n}{n-2}^p}\right) + \left(\frac{1}{\binom{n}{3}^p} + \cdots + \frac{1}{\binom{n}{n-3}^p}\right)$$

$$\leq 1 + \frac{2}{n^p} + \frac{2^{p+1}}{n^p(n-1)^p} + \frac{n-5}{\binom{n}{3}^p},$$

since $\binom{n}{k} \geq \binom{n}{3}$, for $3 \leq k \leq n-3$. Thus,

$$1 + \frac{2}{n^p} \leq a_n \leq 1 + \frac{2}{n^p} + \frac{2^{p+1}}{n^p(n-1)^p} + \frac{6^p(n-5)}{(n(n-1)(n-2))^p}.$$

Hence,

$$\left(1 + \frac{2}{n^p}\right)^{n^p} \leq a_n^{n^p} \leq \left(1 + \frac{2}{n^p} + \frac{2^{p+1}}{n^p(n-1)^p} + \frac{6^p(n-5)}{(n(n-1)(n-2))^p}\right)^{n^p},$$

and the result follows by passing to the limit, as $n \to \infty$, in the preceding inequality.

1.7. The limit equals $\ln^2 2$ (see [28]).

1.8. We prove that the limit equals 0 if $0 < \alpha < 2$ and $-\infty$ if $\alpha \geq 2$. Let $x_n = \sqrt[n]{n!}/n$ and let $y_n = \ln x_n$. Using Stirling's formula, $\ln n! = \ln\sqrt{2\pi} + (n+1/2)\ln n - n + O(1/n)$, we get that $y_n = \frac{1}{n}\ln n! - \ln n = \frac{\ln 2\pi n}{2n} - 1 + O(1/n^2)$, and hence,

$$\ln x_{n+1} - \ln x_n = -\frac{\ln 2\pi}{2n(n+1)} + \frac{n\ln\frac{n+1}{n} - \ln n}{2n(n+1)} + O\left(\frac{1}{n^2}\right).$$

Lagrange's Mean Value Theorem implies that $\ln x_{n+1} - \ln x_n = (x_{n+1} - x_n)1/c_n$, where c_n is between x_n and x_{n+1}. Also, we have that $\lim_{n\to\infty} c_n = 1/e$. It follows that $n^\alpha(\ln x_{n+1} - \ln x_n) = n^\alpha(x_{n+1} - x_n)\cdot 1/c_n$, and hence, $\lim_{n\to\infty} n^\alpha(x_{n+1} - x_n) = 1/e \cdot \lim_{n\to\infty} n^\alpha(\ln x_{n+1} - \ln x_n)$. On the other hand,

$$n^\alpha(\ln x_{n+1} - \ln x_n) = -\ln(2\pi)\frac{n^\alpha}{2n(n+1)} + \frac{n^\alpha}{2n(n+1)}\ln\left(\frac{n+1}{n}\right)^n - \frac{n^\alpha \ln n}{2n(n+1)}$$

$$+ n^\alpha \cdot O\left(\frac{1}{n^2}\right). \tag{1.2}$$

When $\alpha = 2$, we get, from (1.2), that $\lim_{n\to\infty} n^2(\ln x_{n+1} - \ln x_n) = -\infty$.
When $\alpha > 2$, we have

$$\lim_{n\to\infty} n^\alpha(\ln x_{n+1} - \ln x_n) = \lim_{n\to\infty} n^{\alpha-2} \cdot \lim_{n\to\infty} n^2(\ln x_{n+1} - \ln x_n) = -\infty.$$

When $0 < \alpha < 2$, we have, based on (1.2), that $\lim_{n\to\infty} n^\alpha(\ln x_{n+1} - \ln x_n) = 0$.

Remark. The problem implies that for $\alpha \in (0,2)$,

$$\frac{\sqrt[n+1]{(n+1)!}}{n+1} - \frac{\sqrt[n]{n!}}{n} = o\left(\frac{1}{n^\alpha}\right).$$

1.9. The limit equals $e^{\gamma(a)}/a$, where

$$\gamma(a) = \lim_{n\to\infty}\left(\frac{1}{a} + \frac{1}{a+1} + \cdots + \frac{1}{a+n-1} - \ln\frac{a+n-1}{a}\right).$$

We have that

$$e^{\frac{1}{a}+\frac{1}{a+1}+\cdots+\frac{1}{a+n}} - e^{\frac{1}{a}+\frac{1}{a+1}+\cdots+\frac{1}{a+n-1}} = e^{\frac{1}{a}+\frac{1}{a+1}+\cdots+\frac{1}{a+n-1}}\left(e^{\frac{1}{a+n}}-1\right)$$

$$= e^{\frac{1}{a}+\frac{1}{a+1}+\cdots+\frac{1}{a+n-1}-\ln\frac{a+n-1}{a}} \cdot \frac{e^{1/(a+n)}-1}{1/(a+n)} \cdot \frac{a+n-1}{a(n+a)}, \qquad (1.3)$$

and the result follows by passing to the limit, as $n \to \infty$, in (1.3).

Remark. When $a = 1$, one has that

$$\lim_{n\to\infty}\left(e^{1+\frac{1}{2}+\cdots+\frac{1}{n+1}} - e^{1+\frac{1}{2}+\cdots+\frac{1}{n}}\right) = e^{\gamma(1)} = e^\gamma.$$

Interesting properties of the sequence $(\gamma_n(a))_{n\in\mathbb{N}}$, defined by

$$\gamma_n(a) = \frac{1}{a} + \frac{1}{a+1} + \cdots + \frac{1}{a+n-1} - \ln\frac{a+n-1}{a},$$

are studied in [119].

1.10. For a solution, see [113].

1.11. The limit equals $(e^\pi + e^{-\pi})/2\pi$. We need Euler's product formula for the sine function (see [109, Sect. 3, p. 12])

$$\sin \pi z = \pi z \prod_{n=1}^{\infty}\left(1 - \frac{z^2}{n^2}\right), \quad z \in \mathbb{C}.$$

Replacing z by iz and using Euler's formula, $\sin\alpha = \frac{e^{i\alpha}-e^{-i\alpha}}{2i}$, we obtain that

$$\frac{\sin \pi iz}{\pi iz} = \frac{e^{\pi z} - e^{-\pi z}}{2\pi z} = \prod_{n=1}^{\infty}\left(1 + \frac{z^2}{n^2}\right), \quad z \in \mathbb{C}. \qquad (1.4)$$

When $z = 1$ and $z = 2$, we obtain, based on (1.4), that

$$\frac{e^\pi - e^{-\pi}}{2\pi} = \prod_{n=1}^{\infty}\left(1 + \frac{1}{n^2}\right) \quad \text{and} \quad \frac{e^{2\pi} - e^{-2\pi}}{4\pi} = \prod_{n=1}^{\infty}\left(1 + \frac{4}{n^2}\right). \qquad (1.5)$$

Let

$$x_n = \prod_{m=1}^{n} \left(1 - \frac{1}{m} + \frac{5}{4m^2} \right) = \prod_{m=1}^{n} \frac{(2m-1)^2 + 4}{4m^2}.$$

We have, based on (1.5), that

$$\lim_{n \to \infty} \prod_{k=1}^{2n} \left(1 + \frac{4}{k^2} \right) = \frac{e^{2\pi} - e^{-2\pi}}{4\pi} \quad \text{and} \quad \lim_{n \to \infty} \prod_{k=1}^{n} \left(1 + \frac{1}{k^2} \right) = \frac{e^{\pi} - e^{-\pi}}{2\pi}. \quad (1.6)$$

On the other hand,

$$\prod_{k=1}^{2n} \left(1 + \frac{4}{k^2} \right) = \prod_{m=1}^{n} \left(1 + \frac{4}{4m^2} \right) \cdot \prod_{m=1}^{n} \left(1 + \frac{4}{(2m-1)^2} \right)$$

$$= \prod_{m=1}^{n} \frac{4m^2 + 4}{4m^2} \cdot \prod_{m=1}^{n} \frac{(2m-1)^2 + 4}{(2m-1)^2}$$

$$= \prod_{m=1}^{n} \frac{4(m^2+1)}{(2m-1)^2} \cdot \prod_{m=1}^{n} \frac{(2m-1)^2 + 4}{4m^2}.$$

We have,

$$\prod_{m=1}^{n} \frac{4(m^2+1)}{(2m-1)^2} = 4^n \prod_{m=1}^{n} \frac{m^2+1}{m^2} \cdot \prod_{m=1}^{n} \frac{m^2}{(2m-1)^2}$$

$$= \left(\prod_{m=1}^{n} \frac{m^2+1}{m^2} \right) \cdot \frac{4^{2n}(n!)^4}{((2n)!)^2}.$$

It follows that

$$\prod_{k=1}^{2n} \left(1 + \frac{4}{k^2} \right) = \left(\prod_{m=1}^{n} \frac{m^2+1}{m^2} \right) \cdot \frac{4^{2n}(n!)^4}{((2n)!)^2} \cdot x_n.$$

Since $n! = n^n e^{-n} \sqrt{2\pi n}(1 + O(1/n))$, one has that

$$4^{2n}(n!)^4 / ((2n)!)^2 = \pi n (1 + O(1/n)).$$

Thus,

$$\prod_{k=1}^{2n} \left(1 + \frac{4}{k^2} \right) = \left(\prod_{m=1}^{n} \frac{m^2+1}{m^2} \right) \cdot \pi \cdot (n \cdot x_n) \cdot \left(1 + O\left(\frac{1}{n} \right) \right).$$

Letting $n \to \infty$ in the preceding equality, we get, based on (1.6), that

$$\frac{e^{2\pi} - e^{-2\pi}}{4\pi} = \frac{e^{\pi} - e^{-\pi}}{2\pi} \cdot \pi \cdot \lim_{n \to \infty} n x_n.$$

It follows that $\lim_{n \to \infty} n x_n = (e^{\pi} + e^{-\pi})/2\pi$, and the problem is solved.

This problem is due to Raymond Mortini (see [93]). For an alternative solution, see [58].

1.12. The limit equals $e^{\alpha}/(e^{\alpha} - 1)$. A solution, for the case when $\alpha = 1$, is given in [108, p. 56].

1.13. It is worth mentioning the history of this problem. It appears that the case $s = 1$ is an old problem which Bromwich attributed to Joseph Wolstenholme (1829–1891) (see [20, Problem 19, p. 528]). The problem, as Bromwich records it, states that

$$\lim_{n \to \infty} \left[\left(\frac{1}{n}\right)^n + \left(\frac{2}{n}\right)^n + \cdots + \left(\frac{n-1}{n}\right)^n \right] = \frac{1}{e-1},$$

which we call the *Wolstenholme limit*. In 1982, I. J. Schoenberg proposed the problem of proving that $\lim_{n \to \infty} S_n = 1/(e-1)$ where $S_n = (n+1)^{-n} \sum_{k=1}^{n} k^n$ (see [114]). However, it appears that this problem appeared previously in Knopp's book [76, p. 102]. A complete asymptotic expansion of S_n was given by Abel, Ivan, and Lupaş in [2]. In 1996, Dumitru Popa studied a general version of this problem which was published as Problem **C:1789** on page 120, in the renown Romanian journal Gazeta Matematică. His problem, which the reader is invited to solve, is the following one:

C:1789. Let $f : (0,1] \to (0,\infty)$ be a differentiable function on $(0,1]$ with $f'(1) > 0$ and $\ln f$ having decreasing derivative. Let $(x_n)_{n \in \mathbb{N}}$ be the sequence defined by

$$x_n = \sum_{k=1}^{n} \left[f\left(\frac{k}{n}\right) \right]^n.$$

Then,

$$\lim_{n \to \infty} x_n = \begin{cases} 0 & \text{if } f(1) < 1, \\ \frac{1}{1 - e^{-f'(1)}} & \text{if } f(1) = 1, \\ \infty & \text{if } f(1) > 1. \end{cases}$$

Recently, Michael Spivey [121] illustrates the importance of Euler–Maclaurin summation formula in analyzing the behavior of the sequence $1^m + 2^m + \cdots + m^m$ as $m \to \infty$, and he rediscovers, as a consequence of his results, the Wolstenholme limit. The problem is revived and brought into light by Finbarr Holland [42] who *attributes* the problem to Spivey and provides two alternative ways to determine the limit, one elementary and the second by using the elegance and the power of Lebesgue integral.

It is clear that this beautiful problem has attracted the interest of lots of mathematicians who provided elegant and nice solutions for it. However, taking into account that the first edition of Bromwich's book *An Introduction to the Theory of Infinite Series* was published in 1908, we consider, as Bromwich did, that this problem, whose history appears to be lost with the passing of time, is due to Wolstenholme.

1.14. We need the following result (see [65, Entry 5, p. 109]):

$$\lim_{n \to \infty} \left(n - \Gamma\left(\frac{1}{n}\right) \right) = \gamma. \tag{1.7}$$

First, we note we have an indeterminate form of type $\infty - \infty$ since the sequence $\sqrt[n]{\Gamma(1/1)\Gamma(1/2)\cdots\Gamma(1/n)}$ is the geometric mean of $\Gamma(1/n)$ which, based on (1.7), tends to ∞. On the other hand, we prove that

$$\lim_{n \to \infty} \frac{\sqrt[n]{\Gamma\left(\frac{1}{1}\right)\Gamma\left(\frac{1}{2}\right)\cdots\Gamma\left(\frac{1}{n}\right)}}{n} = \frac{1}{e}. \tag{1.8}$$

Let $x_n = \Gamma(1/1)\Gamma(1/2)\cdots\Gamma(1/n)/n^n$. A calculation shows that

$$\frac{x_{n+1}}{x_n} = \frac{\Gamma\left(\frac{1}{n+1}\right)}{n+1} \cdot \left(\frac{n}{n+1}\right)^n.$$

Using (1.7) we get that $x_{n+1}/x_n \to 1/e$, and hence, the Cauchy–d'Alembert criteria imply that (1.8) holds. Similarly one can prove that

$$\lim_{n \to \infty} \frac{\sqrt[n+1]{\Gamma\left(\frac{1}{1}\right)\Gamma\left(\frac{1}{2}\right)\cdots\Gamma\left(\frac{1}{n+1}\right)}}{n} = \frac{1}{e}. \tag{1.9}$$

Let $I = \left[\sqrt[n+1]{\Gamma(1/1)\Gamma(1/2)\cdots\Gamma(1/n+1)}/n, \sqrt[n]{\Gamma(1/1)\Gamma(1/2)\cdots\Gamma(1/n)}/n \right]$. The order of the two endpoints of the interval is not important. We apply Lagrange's Mean Value Theorem to the function $f(x) = \ln x$, on the interval I, and we get that

$$\frac{n \ln \left(\dfrac{\sqrt[n+1]{\Gamma\left(\frac{1}{1}\right)\Gamma\left(\frac{1}{2}\right)\cdots\Gamma\left(\frac{1}{n+1}\right)}}{\sqrt[n]{\Gamma\left(\frac{1}{1}\right)\Gamma\left(\frac{1}{2}\right)\cdots\Gamma\left(\frac{1}{n}\right)}} \right)}{\sqrt[n+1]{\Gamma\left(\frac{1}{1}\right)\Gamma\left(\frac{1}{2}\right)\cdots\Gamma\left(\frac{1}{n+1}\right)} - \sqrt[n]{\Gamma\left(\frac{1}{1}\right)\Gamma\left(\frac{1}{2}\right)\cdots\Gamma\left(\frac{1}{n}\right)}} = \frac{1}{c_n}, \tag{1.10}$$

where $c_n \in I$. Combining (1.8) and (1.9), we get that

$$\lim_{n \to \infty} c_n = \frac{1}{e}. \tag{1.11}$$

On the other hand, a calculation shows that

$$n \ln \frac{\sqrt[n+1]{\Gamma\left(\frac{1}{1}\right)\Gamma\left(\frac{1}{2}\right)\cdots\Gamma\left(\frac{1}{n+1}\right)}}{\sqrt[n]{\Gamma\left(\frac{1}{1}\right)\Gamma\left(\frac{1}{2}\right)\cdots\Gamma\left(\frac{1}{n}\right)}} = \ln \sqrt[n+1]{\frac{\Gamma^n\left(\frac{1}{n+1}\right)}{\Gamma\left(\frac{1}{1}\right)\Gamma\left(\frac{1}{2}\right)\cdots\Gamma\left(\frac{1}{n}\right)}}.$$

Let $y_n = \frac{\Gamma^n\left(\frac{1}{n+1}\right)}{\Gamma\left(\frac{1}{1}\right)\Gamma\left(\frac{1}{2}\right)\cdots\Gamma\left(\frac{1}{n}\right)}$. Using (1.7), one has that

$$\frac{y_{n+1}}{y_n} = \left(\frac{\Gamma\left(\frac{1}{n+2}\right)}{\Gamma\left(\frac{1}{n+1}\right)}\right)^{n+1} \to e,$$

and hence, based on the Cauchy–d'Alembert criteria, we have that

$$n \ln \frac{\sqrt[n+1]{\Gamma\left(\frac{1}{1}\right)\Gamma\left(\frac{1}{2}\right)\cdots\Gamma\left(\frac{1}{n+1}\right)}}{\sqrt[n]{\Gamma\left(\frac{1}{1}\right)\Gamma\left(\frac{1}{2}\right)\cdots\Gamma\left(\frac{1}{n}\right)}} \to 1. \tag{1.12}$$

Combining (1.10)–(1.12), we get that the desired limit holds and the problem is solved.

For an alternative solution, see [85].

Remark. It is worth mentioning that one can also prove that if

$$x_n = \sqrt[n]{\zeta(1+1/1)\zeta(1+1/2)\cdots\zeta(1+1/n)},$$

then $\lim_{n \to \infty}(x_{n+1} - x_n) = 1/e$. This can be proved by using the same technique as above combined with the limit $\lim_{n \to \infty} \zeta(1+1/n) - n = \gamma$, which is a particular case of Dirichlet's Theorem, $\lim_{x \to 0} \zeta(1+x) - 1/x = \gamma$ (see Appendix A, Theorem A.3).

1.15. The limit equals

$$\begin{cases} e^{\frac{1}{\alpha+1}} & \text{if } \alpha+1=\beta, \\ 1 & \text{if } \alpha+1<\beta, \\ \infty & \text{if } \alpha+1>\beta. \end{cases}$$

Let

$$x_n = \sum_{k=1}^{n} \ln\left(1 + \frac{k^\alpha}{n^\beta}\right) = \sum_{k=1}^{n} \frac{\ln\left(1 + \frac{k^\alpha}{n^\beta}\right)}{\frac{k^\alpha}{n^\beta}} \cdot \frac{k^\alpha}{n^\beta}.$$

It follows that

$$\min_{k=1,\ldots,n}\left(\frac{\ln\left(1+\frac{k^\alpha}{n^\beta}\right)}{\frac{k^\alpha}{n^\beta}}\right)\sum_{k=1}^{n}\frac{k^\alpha}{n^\beta}\le x_n\le\left(\max_{k=1,\ldots,n}\frac{\ln\left(1+\frac{k^\alpha}{n^\beta}\right)}{\frac{k^\alpha}{n^\beta}}\right)\sum_{k=1}^{n}\frac{k^\alpha}{n^\beta}.$$

We distinguish here the following cases:

Case 1. $\alpha+1=\beta$. We have, since $\lim_{x\to0}\ln(1+x)/x=1$, that

$$\lim_{n\to\infty}\left(\min_{k=1,\ldots,n}\frac{\ln\left(1+\frac{k^\alpha}{n^\beta}\right)}{\frac{k^\alpha}{n^\beta}}\right)=1\quad\text{and}\quad\lim_{n\to\infty}\left(\max_{k=1,\ldots,n}\frac{\ln\left(1+\frac{k^\alpha}{n^\beta}\right)}{\frac{k^\alpha}{n^\beta}}\right)=1.$$

$$(1.13)$$

Also,

$$\lim_{n\to\infty}\sum_{k=1}^{n}\frac{k^\alpha}{n^\beta}=\lim_{n\to\infty}\frac{1}{n}\sum_{k=1}^{n}\left(\frac{k}{n}\right)^\alpha=\int_0^1 x^\alpha dx=\frac{1}{\alpha+1}.$$

Case 2. $\alpha+1<\beta$. In this case we still have that (1.13) holds. However,

$$\lim_{n\to\infty}\sum_{k=1}^{n}\frac{k^\alpha}{n^\beta}=\left(\lim_{n\to\infty}\frac{1}{n}\sum_{k=1}^{n}\frac{k^\alpha}{n^\alpha}\right)\cdot\left(\lim_{n\to\infty}\frac{1}{n^{\beta-\alpha-1}}\right)=0.$$

Case 3. $\alpha+1>\beta$. We have, since $\ln(1+x)\ge x/2$ for $x\in[0,1]$, that

$$x_n>\frac{1}{2}\sum_{k=1}^{n}\frac{k^\alpha}{n^\beta}>\frac{n}{2n^\beta}\cdot\sqrt[n]{(n!)^\alpha},$$

where the last inequality follows from the Arithmetic Mean–Geometric Mean Inequality. Using Stirling's formula, $n!\sim\sqrt{2\pi n}(n/e)^n$, we get that $\lim_{n\to\infty}x_n=\infty$, and the problem is solved.

1.16. The limit equals $e^{\alpha/p}$. Let $(x_n)_{n\in\mathbb{N}}$ be the sequence defined by

$$x_n=\sum_{k=1}^{n}\ln\left(1+\frac{\alpha k^{p-1}}{n^p-k^{p-1}}\right).$$

We have that

$$\min_{k=1,\ldots,n}\left(\frac{\ln\left(1+\frac{\alpha k^{p-1}}{n^p-k^{p-1}}\right)}{\frac{\alpha k^{p-1}}{n^p-k^{p-1}}}\right)\sum_{k=1}^{n}\frac{\alpha k^{p-1}}{n^p-k^{p-1}}\le x_n$$

$$\le\max_{k=1,\ldots,n}\left(\frac{\ln\left(1+\frac{\alpha k^{p-1}}{n^p-k^{p-1}}\right)}{\frac{\alpha k^{p-1}}{n^p-k^{p-1}}}\right)\sum_{k=1}^{n}\frac{\alpha k^{p-1}}{n^p-k^{p-1}}.$$

Since $\lim_{x\to 0}\ln(1+x)/x=1$, we get that

$$\lim_{n\to\infty}\left(\min_{k=1,\dots,n}\frac{\ln\left(1+\frac{\alpha k^{p-1}}{n^p-k^{p-1}}\right)}{\frac{\alpha k^{p-1}}{n^p-k^{p-1}}}\right)=1 \quad\text{and}\quad \lim_{n\to\infty}\left(\max_{k=1,\dots,n}\frac{\ln\left(1+\frac{\alpha k^{p-1}}{n^p-k^{p-1}}\right)}{\frac{\alpha k^{p-1}}{n^p-k^{p-1}}}\right)=1.$$

Now, we calculate

$$\lim_{n\to\infty}\sum_{k=1}^{n}\frac{k^{p-1}}{n^p-k^{p-1}}.$$

We note that $n^p-n^{p-1}\le n^p-k^{p-1}\le n^p$, and hence,

$$\sum_{k=1}^{n}\frac{k^{p-1}}{n^p-n^{p-1}}\ge\sum_{k=1}^{n}\frac{k^{p-1}}{n^p-k^{p-1}}\ge\sum_{k=1}^{n}\frac{k^{p-1}}{n^p}.$$

We have

$$\lim_{n\to\infty}\sum_{k=1}^{n}\frac{k^{p-1}}{n^p}=\lim_{n\to\infty}\frac{1}{n}\sum_{k=1}^{n}\left(\frac{k}{n}\right)^{p-1}=\int_0^1 x^{p-1}\mathrm{d}x=\frac{1}{p}.$$

On the other hand,

$$\lim_{n\to\infty}\sum_{k=1}^{n}\frac{k^{p-1}}{n^p-n^{p-1}}=\lim_{n\to\infty}\frac{1}{n}\sum_{k=1}^{n}\left(\frac{k}{n}\right)^{p-1}\cdot\lim_{n\to\infty}\left(\frac{n^p}{n^p-n^{p-1}}\right)=\frac{1}{p}.$$

Putting all these together we get that $x_n\to\alpha/p$.

Remark. When $\alpha=p$, respectively, when $\alpha=p=2$, we have

$$\lim_{n\to\infty}\prod_{k=1}^{n}\frac{n^p+(p-1)k^{p-1}}{n^p-k^{p-1}}=e \quad\text{and}\quad \lim_{n\to\infty}\prod_{k=1}^{n}\frac{n^2+k}{n^2-k}=e,$$

and by the second limit, we rediscover a problem of Pólya and Szegö (see [104, Problem 55, p. 45]).

1.17. The limit equals $2\alpha/k$. We have, based on Lagrange's Mean Value Theorem, that

$$1-\prod_{j=n}^{\infty}\frac{j^{k+1}-\alpha}{j^{k+1}+\alpha}=1-e^{\sum_{j=n}^{\infty}\ln\left(1-\frac{2\alpha}{j^{k+1}+\alpha}\right)}=-\sum_{j=n}^{\infty}\ln\left(1-\frac{2\alpha}{j^{k+1}+\alpha}\right)\cdot e^{\theta_n},$$

where $\theta_n\in\left(\sum_{j=n}^{\infty}\ln\left(1-\frac{2\alpha}{j^{k+1}+\alpha}\right),0\right)$, and we note that $\theta_n\to 0$. On the other hand, we have that, for $x\in[0,1)$, the following inequalities hold

$$-\frac{x}{1-x}\le\ln(1-x)\le -x.$$

It follows, based on the preceding inequalities, that

$$-2\alpha \sum_{j=n}^{\infty} \frac{1}{j^{k+1} - \alpha} \le \sum_{j=n}^{\infty} \ln\left(1 - \frac{2\alpha}{j^{k+1} + \alpha}\right) \le -2\alpha \sum_{j=n}^{\infty} \frac{1}{j^{k+1} + \alpha}.$$

A calculation, based on Stolz–Cesàro lemma (the $0/0$ case), shows that

$$\lim_{n \to \infty} n^k \sum_{j=n}^{\infty} \frac{1}{j^{k+1} - \alpha} = \frac{1}{k} \quad \text{and} \quad \lim_{n \to \infty} n^k \sum_{j=n}^{\infty} \frac{1}{j^{k+1} + \alpha} = \frac{1}{k}.$$

Putting all these together we get that the limit equals $2\alpha/k$.

1.18. The limit equals $\frac{p}{p+1} \cdot \left(k^{(p+1)/p} - 1\right)$. Since $L = \lim_{n \to \infty} x_n / \sqrt[p]{n}$ we have that, for $0 < \varepsilon < L$, there is $n_0 \in \mathbb{N}$ such that for all $n \ge n_0$ one has the following inequalities $(L - \varepsilon) \sqrt[p]{n} < x_n < (L + \varepsilon) \sqrt[p]{n}$. It follows that, for all $n \ge n_0$, one has that

$$\frac{L - \varepsilon}{L + \varepsilon} \cdot \frac{\sqrt[p]{n} + \sqrt[p]{n+1} + \cdots + \sqrt[p]{kn}}{n \sqrt[p]{n}} < \frac{x_n + x_{n+1} + \cdots + x_{kn}}{nx_n}$$

$$< \frac{L + \varepsilon}{L - \varepsilon} \cdot \frac{\sqrt[p]{n} + \sqrt[p]{n+1} + \cdots + \sqrt[p]{kn}}{n \sqrt[p]{n}}. \quad (1.14)$$

On the other hand,

$$\frac{\sqrt[p]{n} + \sqrt[p]{n+1} + \cdots + \sqrt[p]{kn}}{n \sqrt[p]{n}} = \frac{1}{n}\left(1 + \sqrt[p]{1 + \frac{1}{n}} + \sqrt[p]{1 + \frac{2}{n}} + \cdots + \sqrt[p]{1 + \frac{(k-1)n}{n}}\right).$$

Let $f : [0, k-1] \to \mathbb{R}$ be the function defined by $f(x) = \sqrt[p]{1+x}$, and consider the partition $0, \frac{1}{n}, \frac{2}{n}, \ldots, 1, 1 + \frac{1}{n}, \ldots, k-2, k-2 + \frac{1}{n}, \ldots, k-2 + \frac{n-1}{n}, k-1$, and the sequence of intermediary points given by $\frac{1}{n}, \frac{2}{n}, \ldots, 1, 1 + \frac{1}{n}, \ldots, k-2, k-2 + \frac{1}{n}, \ldots, k-2 + \frac{n-1}{n}, k-1$. Since f is Riemann integrable we obtain that

$$\lim_{n \to \infty} \frac{\sqrt[p]{n} + \sqrt[p]{n+1} + \cdots + \sqrt[p]{kn}}{n \sqrt[p]{n}} = \int_0^{k-1} \sqrt[p]{1+x}\,dx = \frac{p}{p+1} \cdot \left(k^{(p+1)/p} - 1\right).$$

$$(1.15)$$

Letting $n \to \infty$ in (1.14) and using (1.15), we get that

$$\frac{L - \varepsilon}{L + \varepsilon} \cdot \frac{p}{p+1} \cdot \left(k^{(p+1)/p} - 1\right) \le \lim_{n \to \infty} \frac{x_n + x_{n+1} + \cdots + x_{kn}}{nx_n}$$

$$\le \frac{L + \varepsilon}{L - \varepsilon} \cdot \frac{p}{p+1} \cdot \left(k^{(p+1)/p} - 1\right).$$

Since $\varepsilon > 0$ is arbitrary taken the result follows.

1.19. The problem, which is recorded as a lemma (see [91, Lemma, p. 434]), is used for introducing a new approach for determining asymptotic evaluations of products of the form $f(n) = c_1 c_2 \cdots c_n$.

For an alternative solution, apply Stolz–Cesàro lemma (the $0/0$ case).

1.20. The limit equals $\int_0^1 \frac{dx}{1+x+\sqrt{1+x}} = 2\ln \frac{1+\sqrt{2}}{2}$. We have, since

$$n + k + b + \sqrt{n^2 + kn + a} \geq n + k + \sqrt{n^2 + kn},$$

that

$$x_n = \sum_{k=1}^n \frac{1}{n+k+b+\sqrt{n^2+kn+a}} \leq \frac{1}{n} \sum_{k=1}^n \frac{1}{1+k/n+\sqrt{1+k/n}},$$

and hence, $\lim_{n \to \infty} x_n \leq \int_0^1 \frac{dx}{1+x+\sqrt{1+x}}$. Let p be the integer part of b, that is, $p = \lfloor b \rfloor$. We note that for all $n \geq a/(p+1-b)$ and all $k = 1, \ldots, n-p-1$, the following inequalities hold

$$\frac{k+p}{n} \leq \frac{k}{n} + \frac{b}{n} + \frac{a}{n^2} \leq \frac{k+p+1}{n}.$$

We have

$$x_n = \sum_{k=1}^{n-p-1} \frac{1}{n+k+b+\sqrt{n^2+kn+a}} + \sum_{k=n-p}^n \frac{1}{n+k+b+\sqrt{n^2+kn+a}}$$

$$= \left(\sum_{k=1}^{n-p-1} \frac{1}{n+k+b+\sqrt{n^2+kn+a}} + \sum_{k=1}^{p+1} \frac{1}{k+n+\sqrt{n^2+kn}} \right)$$

$$- \sum_{k=1}^{p+1} \frac{1}{k+n+\sqrt{n^2+kn}} + \sum_{k=n-p}^n \frac{1}{n+k+b+\sqrt{n^2+kn+a}}$$

$$= S_1 - S_2 + S_3. \tag{1.16}$$

We note that

$$S_2 = \sum_{k=1}^{p+1} \frac{1}{k+n+\sqrt{n^2+kn}} \leq \frac{p+1}{n}$$

$$S_3 = \sum_{k=n-p}^n \frac{1}{n+k+b+\sqrt{n^2+kn+a}} < \frac{p+1}{n}$$

and hence, $\lim_{n \to \infty} S_2 = \lim_{n \to \infty} S_3 = 0$.

Since $n + k + b + \sqrt{n^2+kn+a} \leq n + k + b + a/n + \sqrt{n^2+nk+bn+a}$, we have

$$S_1 = \sum_{k=1}^{n-p-1} \frac{1}{n+k+b+\sqrt{n^2+kn+a}} + \sum_{k=1}^{p+1} \frac{1}{k+n+\sqrt{n^2+kn}}$$

$$\geq \frac{1}{n}\sum_{k=1}^{n-p-1} \frac{1}{1+k/n+b/n+a/n^2+\sqrt{1+k/n+b/n+a/n^2}}$$

$$+\frac{1}{n}\sum_{k=1}^{p+1} \frac{1}{1+k/n+\sqrt{1+k/n}}$$

$$= \frac{1}{n}\left(\sum_{k=1}^{n-p-1} \frac{1}{1+\frac{k}{n}+\frac{b}{n}+\frac{a}{n^2}+\sqrt{1+\frac{k}{n}+\frac{b}{n}+\frac{a}{n^2}}} + \sum_{k=1}^{p+1} \frac{1}{1+\frac{k}{n}+\sqrt{1+\frac{k}{n}}}\right)$$

$$= S_1'. \tag{1.17}$$

We note that S_1' is the Riemann sum of $f(x) = 1/(1+x+\sqrt{1+x})$ on the interval $[0,1]$ associated with the partition $0, \frac{1}{n}, \ldots, \frac{n-1}{n}, 1$ and the system of intermediary points

$$\frac{1}{n}, \frac{2}{n}, \ldots, \frac{p}{n}, \frac{p+1}{n}, \frac{1+b}{n}+\frac{a}{n^2}, \frac{2+b}{n}+\frac{a}{n^2}, \ldots, \frac{n-p-1+b}{n}+\frac{a}{n^2}.$$

Thus, $\lim_{n\to\infty} S_1 \geq \int_0^1 \frac{dx}{1+x+\sqrt{1+x}}$, and we get, based on (1.16) and (1.17), that $\lim_{n\to\infty} x_n \geq \int_0^1 \frac{dx}{1+x+\sqrt{1+x}} = 2\ln\frac{1+\sqrt{2}}{2}$.

For an alternative solution, see [100].

1.21. The limit equals $1 - \ln 2$. We need the following lemma.

Lemma 1.1. *Let α and β be nonnegative real numbers. Then,*

$$\lim_{n\to\infty} \sum_{k=1}^{n} \frac{k+\alpha}{n^2+kn+\beta} = \int_0^1 \frac{x}{x+1}dx = 1 - \ln 2.$$

Proof. A calculation shows that

$$x_n = \sum_{k=1}^{n} \frac{k+\alpha}{n^2+kn+\beta} = \frac{1}{n}\sum_{k=1}^{n} \frac{\frac{k}{n}+\frac{\beta}{n^2}}{1+\frac{k}{n}+\frac{\beta}{n^2}} + \frac{\alpha n-\beta}{n}\sum_{k=1}^{n} \frac{1}{n^2+kn+\beta}.$$

On the other hand, we have that

$$\frac{\alpha n-\beta}{2n^2+\beta} \leq \frac{\alpha n-\beta}{n}\sum_{k=1}^{n} \frac{1}{n^2+kn+\beta} \leq \frac{\alpha n-\beta}{n^2+n+\beta},$$

and hence,

$$\lim_{n\to\infty} \frac{\alpha n - \beta}{n} \sum_{k=1}^{n} \frac{1}{n^2 + kn + \beta} = 0.$$

We prove that

$$\lim_{n\to\infty} \sum_{k=1}^{n} \frac{k+\alpha}{n^2+kn+\beta} = \lim_{n\to\infty} \frac{1}{n} \sum_{k=1}^{n} \frac{\frac{k}{n}+\frac{\beta}{n^2}}{1+\frac{k}{n}+\frac{\beta}{n^2}} = \lim_{n\to\infty} \frac{1}{n} \sum_{k=0}^{n-1} \frac{\frac{k}{n}+\frac{\beta}{n^2}}{1+\frac{k}{n}+\frac{\beta}{n^2}} = 1 - \ln 2.$$

To see this, we note that for all $k = 0, 1, \ldots, n-1$, and for large n, i.e., for $n \geq \beta$, one has

$$\frac{k}{n} \leq \frac{k}{n} + \frac{\beta}{n^2} \leq \frac{k+1}{n}.$$

Thus, the preceding sum is the Riemann sum of $f(x) = x/(1+x)$ on $[0,1]$ associated with the partition $0, \frac{1}{n}, \ldots, \frac{n-1}{n}, 1$, and the system of intermediary points $\frac{k}{n} + \frac{\beta}{n^2}$, $k = 0, 1, \ldots, n-1$. The lemma is proved.

Now we are ready to solve the problem. Without losing generality let $b \leq c$. We have that

$$\frac{k+a}{n^2+kn+c} \leq \frac{k+a}{\sqrt{n^2+kn+b} \cdot \sqrt{n^2+kn+c}} \leq \frac{k+a}{n^2+kn+b},$$

and the problem is solved by using the Squeeze Theorem and Lemma 1.1.

For an alternative solution, see [101].

1.22. The limit equals $2\sqrt{2} - 2$. We need the following lemma.

Lemma 1.2. *Let α and β be nonnegative real numbers. Then,*

$$\lim_{n\to\infty} \sum_{k=1}^{n} \frac{\sqrt{n^2+kn+\alpha}}{n^2+kn+\beta} = 2\sqrt{2} - 2.$$

Proof. We distinguish the following two cases:

Case 1. $\alpha \leq 2\beta$. Let $\lambda > 2\beta - \alpha$ be fixed and let n be an integer such that $n \geq \frac{\beta^2 - \lambda\alpha}{\alpha + \lambda - 2\beta}$. We have

$$\frac{1}{\sqrt{n^2+kn+\lambda}} \leq \frac{\sqrt{n^2+kn+\alpha}}{n^2+kn+\beta} \leq \frac{1}{\sqrt{n^2+kn}}$$

$$\lim_{n\to\infty} \frac{1}{n} \sum_{k=1}^{n} \frac{1}{\sqrt{1+\frac{k}{n}+\frac{\lambda}{n^2}}} \leq \lim_{n\to\infty} \sum_{k=1}^{n} \frac{\sqrt{n^2+kn+\alpha}}{n^2+kn+\beta} \leq \lim_{n\to\infty} \frac{1}{n} \sum_{k=1}^{n} \frac{1}{\sqrt{1+\frac{k}{n}}}.$$

We prove that

$$\lim_{n\to\infty}\frac{1}{n}\sum_{k=1}^{n}\frac{1}{\sqrt{1+\frac{k}{n}+\frac{\lambda}{n^2}}}=\lim_{n\to\infty}\frac{1}{n}\sum_{k=0}^{n-1}\frac{1}{\sqrt{1+\frac{k}{n}+\frac{\lambda}{n^2}}}=2\sqrt{2}-2.$$

To see this, we note that for $n\geq\lambda$ and $k=0,\ldots,n-1$, we have

$$\frac{k}{n}\leq\frac{k}{n}+\frac{\lambda}{n^2}\leq\frac{k+1}{n}.$$

Thus, the preceding sum is the Riemann sum of $f(x)=1/\sqrt{1+x}$ on the interval $[0,1]$ associated with the partition $0,\frac{1}{n},\frac{2}{n},\ldots,\frac{k-1}{n},\frac{k}{n},\ldots,\frac{n-1}{n},1$, and the system of intermediary points $\frac{k}{n}+\frac{\lambda}{n^2}$, $k=0,\ldots,n-1$, and the result follows.

Case 2. $\alpha>2\beta$. Let λ be such that $\lambda>\alpha-2\beta$ and $n\geq\beta^2/(\alpha-2\beta)$. We have

$$\frac{1}{\sqrt{n^2+kn}}\leq\frac{\sqrt{n^2+kn+\alpha}}{n^2+kn+\beta}\leq\frac{1}{\sqrt{n^2+kn-\lambda}}.$$

Thus,

$$\lim_{n\to\infty}\frac{1}{n}\sum_{k=1}^{n}\frac{1}{\sqrt{1+\frac{k}{n}}}\leq\lim_{n\to\infty}\sum_{k=1}^{n}\frac{\sqrt{n^2+kn+\alpha}}{n^2+kn+\beta}\leq\lim_{n\to\infty}\frac{1}{n}\sum_{k=1}^{n}\frac{1}{\sqrt{1+\frac{k}{n}-\frac{\lambda}{n^2}}}.$$

On the other hand,

$$\lim_{n\to\infty}\frac{1}{n}\sum_{k=1}^{n}\frac{1}{\sqrt{1+\frac{k}{n}-\frac{\lambda}{n^2}}}=\int_0^1\frac{1}{\sqrt{x+1}}dx=2\sqrt{2}-2.$$

This follows by observing that the preceding sum is the Riemann sum of $f(x)=1/\sqrt{x+1}$ on $[0,1]$ associated with the partition $0,\frac{1}{n},\ldots,\frac{n-1}{n},1$, and the system of intermediary points $\frac{k}{n}-\frac{\lambda}{n^2}$, $k=1,\ldots,n$. The inequalities $\frac{k-1}{n}<\frac{k}{n}-\frac{\lambda}{n^2}<\frac{k}{n}$ hold for all $k=1,\ldots,n$ and $n\geq\lambda$.

Without losing the generality, we consider $b\leq c$. We have

$$\frac{\sqrt{n^2+kn+a}}{n^2+kn+c}\leq\frac{\sqrt{n^2+kn+a}}{\sqrt{n^2+kn+b}\cdot\sqrt{n^2+kn+c}}\leq\frac{\sqrt{n^2+kn+a}}{n^2+kn+b},$$

and the problem is solved by using the Squeeze Theorem and Lemma 1.2. For an alternative solution, see [96].

1.23. The limit equals $2(\sqrt{2}-1)/3$. We need the following lemma.

Lemma 1.3. *Let $k\geq0$ be a real number. Then,*

$$\lim_{n\to\infty}\frac{1}{n^3}\sum_{i=1}^{n}\sum_{j=1}^{n}\frac{ij}{\sqrt{(i+k)^2+(j+k)^2}}=\frac{2(\sqrt{2}-1)}{3}.$$

Proof. First we consider the case when $k = 0$. Let $y_n = 1/n^3 \sum_{i=1}^{n} \sum_{j=1}^{n} ij/\sqrt{i^2 + j^2}$.
Using Stolz–Cesàro lemma (the ∞/∞ case) we get that

$$\lim_{n\to\infty} y_n = \lim_{n\to\infty} \frac{1}{3n^2 + 3n + 1} \left(2\sum_{i=1}^{n} \frac{i(n+1)}{\sqrt{i^2 + (n+1)^2}} + \frac{(n+1)^2}{\sqrt{2(n+1)^2}} \right)$$

$$= \lim_{n\to\infty} \frac{1}{3n^2 + 3n + 1} \left(2(n+1)\sum_{i=1}^{n} \frac{i/(n+1)}{\sqrt{1 + (i/(n+1))^2}} + \frac{n+1}{\sqrt{2}} \right)$$

$$= \lim_{n\to\infty} \frac{2(n+1)^2}{3n^2 + 3n + 1} \cdot \lim_{n\to\infty} \frac{1}{n+1} \sum_{i=1}^{n} \frac{i/(n+1)}{\sqrt{1 + (i/(n+1))^2}}$$

$$= \frac{2}{3} \int_0^1 \frac{x}{\sqrt{1+x^2}} dx$$

$$= \frac{2}{3}(\sqrt{2} - 1).$$

Now we consider the case when $k > 0$. Let

$$z_n = \frac{1}{n^3} \sum_{i=1}^{n} \sum_{j=1}^{n} \frac{ij}{\sqrt{(i+k)^2 + (j+k)^2}},$$

and we note that it suffices to prove

$$\lim_{n\to\infty} \left(\frac{1}{n^3} \sum_{i=1}^{n} \sum_{j=1}^{n} \frac{ij}{\sqrt{(i+k)^2 + (j+k)^2}} - \frac{1}{n^3} \sum_{i=1}^{n} \sum_{j=1}^{n} \frac{ij}{\sqrt{i^2 + j^2}} \right) = 0.$$

We have

$$d_n = \frac{1}{n^3} \sum_{i=1}^{n} \sum_{j=1}^{n} \frac{ij}{\sqrt{(i+k)^2 + (j+k)^2}} - \frac{1}{n^3} \sum_{i=1}^{n} \sum_{j=1}^{n} \frac{ij}{\sqrt{i^2 + j^2}}$$

$$= \frac{-1}{n^3} \sum_{i=1}^{n} \sum_{j=1}^{n} \frac{(2k^2 + 2ki + 2kj)ij}{\sqrt{(i+k)^2 + (j+k)^2}\sqrt{i^2 + j^2} \left(\sqrt{(i+k)^2 + (j+k)^2} + \sqrt{i^2 + j^2} \right)}.$$

It follows, since $\sqrt{i^2 + j^2} \geq \sqrt{2ij}$, that

$$|d_n| \leq \frac{1}{n^3} \sum_{i=1}^{n} \sum_{j=1}^{n} \frac{(2k^2 + 2ki + 2kj)ij}{\left(\sqrt{i^2 + j^2} \right)^3} \leq \frac{1}{\sqrt{2}n^3} \sum_{i=1}^{n} \sum_{j=1}^{n} \frac{k^2 + ki + kj}{\sqrt{i}\sqrt{j}}$$

$$= \frac{k^2}{\sqrt{2}n^3} \left(\sum_{i=1}^{n} \frac{1}{\sqrt{i}} \right) \left(\sum_{j=1}^{n} \frac{1}{\sqrt{j}} \right) + \frac{k}{\sqrt{2}} \left(\frac{1}{n} \sum_{j=1}^{n} \frac{1}{\sqrt{j}} \right) \left(\frac{1}{n^2} \sum_{i=1}^{n} \sqrt{i} \right)$$

$$+ \frac{k}{\sqrt{2}} \left(\frac{1}{n} \sum_{i=1}^{n} \frac{1}{\sqrt{i}} \right) \left(\frac{1}{n^2} \sum_{j=1}^{n} \sqrt{j} \right).$$

Letting $n \to \infty$ in the preceding inequality and using that

$$\lim_{n \to \infty} \frac{1}{n} \sum_{j=1}^{n} \frac{1}{\sqrt{j}} = 0 \quad \text{and} \quad \lim_{n \to \infty} \frac{1}{n^2} \sum_{j=1}^{n} \sqrt{j} = 0,$$

we obtain that $\lim_{n \to \infty} d_n = 0$ and the lemma is proved.

Now we are ready to solve the problem. Let

$$x_n = \frac{1}{n^3} \sum_{i=1}^{n} \sum_{j=1}^{n} \frac{ij}{\sqrt{i^2 + j^2 + ai + bj + c}},$$

and let $k = \max\left\{a/2, b/2, \sqrt{c/2}\right\}$. We get, since

$$i^2 + j^2 \le i^2 + j^2 + ai + bj + c \le (i+k)^2 + (j+k)^2,$$

that

$$\frac{1}{n^3} \sum_{i=1}^{n} \sum_{j=1}^{n} \frac{ij}{\sqrt{(i+k)^2 + (j+k)^2}} \le x_n \le \frac{1}{n^3} \sum_{i=1}^{n} \sum_{j=1}^{n} \frac{ij}{\sqrt{i^2 + j^2}}. \tag{1.18}$$

Letting $n \to \infty$ in (1.18) and using Lemma 1.3, we get that $\lim_{n \to \infty} x_n = 2(\sqrt{2}-1)/3$.

1.24. We have, since $\sin^2(nx) + \cos^2(nx) = 1$, that $\lim_{n \to \infty} \sin(nx) = 0$. On the other hand, $\sin(nx)\sin x = \cos(nx)\cos x - \cos(n+1)x$. Letting $n \to \infty$ in the preceding equality, we get that $0 = \cos x - 1$, which implies that $x = 2k\pi$.

1.25. We prove that the limit, denoted by $L(\alpha, \beta)$, equals

$$L(\alpha, \beta) = \begin{cases} \infty & \text{if} \quad 2\alpha > 2\beta - 1, \\ 0 & \text{if} \quad 2\alpha < 2\beta - 1, \\ \int_0^1 \frac{1}{\sqrt{1+x}} dx = 2\sqrt{2} - 2 & \text{if} \quad 2\alpha = 2\beta - 1. \end{cases}$$

We need the following lemma.

Lemma 1.4. *Let $a, b, u \ge 0$ be real numbers and let $v > 0$. Then*

$$\lim_{n \to \infty} \frac{1}{n} \sum_{k=1}^{n} \frac{\left(1 + \frac{k}{n} + \frac{a}{n^2}\right)^u}{\left(1 + \frac{k}{n} + \frac{b}{n^2}\right)^v} = \int_0^1 (1+x)^{u-v} dx.$$

Let

$$x_n = \sum_{k=1}^{n} \frac{(n^2 + kn + a)^\alpha}{(n^2 + kn + b)^\beta} = \frac{n^{2\alpha}}{n^{2\beta-1}} \cdot \frac{1}{n} \sum_{k=1}^{n} \frac{\left(1 + \frac{k}{n} + \frac{a}{n^2}\right)^\alpha}{\left(1 + \frac{k}{n} + \frac{b}{n^2}\right)^\beta}.$$

An application of Lemma 1.4, with $u = \alpha$, $v = \beta$, shows that the desired result holds. Now we prove the lemma.

Proof. Let

$$y_n = \frac{1}{n} \sum_{k=1}^{n} \frac{\left(1 + \frac{k}{n} + \frac{a}{n^2}\right)^u}{\left(1 + \frac{k}{n} + \frac{b}{n^2}\right)^v}.$$

We have

$$y_n = \frac{1}{n} \sum_{k=1}^{n} \frac{\left(1 + \frac{k}{n} + \frac{a}{n^2}\right)^u - \left(1 + \frac{k}{n} + \frac{b}{n^2}\right)^u}{\left(1 + \frac{k}{n} + \frac{b}{n^2}\right)^v} + \frac{1}{n} \sum_{k=1}^{n} \frac{\left(1 + \frac{k}{n} + \frac{b}{n^2}\right)^u}{\left(1 + \frac{k}{n} + \frac{b}{n^2}\right)^v}. \qquad (1.19)$$

Without losing the generality, we assume that $a \geq b$. Thus,

$$0 \leq \frac{1}{n} \sum_{k=1}^{n} \frac{\left(1 + \frac{k}{n} + \frac{a}{n^2}\right)^u - \left(1 + \frac{k}{n} + \frac{b}{n^2}\right)^u}{\left(1 + \frac{k}{n} + \frac{b}{n^2}\right)^v}$$

$$\leq \frac{1}{n} \sum_{k=1}^{n} \left(\left(1 + \frac{k}{n} + \frac{a}{n^2}\right)^u - \left(1 + \frac{k}{n} + \frac{b}{n^2}\right)^u \right)$$

$$= \frac{1}{n} \sum_{k=1}^{n} \left(1 + \frac{k}{n} + \frac{b}{n^2}\right)^u \left(\left(1 + \frac{a-b}{n^2 + kn + b}\right)^u - 1 \right)$$

$$\leq \left(\left(1 + \frac{a-b}{n^2 + b}\right)^u - 1 \right) \frac{1}{n} \sum_{k=1}^{n} \left(1 + \frac{k}{n} + \frac{b}{n^2}\right)^u$$

$$\leq \left(\left(1 + \frac{a-b}{n^2 + b}\right)^u - 1 \right) \frac{1}{n} \sum_{k=1}^{n} \left(1 + \frac{k+1}{n}\right)^u,$$

where the last inequality follows from the fact that for large n one has $n > b$. We have, based on the preceding inequality, that

$$\lim_{n \to \infty} \frac{1}{n} \sum_{k=1}^{n} \frac{\left(1 + \frac{k}{n} + \frac{a}{n^2}\right)^u - \left(1 + \frac{k}{n} + \frac{b}{n^2}\right)^u}{\left(1 + \frac{k}{n} + \frac{b}{n^2}\right)^v} = 0. \qquad (1.20)$$

Also, for large n and for $k = 1, 2, \ldots, n - 1$, we have

$$\frac{k}{n} < \frac{k}{n} + \frac{b}{n^2} < \frac{k+1}{n}.$$

On the other hand, we note that

$$\lim_{n \to \infty} \frac{1}{n} \sum_{k=1}^{n} \frac{\left(1 + \frac{k}{n} + \frac{b}{n^2}\right)^u}{\left(1 + \frac{k}{n} + \frac{b}{n^2}\right)^v} = \int_0^1 \frac{(1+x)^u}{(1+x)^v} dx = \int_0^1 (1+x)^{u-v} dx, \qquad (1.21)$$

since the preceding sum is the Riemann sum of the function $f(x) = (1+x)^{u-v}$ on the interval $[0,1]$ associated with the partition $0, \frac{1}{n}, \ldots, \frac{n-1}{n}, \frac{n}{n}$ and the system of intermediary points $\frac{1}{n} + \frac{b}{n^2}, \frac{2}{n} + \frac{b}{n^2}, \ldots, \frac{n-1}{n} + \frac{b}{n^2}$. Combining (1.19)–(1.21), we get that the lemma is proved.

Remark. The following limits hold

$$L\left(0, \frac{1}{2}\right) = \lim_{n \to \infty} \sum_{k=1}^{n} \frac{1}{\sqrt{n^2 + kn + b}} = 2\sqrt{2} - 2,$$

$$L\left(\frac{1}{2}, 1\right) = \lim_{n \to \infty} \sum_{k=1}^{n} \frac{\sqrt{n^2 + kn + a}}{n^2 + kn + b} = 2\sqrt{2} - 2.$$

1.26. The limit equals $(-1)^n$. Let L_n be the value of the limit. We have, based on Lagrange's Mean Value Theorem, that

$$f_n(x) - f_{n-1}(x) = \exp(\ln f_n(x)) - \exp(\ln f_{n-1}(x))$$
$$= \exp(f_{n-1}(x) \ln x) - \exp(f_{n-2}(x) \ln x)$$
$$= (f_{n-1}(x) - f_{n-2}(x)) \cdot \ln x \cdot \exp(\theta_n(x)),$$

where $\theta_n(x)$ is between $f_{n-1}(x) \ln x$ and $f_{n-2}(x) \ln x$. This implies that $\lim_{x \to 1} \theta_n(x) = 0$. Thus,

$$L_n = \lim_{x \to 1} \frac{f_n(x) - f_{n-1}(x)}{(1-x)^n} = \lim_{x \to 1} \left(\frac{f_{n-1}(x) - f_{n-2}(x)}{(1-x)^{n-1}} \cdot \frac{\ln x}{1-x} \cdot \exp(\theta_n(x)) \right) = -L_{n-1}.$$

It follows, since

$$L_2 = \lim_{x \to 1} \frac{x^x - x}{(1-x)^2} = 1,$$

that $L_n = (-1)^{n-2} L_2 = (-1)^n$, and the problem is solved.

1.27. (a) We need the following lemma.

Lemma 1.5. *Let* $0 < x < \frac{1}{e}$. *The following inequalities hold*

$$x < \frac{1}{\ln \frac{1}{x}} \quad and \quad x^{\frac{1}{e}} = x^{x^{\frac{1}{\ln \frac{1}{x}}}} < \frac{1}{\ln \frac{1}{x}}.$$

Proof. The inequalities can be proved by straightforward calculations.

Now we are ready to solve the problem. Let $(h_k)_{k\geq 1}$ be the sequence of functions defined on $(0,1)$ by the recurrence relation $h_{k+1}(x) = x^{h_k(x)}$, with $h_1(x) = x$. It is easy to check that $h_1(x) = x < x^x = h_2(x) < 1$. It follows that $h_2(x) = x^x > x^{x^x} = h_3(x)$. Continuing this process, we obtain that the following sequence of inequalities hold

$$h_1(x)<h_3(x)<\cdots<h_{2k-1}(x)<h_{2k+1}(x)<\cdots<h_{2k}(x)<h_{2k-2}(x)<\cdots<h_2(x).$$
(1.22)

Thus, the sequence (h_{2k+1}) increases. We show that for each $k \geq 1$, one has that

$$x < h_{2k+1}(x) < \frac{1}{\ln\frac{1}{x}}, \quad 0 < x < \frac{1}{e}.$$
(1.23)

The left inequality holds, based on (1.22), for all $x \in (0,1)$. We prove that the right inequality holds by induction on k. We know, based on the first inequality of Lemma 1.5, that $x < \frac{1}{\ln\frac{1}{x}}$. This implies that $x^x > x^{\frac{1}{\ln\frac{1}{x}}}$, which in turn implies that

$$h_3(x) = x^{x^x} < x^{x^{\frac{1}{\ln\frac{1}{x}}}} = x^{\frac{1}{e}} < \frac{1}{\ln\frac{1}{x}},$$

where the last inequality follows from the second inequality of Lemma 1.5. Thus, (1.23) is verified when $k = 1$. For proving the inductive step, we assume that (1.23) holds for fixed k, i.e.,

$$h_{2k+1}(x) < \frac{1}{\ln\frac{1}{x}}$$
(1.24)

and we show that $h_{2k+3}(x) < \frac{1}{\ln\frac{1}{x}}$. Inequality (1.24) implies that

$$x^{h_{2k+1}(x)} > x^{\frac{1}{\ln\frac{1}{x}}} \Rightarrow h_{2k+3}(x) = x^{x^{h_{2k+1}(x)}} < x^{x^{\frac{1}{\ln\frac{1}{x}}}} = x^{\frac{1}{e}} < \frac{1}{\ln\frac{1}{x}},$$

where the last inequality follows from the lemma, and the induction is completed. Since $\frac{1}{2k+1} < \frac{1}{e}$ for all $k \geq 1$ we obtain from (1.23), with $x = \frac{1}{2k+1}$, that

$$\frac{1}{2k+1} < x_k = h_{2k+1}\left(\frac{1}{2k+1}\right) < \frac{1}{\ln(2k+1)},$$

and the first part of the problem is solved.

(b) We have, based on (1.23), that for all $k \geq 1$, one has that $x < h_{2k-1}(x) < \frac{1}{\ln\frac{1}{x}}$ for all $x \in (0,\frac{1}{e})$. It follows that

$$x^x > x^{h_{2k-1}(x)} > x^{\frac{1}{\ln\frac{1}{x}}}, \quad x \in \left(0,\frac{1}{e}\right).$$
(1.25)

Letting $x = \frac{1}{2k}$ in (1.25), we get, since $\frac{1}{2k} < \frac{1}{e}$ for $k \geq 2$, that

$$\frac{1}{2k}^{\frac{1}{2k}} > y_k = \frac{1}{2k} h_{2k-1}\left(\frac{1}{2k}\right) > \frac{1}{2k}^{\frac{1}{\ln(2k)}}, \quad k \geq 2.$$

On the other hand, it is elementary to prove that $1 - x > x^x$ for all $x \in (0, 1/4)$, and hence, $y_k < \frac{1}{2k}^{\frac{1}{2k}} < 1 - \frac{1}{2k}$.

Now we prove[5] that $y_k > 1 - 1/\ln(2k)$. One can check that the inequality holds for $k = 10$ and we prove it for $k \geq 11$. To do this we show that

$$f(x) = \frac{\ln\left(1 + \frac{1}{\ln x}\right)}{x \ln x} > e, \quad 0 < x \leq \frac{1}{22}. \tag{1.26}$$

A calculation shows that

$$f'(x) = -\frac{1 + (1 + \ln x)^2 \ln\left(1 + \frac{1}{\ln x}\right)}{(1 + \ln x)(x \ln x)^2}.$$

Since $\ln(1 + x) + x^2/(x + 1)^2 < 0$ for $-1/\ln 22 \leq x < 0$ one has that $1 + (1/x + 1)^2 \ln(1 + x) < 0$ for $-1/\ln 22 \leq x < 0$. Thus, $f'(x) < 0$ for $0 < x \leq 1/22$. This implies that $f(x) > f(1/22) > e$ and (1.26) is proved. Let $g_x(a) = x^{x^a}$ and let $h_x(a) = g_x(a) - a$, where $0 < x \leq 1/22$ and $-\infty < a < \infty$. A calculation shows

$$g'_x(a) = x^a \ln^2 x g_x(a) > 0 \quad \text{and} \quad g''_x(a) = x^a \ln^3 x(1 + x^a \ln x) g_x(a).$$

Since $1 + x^a \ln x \geq 1 + x^{1 + 1/\ln x} \ln x = 1 + e x \ln x$, for $a \geq 1 + 1/\ln x$, one has that $g''_x(a) = h''_x(a) < 0$, for $a \geq 1 + 1/\ln x$. Inequality (1.26) is equivalent to $h_x(1 + 1/\ln x) > 0$. On the other hand, $h_x(1) < 0$, and it follows, since $h''_x(a) < 0$, that there is a unique $\alpha_x \in (1 + 1/\ln x, 1)$ such that $g_x(\alpha_x) = \alpha_x$. We note that $\alpha_{1/(2k)} < 1$ and $g'_{1/(2k)}(a) > 0$, which implies that $\alpha_{1/(2k)} = g_{1/(2k)}(\alpha_{1/(2k)}) < g_{1/(2k)}(1)$. Since $\alpha_{1/(2k)} < g_{1/(2k)}(1)$ and $g'_{1/(2k)}(a) > 0$, we get that $\alpha_{1/(2k)} = g_{1/(2k)}(\alpha_{1/(2k)}) < g_{1/(2k)}(g_{1/(2k)}(1))$. Continuing in this way we get the result $1 - 1/\ln(2k) = 1 + 1/\ln(1/(2k)) < \alpha_{1/(2k)} < y_k$.

(c) The limit does not exist (see parts (a) and (b)).

For an alternative solution, of parts (a) and (b), the reader is referred to [37].

Remark. This problem is about studying the behavior of a very special sequence involving iterated exponentials. However, the problem differs significantly, in terms of the results, by the classical problem involving iterated exponentials. Recall that if $a > 0$ is a real number, then the sequence

$$a, \quad a^a, \quad a^{a^a}, \quad a^{a^{a^a}}, \ldots$$

---

[5]This part of the proof is due to Manyama (see [87]).

is called the sequence of *iterated exponentials* generated by a. The question of determining the values of a for which this sequence converges as well as a nice discussion of the history of this problem, who solved it first and how it appeared in the mathematical literature, is given in [4].

1.28. (a) We have

$$x_n = \sum_{i=1}^{n} \sum_{j=1}^{n} \frac{1}{i+j} = \sum_{i=1}^{n} \sum_{j=1}^{n} \int_0^1 x^{i+j-1} dx$$

$$= \int_0^1 x \left(\sum_{j=1}^{n} x^{j-1} \right)^2 dx = \int_0^1 x \left(\frac{1-x^n}{1-x} \right)^2 dx$$

$$= \int_0^1 x(1-x^n)^2 \left(\sum_{m=0}^{\infty} (m+1)x^m \right) dx$$

$$= \sum_{m=0}^{\infty} (m+1) \int_0^1 x^{m+1}(1-2x^n+x^{2n})dx$$

$$= \sum_{m=0}^{\infty} (m+1) \left(\frac{1}{m+2} - \frac{2}{n+m+2} + \frac{1}{2n+m+2} \right)$$

$$= 1 + (2n+1)H_{2n+1} - (2n+2)H_{n+1}.$$

The last equality follows by calculating the limit of the mth partial sum of the series.
(b) We note that

$$x_n - (2n+1)\ln 2 + \ln n = 1 - \gamma + (2n+1)\left(H_{2n+1} - \ln(2n+1) - \gamma\right)$$

$$- (2n+2)\left(H_{n+1} - \ln(n+1) - \gamma\right)$$

$$+ (2n+1)\ln \frac{2n+1}{2n+2} + \ln \frac{n}{n+1},$$

and the result follows since $\lim_{n\to\infty} n(H_n - \ln n - \gamma) = 1/2$.
(c) Part (c) follows from part (a) of the problem (see also the solution of Problem **1.29**, the case when $k = 2$).

1.29. The limit equals

$$\int_0^1 \cdots \int_0^1 \frac{dx_1 dx_2 \cdots dx_k}{x_1 + x_2 + \cdots + x_k}.$$

We have, since $1/a = \int_0^\infty e^{-at}\,dt$, that

$$\int_0^1 \cdots \int_0^1 \frac{dx_1\,dx_2\cdots dx_k}{x_1 + x_2 + \cdots + x_k} = \int_0^1 \cdots \int_0^1 \left(\int_0^\infty e^{-(x_1+x_2+\cdots+x_k)t}\,dt \right) dx_1\,dx_2\cdots dx_k$$

$$= \int_0^\infty \left(\left(\int_0^1 e^{-x_1 t}\,dx_1 \right) \left(\int_0^1 e^{-x_2 t}\,dx_2 \right) \cdots \left(\int_0^1 e^{-x_k t}\,dx_k \right) \right) dt$$

$$= \int_0^\infty \left(\frac{1-e^{-x}}{x} \right)^k dx.$$

Now, since $\frac{1}{x^k} = \frac{1}{(k-1)!} \int_0^\infty e^{-xt} t^{k-1}\,dt$, we obtain that

$$\int_0^\infty \frac{(1-e^{-x})^k}{x^k}\,dx = \frac{1}{(k-1)!} \int_0^\infty (1-e^{-x})^k \left(\int_0^\infty e^{-xt} t^{k-1}\,dt \right) dx$$

$$= \frac{1}{(k-1)!} \int_0^\infty t^{k-1} \left(\int_0^\infty (1-e^{-x})^k e^{-xt}\,dx \right) dt$$

$$= \frac{(-1)^k}{(k-1)!} \int_0^\infty t^{k-1} \left(\int_0^\infty (e^{-x}-1)^k e^{-xt}\,dx \right) dt$$

$$= \frac{(-1)^k}{(k-1)!} \int_0^\infty t^{k-1} \left(\int_0^\infty \sum_{j=0}^k \binom{k}{j} e^{-x(j+t)} (-1)^{k-j}\,dx \right) dt.$$

Since $(-1)^{k-j} = (-1)^{k+j}$, we get that

$$\int_0^\infty \frac{(1-e^{-x})^k}{x^k}\,dx = \frac{1}{(k-1)!} \int_0^\infty t^{k-1} \left(\sum_{j=0}^k (-1)^j \binom{k}{j} \frac{1}{t+j} \right) dt.$$

A calculation shows that

$$\frac{1}{t(t+1)\cdots(t+k)} = \sum_{j=0}^k \frac{a_j}{t+j}, \quad \text{where} \quad a_j = \frac{(-1)^j}{k!} \binom{k}{j},$$

and hence,

$$\int_0^\infty \frac{(1-e^{-x})^k}{x^k}\,dx = k \int_0^\infty \frac{t^{k-2}}{(t+1)(t+2)\cdots(t+k)}\,dt.$$

On the other hand,

$$\frac{t^{k-2}}{(t+1)\cdots(t+k)} = \sum_{j=1}^{k} \frac{b_j}{t+j}, \quad \text{where} \quad b_j = \frac{(-1)^{k-j-1} j^{k-2}}{(k-1)!} \binom{k-1}{j-1},$$

and note that $b_1 + b_2 + \cdots + b_k = 0$. Thus,

$$\int_0^\infty \frac{t^{k-2}}{(t+1)(t+2)\cdots(t+k)} dt = \int_0^\infty \left(\sum_{j=1}^{k} \frac{b_j}{t+j} \right) dt$$

$$= \left. \left(\sum_{j=1}^{k} b_j \ln(t+j) \right) \right|_{t=0}^{t=\infty}$$

$$= - \sum_{j=1}^{k} b_j \ln j,$$

since $\lim_{t\to\infty} \sum_{j=1}^{k} b_j \ln(t+j) = 0$. To prove this, we have, since $b_1 = -b_2 - \cdots - b_k$, that $\lim_{t\to\infty} \sum_{j=1}^{k} b_j \ln(t+j) = \lim_{t\to\infty} \sum_{j=2}^{k} b_j \ln \frac{t+j}{t+1} = 0$. It follows that the limit equals

$$\frac{k}{(k-1)!} \left(\sum_{j=2}^{k} (-1)^{k-j} j^{k-2} \binom{k-1}{j-1} \ln j \right),$$

and the problem is solved.

1.30. Let $S_n = \sum_{l=0}^{n} 1/(a^l + x^k)$ and let $F(n) = \int_0^n dt/(a^t + x^k)$. Using Lagrange's Mean Value Theorem, we get that

$$\frac{1}{a^{n+1} + x^k} \le F(n+1) - F(n) \le \frac{1}{a^n + x^k},$$

and it follows that

$$F(n+1) \le S_n < F(n+1) + \frac{1}{1+x^k} - \frac{1}{a^{n+1} + x^k}.$$

Letting $n \to \infty$ we get that

$$\int_0^\infty \frac{dt}{a^t + x^k} \le \sum_{n=0}^{\infty} \frac{1}{a^n + x^k} \le \int_0^\infty \frac{dt}{a^t + x^k} + \frac{1}{1+x^k}.$$

A calculation, based on the substitution $a^t = y$, shows that

$$\int_0^\infty \frac{dt}{a^t + x^k} = \frac{\ln(1+x^k)}{x^k \ln a}.$$

Therefore, we get that

$$\frac{\ln(1+x^k)}{\ln a \ln x} \le \frac{x^k}{\ln x} \cdot \sum_{n=0}^{\infty} \frac{1}{a^n + x^k} \le \frac{\ln(1+x^k)}{\ln a \ln x} + \frac{x^k}{1+x^k} \cdot \frac{1}{\ln x}. \qquad (1.27)$$

Letting $x \to \infty$ in (1.27), we get that the desired limit holds. Also, it follows from (1.27) that

$$\ln \ln x \left(\frac{\ln(1+x^k)}{\ln a \ln x} - \frac{k}{\ln a} \right) \le \ln \ln x \left(\frac{x^k}{\ln x} \cdot \sum_{n=0}^{\infty} \frac{1}{a^n + x^k} - \frac{k}{\ln a} \right)$$

$$\le \ln \ln x \left(\frac{\ln(1+x^k)}{\ln a \ln x} - \frac{k}{\ln a} \right) + \frac{\ln \ln x}{\ln x} \cdot \frac{x^k}{1+x^k}.$$

Since

$$\lim_{x \to \infty} \ln \ln x \left(\frac{\ln(1+x^k)}{\ln a \ln x} - \frac{k}{\ln a} \right) = \lim_{x \to \infty} \frac{\ln \frac{1+x^k}{x^k} \ln \ln x}{\ln a \ln x} = 0,$$

we get that part (b) of the problem is solved.

1.31. (a) The limit equals 1 when $a = 2$, 0 when $a \in (0,2)$, and ∞ when $a > 2$. We need the following lemma.

Lemma 1.6. *The following limit holds* $\lim_{x \to \infty} 2^x (\zeta(x) - 1) = 1$.

Proof. We note that if $x > 1$, we have $\zeta(x) = \sum_{n=1}^{\infty} \frac{1}{n^x} = 1 + \frac{1}{2^x} + \sum_{n=3}^{\infty} \frac{1}{n^x} > 1 + \frac{1}{2^x}$, and hence,

$$1 < 2^x (\zeta(x) - 1). \qquad (1.28)$$

On the other hand, $\sum_{n=3}^{\infty} \frac{1}{n^x} < \int_2^{\infty} \frac{1}{t^x} dt = \frac{2^{1-x}}{x-1}$, from which it follows that

$$2^x (\zeta(x) - 1) < \frac{x+1}{x-1}. \qquad (1.29)$$

From (1.28) and (1.29) we get that $\lim_{x \to \infty} 2^x (\zeta(x) - 1) = 1$.

Now we are ready to solve the problem. First, we prove that $S_n = \sum_{k=1}^{\infty} \frac{1}{k(k+1)^n} = n - \zeta(2) - \zeta(3) - \cdots - \zeta(n)$. We have, since

$$\frac{1}{k(k+1)^n} = \frac{1}{k(k+1)^{n-1}} - \frac{1}{(k+1)^n},$$

that

$$\sum_{k=1}^{\infty} \frac{1}{k(k+1)^n} = \sum_{k=1}^{\infty} \frac{1}{k(k+1)^{n-1}} - \sum_{k=1}^{\infty} \frac{1}{(k+1)^n},$$

and hence, $S_n = S_{n-1} - (\zeta(n) - 1)$. Iterating this equality we obtain that $S_n = S_1 - (\zeta(2) + \zeta(3) + \cdots + \zeta(n) - (n-1))$ and since $S_1 = \sum_{k=1}^{\infty} 1/(k(k+1)) = 1$, we obtain that $S_n = n - \zeta(2) - \zeta(3) - \cdots - \zeta(n)$.

First we consider the case when $a = 2$. Let $L = \lim_{n \to \infty} 2^n S_n$. Since S_n verifies the recurrence formula $S_n = S_{n-1} - (\zeta(n) - 1)$, it follows that $2^n S_n = 2 \cdot 2^{n-1} S_{n-1} - 2^n(\zeta(n) - 1)$. Letting n tend to ∞ in the preceding equality and using the lemma, we get that $L = 2L - 1$, from which it follows that $L = 1$. If $a < 2$, we have that $L = \lim_{n \to \infty} a^n S_n = \lim_{n \to \infty} 2^n S_n \cdot \lim_{n \to \infty} (a/2)^n = 0$, and if $a > 2$, we get, based on the same reasoning, that $L = \infty$.

(b) The limit equals $1/2$. Use that $\lim_{n \to \infty} 2^n (n - \zeta(2) - \zeta(3) - \cdots - \zeta(n)) = 1$.

1.32. (a) The limit equal $\int_0^1 f(x)dx$. First we note that for all $k = 1, \ldots, n-1$, one has that $\frac{k-1}{n} < \frac{k}{n+1} < \frac{k}{n}$. We have, based on Lagrange's Mean Value Theorem, that

$$
\begin{aligned}
x_{n+1} - x_n &= \sum_{k=1}^{n-1} \left(f\left(\frac{k}{n+1}\right) - f\left(\frac{k}{n}\right) \right) + f\left(\frac{n}{n+1}\right) \\
&= \sum_{k=1}^{n-1} \left(\frac{k}{n+1} - \frac{k}{n} \right) f'(\theta_{k,n}) + f\left(\frac{n}{n+1}\right) \\
&= -\frac{1}{n+1} \sum_{k=1}^{n-1} \frac{k}{n} f'(\theta_{k,n}) + f\left(\frac{n}{n+1}\right) \\
&= -\frac{1}{n+1} \sum_{k=1}^{n-1} \frac{k}{n} \left(f'(\theta_{k,n}) - f'\left(\frac{k}{n}\right) \right) \\
&\quad - \frac{1}{n+1} \sum_{k=1}^{n-1} \frac{k}{n} f'\left(\frac{k}{n}\right) + f\left(\frac{n}{n+1}\right),
\end{aligned}
\tag{1.30}
$$

where $\frac{k}{n+1} < \theta_{k,n} < \frac{k}{n}$. On the other hand, since f' is uniformly continuous, we get that for $\varepsilon > 0$ one has that

$$
\left| \frac{1}{n+1} \sum_{k=1}^{n-1} \frac{k}{n} \left(f'(\theta_{k,n}) - f'\left(\frac{k}{n}\right) \right) \right| \le \frac{1}{n+1} \sum_{k=1}^{n-1} \varepsilon \cdot \frac{k}{n} = \varepsilon \cdot \frac{n-1}{2(n+1)},
$$

and since ε is arbitrary fixed, we get that

$$
\lim_{n \to \infty} \frac{1}{n+1} \sum_{k=1}^{n-1} \frac{k}{n} \left(f'(\theta_{k,n}) - f'(k/n) \right) = 0.
$$

Letting n tend to ∞ in (1.30), we get that

$$
\lim_{n \to \infty} (x_{n+1} - x_n) = -\int_0^1 x f'(x)dx + f(1) = \int_0^1 f(x)dx.
$$

(b) This part is *open*.

1.33. The limit equals

$$\frac{1}{2i} \ln \frac{\Gamma(1-i)}{\Gamma(1+i)} = 0.30164\ 03204\ 67533\ 19788\dots.$$

We have

$$\lim_{n\to\infty}\left(\sum_{k=1}^{n}\arctan 1/k - \ln n\right) = \lim_{n\to\infty}\sum_{k=1}^{n}(\arctan 1/k - 1/k) + \lim_{n\to\infty}\left(\sum_{k=1}^{n}1/k - \ln n\right)$$

$$= \sum_{k=1}^{\infty}(\arctan 1/k - 1/k) + \gamma. \tag{1.31}$$

Since $\arctan x = \sum_{n=0}^{\infty}(-1)^n x^{2n+1}/(2n+1)$ for $-1 < x \le 1$, we get that

$$\sum_{k=1}^{\infty}(\arctan 1/k - 1/k) = \sum_{k=1}^{\infty}\left(\sum_{n=1}^{\infty}\frac{(-1)^n}{2n+1}\left(\frac{1}{k}\right)^{2n+1}\right)$$

$$= \sum_{n=1}^{\infty}\frac{(-1)^n}{2n+1}\left(\sum_{k=1}^{\infty}\left(\frac{1}{k}\right)^{2n+1}\right)$$

$$= \sum_{n=1}^{\infty}\frac{(-1)^n}{2n+1}\zeta(2n+1). \tag{1.32}$$

Now we need the following power series expansion [122, Entry 12, p. 160]:

$$\sum_{k=1}^{\infty}\zeta(2k+1)\frac{t^{2k+1}}{2k+1} = \frac{1}{2}(\ln\Gamma(1-t) - \ln\Gamma(1+t)) - \gamma t, \quad |t| < 1.$$

Letting $t \to i$ in the preceding equality, one has that

$$\sum_{n=1}^{\infty}\frac{(-1)^n}{2n+1}\zeta(2n+1) = \frac{1}{2i}\ln\frac{\Gamma(1-i)}{\Gamma(1+i)} - \gamma. \tag{1.33}$$

Combining (1.31)–(1.33), we get that the problem is solved.
  This problem is due to Mircea Ivan (see [69]).

1.6.2 Limits of Integrals

> *Finally, two days ago, I succeeded - not on account of my hard*
> *efforts, but by the grace of the Lord. Like a sudden flash of*
> *lightning, the riddle was solved. I am unable to say what was*
> *the conducting thread that connected what I previously knew*
> *with what made my success possible.*
>
> Carl Friedrich Gauss (1777–1855)

1.34. The limit equals $\sqrt{2}$. Let $I_n = \int_0^{\pi/2} \sqrt[n]{\sin^n x + \cos^n x}\,dx$. Then,

$$I_n \leq \int_0^{\pi/4} \sqrt[n]{\cos^n x + \cos^n x}\,dx + \int_{\pi/4}^{\pi/2} \sqrt[n]{\sin^n x + \sin^n x}\,dx$$

$$= \sqrt[n]{2}\left(\int_0^{\pi/4} \cos x\,dx + \int_{\pi/4}^{\pi/2} \sin x\,dx\right)$$

$$= \sqrt[n]{2} \cdot \sqrt{2}.$$

On the other hand,

$$I_n \geq \int_0^{\pi/4} \cos x\,dx + \int_{\pi/4}^{\pi/2} \sin x\,dx = \sqrt{2}.$$

Thus, $\sqrt{2} \leq I_n \leq \sqrt[n]{2} \cdot \sqrt{2}$, and the result follows.

Remark. More generally, one can prove that if $f, g : [a,b] \to [0,\infty)$ are continuous functions, then

$$\lim_{n\to\infty} \int_a^b \sqrt[n]{f^n(x) + g^n(x)}\,dx = \int_a^b h(x)\,dx,$$

where $h(x) = \max\{f(x), g(x)\}$.

1.35. The limit equals $2\Gamma(1 + 1/\alpha)$. We need the following lemma.

Lemma 1.7. *Let $k \in (0,1]$ be a fixed real number. Then,*

$$\lim_{n\to\infty} n^{\frac{1}{\alpha}} \int_0^k (1 - y^\alpha)^n\,dy = \Gamma\left(1 + \frac{1}{\alpha}\right).$$

Proof. First we consider the case when $k = 1$. We have, based on the substitution $y^\alpha = t$, that

$$n^{\frac{1}{\alpha}} \int_0^1 (1 - y^\alpha)^n\,dy = n^{\frac{1}{\alpha}} \int_0^1 (1 - t)^n \frac{1}{\alpha} t^{\frac{1}{\alpha}-1}\,dt = \frac{n^{\frac{1}{\alpha}}}{\alpha} B(n+1, 1/\alpha)$$

$$= \frac{n^{\frac{1}{\alpha}}}{\alpha} \cdot \frac{\Gamma(n+1) \cdot \Gamma\left(\frac{1}{\alpha}\right)}{\Gamma\left(n+1+\frac{1}{\alpha}\right)} = \frac{1}{\alpha}\Gamma\left(\frac{1}{\alpha}\right) \cdot n^{\frac{1}{\alpha}} \frac{\Gamma(n+1)}{\Gamma\left(n+1+\frac{1}{\alpha}\right)}.$$

It follows, in view of Stirling's formula, that

$$\lim_{n\to\infty} n^{\frac{1}{\alpha}} \frac{\Gamma(n+1)}{\Gamma\left(n+1+\frac{1}{\alpha}\right)} = 1,$$

and hence,

$$\lim_{n\to\infty} n^{\frac{1}{\alpha}} \int_0^1 (1-y^\alpha)^n \, dy = \frac{1}{\alpha}\Gamma\left(\frac{1}{\alpha}\right) = \Gamma\left(1+\frac{1}{\alpha}\right).$$

Now we consider the case when $k \in (0,1)$. We have that

$$n^{\frac{1}{\alpha}} \int_0^k (1-y^\alpha)^n \, dy = n^{\frac{1}{\alpha}} \int_0^1 (1-y^\alpha)^n \, dy - n^{\frac{1}{\alpha}} \int_k^1 (1-y^\alpha)^n \, dy,$$

and we note that it suffices to prove

$$\lim_{n\to\infty} n^{\frac{1}{\alpha}} \int_k^1 (1-y^\alpha)^n \, dy = 0.$$

We have, since $k \leq y \leq 1$, that

$$0 < n^{\frac{1}{\alpha}} \int_k^1 (1-y^\alpha)^n \, dy < n^{\frac{1}{\alpha}} (1-k)(1-k^\alpha)^n.$$

Since $\lim_{n\to\infty} n^{1/\alpha}(1-k^\alpha)^n = 0$, it follows, in view of the Squeeze Theorem, that

$$\lim_{n\to\infty} n^{\frac{1}{\alpha}} \int_k^1 (1-y^\alpha)^n \, dy = 0,$$

and the lemma is proved.

Now we are ready to solve the problem. Using the substitution $t - x = y$, we obtain that

$$n^{\frac{1}{\alpha}} \int_0^1 (1-(t-x)^\alpha)^n \, dt = n^{\frac{1}{\alpha}} \int_{-x}^{1-x} (1-y^\alpha)^n \, dy$$

$$= n^{\frac{1}{\alpha}} \int_{-x}^0 (1-y^\alpha)^n \, dy + n^{\frac{1}{\alpha}} \int_0^{1-x} (1-y^\alpha)^n \, dy$$

$$= n^{\frac{1}{\alpha}} \int_0^x (1-y^\alpha)^n \, dy + n^{\frac{1}{\alpha}} \int_0^{1-x} (1-y^\alpha)^n \, dy,$$

and the result follows as an application of Lemma 1.7.

1.36. The limit equals $\frac{b-a}{c}$. Since $\frac{1}{c+\sin^2 x \sin^2(x+1)\cdots\sin^2(x+n)} \leq \frac{1}{c}$, we get that

$$\overline{\lim_{n\to\infty}} \int_a^b \frac{dx}{c+\sin^2 x \cdot \sin^2(x+1)\cdots\sin^2(x+n)} \leq \frac{b-a}{c}. \tag{1.34}$$

An application of the Arithmetic Mean–Geometric Mean Inequality shows that

$$\sin^2 x \cdot \sin^2(x+1) \cdots \sin^2(x+n) \leq \left(\frac{\sin^2 x + \sin^2(x+1) + \cdots + \sin^2(x+n)}{n+1} \right)^{n+1}.$$

On the other hand,

$$\sin^2 x + \sin^2(x+1) + \cdots + \sin^2(x+n) = \frac{n+1}{2} - \frac{\sin(n+1)\cos(n+2x)}{2\sin 1}.$$

It follows that

$$\int_a^b \frac{dx}{c + \sin^2 x \cdot \sin^2(x+1) \cdots \sin^2(x+n)} \geq \int_a^b \frac{dx}{c + \left(\frac{1}{2} - \frac{\sin(n+1)\cos(n+2x)}{2(n+1)\sin 1} \right)^{n+1}}.$$

Let

$$f_n(x) = \frac{1}{c + \left(\frac{1}{2} - \frac{\sin(n+1)\cos(n+2x)}{2(n+1)\sin 1} \right)^{n+1}}.$$

We note that, for fixed $x \in [a,b]$, one has $\lim_{n \to \infty} f_n(x) = 1/c$, since

$$0 < \left| \left(\frac{1}{2} - \frac{\sin(n+1)\cos(n+2x)}{2(n+1)\sin 1} \right)^{n+1} \right| \leq \left(\frac{1}{2} + \frac{1}{2(n+1)\sin 1} \right)^{n+1} \to 0.$$

We have, based on Fatou's lemma, that

$$\underline{\lim} \int_a^b \frac{dx}{c + \sin^2 x \cdots \sin^2(x+n)} \geq \underline{\lim} \int_a^b \frac{dx}{c + \left(\frac{1}{2} - \frac{\sin(n+1)\cos(n+2x)}{2(n+1)\sin 1} \right)^{n+1}}$$

$$\geq \int_a^b \left(\lim_{n \to \infty} f_n(x) \right) dx$$

$$= \frac{b-a}{c}. \tag{1.35}$$

Combining (1.34) and (1.35), we get that the problem is solved.

1.37. We prove that $L = 1/(k+1)$. Let $f : [0,1] \to \mathbb{R}$ be given by $f(x) = \ln(1+x)$. Since f is concave we get that $f(x^{n+1}) = f(x \cdot x^n + (1-x) \cdot 0) \geq x f(x^n) + (1-x) f(0)$, and hence, $\ln(1 + x^{n+1}) \geq x \ln(1 + x^n)$. Iterating this inequality we get that $\ln(1 + x^{n+k}) \geq x^k \ln(1 + x^n)$, and hence

$$\int_0^1 \frac{\ln(1+x^{n+k})}{\ln(1+x^n)} dx \geq \int_0^1 x^k dx = \frac{1}{k+1}.$$

On the other hand, since $\ln(1+x) \leq x$, we obtain that

$$\int_0^1 \frac{\ln(1+x^{n+k})}{\ln(1+x^n)} dx \leq \int_0^1 \frac{x^{n+k}}{\ln(1+x^n)} dx.$$

Let $g_n(x) = x^{n+k}/\ln(1+x^n)$, $x \in [0,1]$. Since $\ln(1+x) \geq x/2$ for $x \in [0,1]$, we get that $0 \leq g_n(x) \leq 2x^k \leq 2$ for all $x \in [0,1]$ and all $n \in \mathbb{N}$. On the other hand, since $\lim_{n\to\infty} g_n(x) = x^k$ for $x \in [0,1)$ we get, based on the Bounded Convergence Theorem, that

$$\lim_{n\to\infty} \int_0^1 \frac{x^{n+k}}{\ln(1+x^n)}dx = \int_0^1 x^k dx = \frac{1}{k+1}.$$

The second limit equals 0. We have

$$x_n = n\left(\int_0^1 \frac{\ln(1+x^{n+k})}{\ln(1+x^n)}dx - \frac{1}{k+1}\right) = n\left(\int_0^1 \frac{\ln(1+x^{n+k})}{\ln(1+x^n)}dx - \int_0^1 x^k dx\right)$$

$$= \int_0^1 n\frac{\ln(1+x^{n+k}) - x^k\ln(1+x^n)}{\ln(1+x^n)}dx \stackrel{x^n=y}{=} \int_0^1 \frac{\ln(1+y^{\frac{n+k}{n}}) - y^{\frac{k}{n}}\ln(1+y)}{y\ln(1+y)}y^{\frac{1}{n}}dy.$$

We note that, for $x \in [0,1]$, the following inequalities hold

$$x - \frac{x^2}{2} \leq \ln(1+x) \leq x - \frac{x^2}{2} + \frac{x^3}{3}.$$

Thus,

$$\ln(1+x^{n+k}) - x^k\ln(1+x^n) \leq x^{n+k} - \frac{x^{2n+2k}}{2} + \frac{x^{3n+3k}}{3} - x^k\left(x^n - \frac{x^{2n}}{2}\right)$$

$$= x^{2n+k}\left(\frac{1}{2}(1-x^k) + \frac{x^{n+2k}}{3}\right)$$

$$\leq x^{2n}\left(\frac{1}{2} + \frac{1}{3}\right)$$

$$= \frac{5}{6}x^{2n}.$$

This implies, when $x^n = y$, that $\ln(1+y^{(n+k)/n}) - y^{k/n}\ln(1+y) \leq 5y^2/6$.

Let $f_n(y) = (\ln(1+y^{(n+k)/n}) - y^{k/n}\ln(1+y))y^{1/n}/(y\ln(1+y))$, for $y \in (0,1]$. It follows that $0 < f_n(y) < 5y/(6\ln(1+y))$, $y \in (0,1]$. On the other hand, for $y \in (0,1]$, one has that $\lim_{n\to\infty} f_n(y) = 0$, and the result follows from Lebesgue Convergence Theorem.

Remark. The problem can be solved in a greater generality: If $f : [0,1] \to \mathbb{R}$ is a continuous function, then

$$L(f) = \lim_{n\to\infty} \int_0^1 \frac{\ln(1+x^{n+k})}{\ln(1+x^n)}f(x)dx = \int_0^1 x^k f(x)dx$$

and

$$\lim_{n\to\infty} n\left(\int_0^1 \frac{\ln(1+x^{n+k})}{\ln(1+x^n)}f(x)dx - L(f)\right) = 0.$$

1.38. (a) Let $\varepsilon > 0$. A calculation shows that

$$\int_\varepsilon^1 \frac{f(ax) - f(bx)}{x}dx = \int_\varepsilon^1 \frac{f(ax)}{x}dx - \int_\varepsilon^1 \frac{f(bx)}{x}dx$$

$$= \int_{a\varepsilon}^a \frac{f(t)}{t}dt - \int_{b\varepsilon}^b \frac{f(t)}{t}dt$$

$$= \int_{a\varepsilon}^{b\varepsilon} \frac{f(t)}{t}dt - \int_a^b \frac{f(t)}{t}dt$$

$$\overset{(*)}{=} f(c)\int_{a\varepsilon}^{b\varepsilon} \frac{dt}{t} - \int_a^b \frac{f(t)}{t}dt$$

$$= f(c)\ln\frac{b}{a} - \int_a^b \frac{f(t)}{t}dt,$$

where $c \in [a\varepsilon, b\varepsilon]$ and the equality $(*)$ follow in view of the First Mean Value Theorem for Integral. Thus,

$$\int_0^1 \frac{f(ax) - f(bx)}{x}dx = \lim_{\varepsilon\to 0}\int_\varepsilon^1 \frac{f(ax) - f(bx)}{x}dx = f(0)\ln\frac{b}{a} - \int_a^b \frac{f(t)}{t}dt.$$

(b) The limit equals

$$\begin{cases} \ln b/a - \int_{[a,b]\cap f^{-1}(\{1\})} dx/x & \text{if } f(0) = 1, \\ -\int_{[a,b]\cap f^{-1}(\{1\})} dx/x & \text{if } f(0) < 1. \end{cases}$$

We have, based on part (a), that

$$\int_0^1 \frac{f^n(ax) - f^n(bx)}{x}dx = f^n(0)\ln\frac{b}{a} - \int_a^b \frac{f^n(t)}{t}dt.$$

Let $g_n(x) = f^n(x)/x$, for $x \in [a,b]$, and we note that $|g_n(x)| \le 1/a$. On the other hand, for $x \in [a,b]$, the limit function equals

$$g(x) = \lim_{n\to\infty} g_n(x) = \begin{cases} 0 & \text{for } x \text{ with } -1 < f(x) < 1, \\ 1/x & \text{for } x \text{ with } f(x) = 1. \end{cases}$$

Now the result follows in view of the Bounded Convergence Theorem.

Remark. We note that $\int_{[a,b]\cap f^{-1}(\{1\})} dx/x = \mu([a,b]\cap f^{-1}(\{1\}))$, where μ is the measure defined by $d\mu = dx/x$. Also, if f is a function such that the set $f^{-1}(\{1\})$ is a discrete set, then the limit equals either 0 or $\ln b/a$ according to whether $f(0) < 1$ or $f(0) = 1$.

1.39. (a) For a solution to this part of the problem, see [32].
(b) Use the fact that any continuous function defined on a compact set can be uniformly approximated by a polynomial.

1.40. The limit equals $\frac{\pi}{2}f(0)$. We need the following two lemmas.

Lemma 1.8. *Let n be a nonnegative integer. Then,*

$$\int_0^{\frac{\pi}{2}} \frac{\sin^2 nx}{\sin^2 x}dx = \frac{n\pi}{2}.$$

Proof. Let

$$a_n = \int_0^{\frac{\pi}{2}} \frac{\sin^2 nx}{\sin^2 x}dx.$$

We have

$$a_n - a_{n-1} = \frac{1}{2}\int_0^{\frac{\pi}{2}} \frac{\cos(2n-2)x - \cos 2nx}{\sin^2 x}dx = \int_0^{\frac{\pi}{2}} \frac{\sin(2n-1)x}{\sin x}dx$$

and

$$(a_n - a_{n-1}) - (a_{n-1} - a_{n-2}) = \int_0^{\frac{\pi}{2}} \frac{\sin(2n-1)x - \sin(2n-3)x}{\sin x}dx$$

$$= 2\int_0^{\frac{\pi}{2}} \cos(2n-2)xdx$$

$$= 0.$$

It follows that $a_n - 2a_{n-1} + a_{n-2} = 0$. Thus, $a_n = \alpha + \beta n$, and it follows, since $a_0 = 0$ and $a_1 = \pi/2$, that $\alpha = 0$ and $\beta = \pi/2$. Hence, $a_n = \pi n/2$, and the lemma is proved.

Lemma 1.9. *Let $k \geq 0$ be a nonnegative integer. Then,*

$$\lim_{n\to\infty} \frac{1}{n}\int_0^{\frac{\pi}{2}} \frac{\sin^2 nx}{\sin^2 x}x^k dx = \begin{cases} \frac{\pi}{2} & \text{if } k = 0, \\ 0 & \text{if } k \geq 1. \end{cases}$$

Proof. The case when $k = 0$ follows from Lemma 1.8. We focus on the case when $k \geq 1$. We have, based on Stolz–Cesàro lemma, that

$$\lim_{n\to\infty} \frac{1}{n}\int_0^{\frac{\pi}{2}} \frac{\sin^2 nx}{\sin^2 x}x^k dx = \lim_{n\to\infty} \frac{\int_0^{\frac{\pi}{2}} \frac{\sin^2 nx}{\sin^2 x}x^k dx}{n}$$

$$= \lim_{n\to\infty} \int_0^{\frac{\pi}{2}} \frac{\sin^2(n+1)x - \sin^2 nx}{\sin^2 x}x^k dx$$

$$= \lim_{n\to\infty} \int_0^{\frac{\pi}{2}} \sin(2n+1)x\left(\frac{x^k}{\sin x}\right)dx.$$

When $k \geq 1$, the function $x \to x^k/\sin x$ is integrable over $[0, \pi/2]$, and we get, based on the Riemann–Lebesgue Lemma, that

$$\lim_{n \to \infty} \int_0^{\frac{\pi}{2}} \sin(2n+1)x \left(\frac{x^k}{\sin x}\right) dx = 0.$$

The preceding limit, which can also be calculated by using integration by parts, is a particular case of the following version of the celebrated Riemann–Lebesgue Lemma:

If $f : [a,b] \to \mathbb{R}$ is an integrable function, then

$$\lim_{n \to \infty} \int_a^b f(x) \sin(2n+1)x \, dx = 0.$$

Now we are ready to solve the problem. Let $\varepsilon > 0$ and let $P(x) = \sum_k a_k x^k$ be the polynomial that uniformly approximates f, that is, $|f(x) - P(x)| < \varepsilon$ for all $x \in [0, \pi/2]$. We have

$$\left| \frac{1}{n} \int_0^{\frac{\pi}{2}} \frac{\sin^2 nx}{\sin^2 x} f(x) dx - \frac{1}{n} \int_0^{\frac{\pi}{2}} \frac{\sin^2 nx}{\sin^2 x} P(x) dx \right| \leq \frac{1}{n} \int_0^{\frac{\pi}{2}} \frac{\sin^2 nx}{\sin^2 x} |f(x) - P(x)| dx$$

$$\leq \varepsilon \frac{\pi}{2},$$

and in the limit,

$$\left| \lim_{n \to \infty} \frac{1}{n} \int_0^{\frac{\pi}{2}} \frac{\sin^2 nx}{\sin^2 x} f(x) dx - \lim_{n \to \infty} \frac{1}{n} \int_0^{\frac{\pi}{2}} \frac{\sin^2 nx}{\sin^2 x} P(x) dx \right| \leq \varepsilon \frac{\pi}{2}. \qquad (1.36)$$

On the other hand, we have, based on Lemma 1.9, that

$$\lim_{n \to \infty} \frac{1}{n} \int_0^{\frac{\pi}{2}} \frac{\sin^2 nx}{\sin^2 x} P(x) dx = \lim_{n \to \infty} \frac{1}{n} \int_0^{\frac{\pi}{2}} \frac{\sin^2 nx}{\sin^2 x} \left(\sum_k a_k x^k\right) dx$$

$$= \sum_k a_k \left(\lim_{n \to \infty} \frac{1}{n} \int_0^{\frac{\pi}{2}} \frac{\sin^2 nx}{\sin^2 x} x^k dx \right)$$

$$= a_0 \frac{\pi}{2}. \qquad (1.37)$$

Combining (1.36) and (1.37), we get that

$$\left| \lim_{n \to \infty} \frac{1}{n} \int_0^{\frac{\pi}{2}} \frac{\sin^2 nx}{\sin^2 x} f(x) dx - a_0 \frac{\pi}{2} \right| \leq \varepsilon \frac{\pi}{2}.$$

Letting $\varepsilon \to 0$, we obtain, since $a_0 \to f(0)$, that the desired result follows and the problem is solved.

1.41. The limit equals $f(1)/k$. Let

$$y_n = \int_0^1 \left(x + 2^k x^2 + 3^k x^3 + \cdots + n^k x^n \right) f(x) dx.$$

We have, based on Stolz–Cesàro lemma (the ∞/∞ case), that

$$\lim_{n \to \infty} \frac{1}{n^k} \int_0^1 \left(x + 2^k x^2 + 3^k x^3 + \cdots + n^k x^n \right) f(x) dx = \lim_{n \to \infty} \frac{y_{n+1} - y_n}{(n+1)^k - n^k}$$

$$= \lim_{n \to \infty} \frac{1}{(n+1)^k - n^k} \int_0^1 (n+1)^k x^{n+1} f(x) dx$$

$$= \lim_{n \to \infty} \frac{(n+1)^{k-1}}{(n+1)^k - n^k} \cdot \lim_{n \to \infty} \int_0^1 (n+1) x^{n+1} f(x) dx.$$

It is elementary to prove that, for $k \geq 1$, one has

$$\lim_{n \to \infty} \frac{(n+1)^{k-1}}{(n+1)^k - n^k} = \frac{1}{k}.$$

Now we prove that

$$\lim_{n \to \infty} \int_0^1 n x^n f(x) dx = f(1).$$

Since f is continuous at 1 we get that for $\varepsilon > 0$, there is $\delta \in (0,1)$ such that $f(1) - \varepsilon < f(x) < f(1) + \varepsilon$ for $x \in (\delta, 1]$. We have

$$\int_0^1 n x^n f(x) dx = \int_0^\delta n x^n f(x) dx + \int_\delta^1 n x^n f(x) dx.$$

On the other hand,

$$0 < \left| \int_0^\delta n x^n f(x) dx \right| \leq \delta^{n+1} \cdot \frac{n}{n+1} \cdot ||f||_\infty$$

and

$$\frac{n}{n+1} (1 - \delta^{n+1})(f(1) - \varepsilon) < \int_\delta^1 n x^n f(x) dx \leq \frac{n}{n+1} (1 - \delta^{n+1})(f(1) + \varepsilon).$$

Passing to the limit, as n tends to ∞, in the preceding relations, we get that $f(1) - \varepsilon \leq \lim_{n \to \infty} \int_0^1 n x^n f(x) dx \leq f(1) + \varepsilon$. Since ε is arbitrary fixed, the result follows.

For an alternative solution, see [63].

1.42. The limit equals $f(1)$. First, we note that since f is an integrable function, it is also bounded; hence, there is $M > 0$ such that $|f(x)| \leq M$, for all $x \in [0,1]$. Also, we note that we have an indeterminate form of type $0 \cdot \infty$ since

$$0 \leq \left| \int_0^1 \left(\sum_{k=n}^\infty \frac{x^k}{k} \right) f(x) dx \right| \leq M \sum_{k=n}^\infty \int_0^1 \frac{x^k}{k} dx = M \sum_{k=n}^\infty \frac{1}{k(k+1)} = \frac{M}{n},$$

and it follows that

$$\lim_{n \to \infty} \int_0^1 \left(\sum_{k=n}^\infty \frac{x^k}{k} \right) f(x)dx = 0.$$

Now we apply Stolz–Cesàro lemma (the $0/0$ case) with

$$a_n = \int_0^1 \left(\sum_{k=n}^\infty \frac{x^k}{k} \right) f(x)dx \quad \text{and} \quad b_n = \frac{1}{n}$$

and we have

$$\lim_{n \to \infty} n \int_0^1 \left(\sum_{k=n}^\infty \frac{x^k}{k} \right) f(x)dx = \lim_{n \to \infty} \frac{\int_0^1 \left(\sum_{k=n}^\infty x^k/k \right) f(x)dx}{1/n}$$

$$= \lim_{n \to \infty} \frac{\int_0^1 \left(\sum_{k=n+1}^\infty x^k/k \right) f(x)dx - \int_0^1 \left(\sum_{k=n}^\infty x^k/k \right) f(x)dx}{1/(n+1) - 1/n}$$

$$= \lim_{n \to \infty} (n+1) \int_0^1 x^n f(x)dx$$

$$= f(1).$$

For proving the last equality, see the solution of Problem **1.41**.

Remark. One can prove, by the same method, that if $p > 0$ is a real number and f is an integrable function which is continuous at 1, then

$$\lim_{n \to \infty} n^p \int_0^1 \left(\sum_{k=n}^\infty \frac{x^k}{k^p} \right) f(x)dx = \frac{f(1)}{p}.$$

1.43. The limit equals $2\ln 2 f(1)$. Use a method similar to the technique used in the previous problem.

1.44. The limit equals $\frac{f(0)\pi}{b\sin(\pi/b)}$. Making the substitution $x = (y/n^a)^{1/b}$, we obtain that

$$n^{\frac{a}{b}} \int_0^1 \frac{f(x)}{1+n^a x^b}dx = \frac{1}{b} \int_0^{n^a} \frac{y^{1/b-1}}{1+y} f\left(\left(\frac{y}{n^a}\right)^{1/b} \right) dy$$

$$= \frac{1}{b} \int_0^\infty \frac{y^{1/b-1}}{1+y} f\left(\left(\frac{y}{n^a}\right)^{1/b} \right) \chi_n(y)dy,$$

where

$$\chi_n(y) = \begin{cases} 1 & \text{if } y \le n^a, \\ 0 & \text{if } y > n^a. \end{cases}$$

Let $f_n : [0, \infty) \to \mathbb{R}$ be the sequence of functions defined by

$$f_n(y) = \frac{y^{1/b-1}}{1+y} f\left(\left(\frac{y}{n^a}\right)^{1/b}\right) \chi_n(y),$$

and note that $\lim_{n\to\infty} f_n(y) = y^{1/b-1}/(1+y) \cdot f(0)$ and $f_n(y)$ is bounded by an integrable function

$$|f_n(y)| \leq \frac{y^{1/b-1}}{1+y} \|f\|_\infty.$$

It follows, based on Lebesgue Convergence Theorem, that

$$\lim_{n\to\infty} \int_0^\infty \frac{y^{1/b-1}}{1+y} f\left(\left(\frac{y}{n^a}\right)^{1/b}\right) \chi_n(y) dy = f(0) \int_0^\infty \frac{y^{1/b-1}}{1+y} dy$$

$$= f(0) \mathrm{B}\left(\frac{1}{b}, 1 - \frac{1}{b}\right)$$

$$= f(0) \frac{\pi}{\sin \pi/b}.$$

1.45. The limit equals $f(0)$. Without losing the generality, we consider that $f(x) \geq 1$ for all $x \in [0, 1]$. Otherwise, we replace f by f/m where $m = \inf_{x\in[0,1]} f(x) \neq 0$. Let L be the value of the limit in question. First, we note that we have an indeterminate form of type 1^∞. To see this, we prove that

$$\lim_{n\to\infty} \int_0^1 g(x) \sqrt[n]{f(x^n)} dx = 1.$$

Let $u_n : [0, 1] \to \mathbb{R}$ be the sequence of functions defined by $u_n(x) = g(x) \sqrt[n]{f(x^n)}$. Let $M_f = \sup_{x\in[0,1]} f(x)$ and $M_g = \sup_{x\in[0,1]} g(x)$ and we observe that $u_n(x) \leq M_g \sqrt[n]{M_f} < M_g \sup_{n\in\mathbb{N}} \sqrt[n]{M_f} < \infty$. It follows that for all $n \in \mathbb{N}$ and all $x \in [0, 1]$, the sequence $(u_n(x))$ is bounded. On the other hand, since f is continuous at 0, we get that the limit function equals

$$u(x) = \lim_{n\to\infty} u_n(x) = \begin{cases} g(1) & \text{if } x = 1, \\ g(x) & \text{if } x \in [0, 1). \end{cases}$$

Using the Bounded Convergence Theorem, we get that

$$\lim_{n\to\infty} \int_0^1 g(x) \sqrt[n]{f(x^n)} dx = \lim_{n\to\infty} \int_0^1 u_n(x) dx = \int_0^1 \left(\lim_{n\to\infty} u_n(x)\right) dx = \int_0^1 g(x) dx = 1.$$

Now we calculate the limit. We have

$$L = \exp\left(\lim_{n\to\infty} n\left(\int_0^1 g(x)e^{\frac{1}{n}\ln f(x^n)}dx - 1\right)\right)$$

$$= \exp\left(\lim_{n\to\infty} n\left(\int_0^1 g(x)e^{\frac{1}{n}\ln f(x^n)}dx - \int_0^1 g(x)dx\right)\right)$$

$$= \exp\left(\lim_{n\to\infty} n\int_0^1 g(x)\left(e^{\frac{1}{n}\ln f(x^n)} - 1\right)dx\right)$$

$$= \exp\left(\lim_{n\to\infty} \int_0^1 g(x)\ln f(x^n)e^{\theta_n(x)}dx\right),$$

since

$$\exp\left(\frac{1}{n}\ln f(x^n)\right) - 1 = \frac{1}{n}\ln f(x^n)\exp\left(\theta_n(x)\right),$$

where $0 \le \theta_n(x) \le (\ln f(x^n))/n$.

Let $h_n : [0,1] \to \mathbb{R}$ be the sequence of functions defined by

$$h_n(x) = g(x)\ln f(x^n)\exp(\theta_n(x)).$$

We note that $0 \le \ln f(x^n) \le \ln M_f$, and for large n, i.e., $n \ge n_0$, one has $0 \le \theta_n(x) \le (\ln M_f)/n \le 1$. Thus, for all $x \in [0,1]$ and all $n \ge n_0$, we have that $|h_n(x)| \le e \cdot M_g \ln M_f$. On the other hand, the limit function equals

$$h(x) = \lim_{n\to\infty} h_n(x) = \begin{cases} g(1)\ln f(1) & \text{if } x = 1, \\ g(x)\ln f(0) & \text{if } x \in [0,1). \end{cases}$$

An application of the Bounded Convergence Theorem shows that

$$\lim_{n\to\infty}\int_0^1 h_n(x)dx = \int_0^1 \left(\lim_{n\to\infty} h_n(x)\right)dx = \int_0^1 (g(x)\ln f(0))dx = \ln f(0),$$

and hence $L = f(0)$.

1.46. (a) The limit equals $f(0)\sqrt{\pi/2}$. We need the following lemma.

Lemma 1.10. *Let* $p \ge 0$ *be an integer. Then,*

$$\lim_{n\to\infty}\sqrt{n}\int_0^{\pi/2}\sin^n x\cos^p x\,dx = \begin{cases} \sqrt{\frac{\pi}{2}} & \text{if } p = 0, \\ 0 & \text{if } p \ge 1. \end{cases}$$

Proof. We have that

$$\sqrt{n}\int_0^{\pi/2}\sin^n x\cos^p x\,dx=\frac{\sqrt{n}}{2}B\left(\frac{n+1}{2},\frac{p+1}{2}\right)=\frac{\sqrt{n}}{2}\cdot\frac{\Gamma\left(\frac{n+1}{2}\right)\Gamma\left(\frac{p+1}{2}\right)}{\Gamma\left(\frac{n+p+2}{2}\right)},$$

and the result follows from Stirling's formula.

Now we are ready to solve the problem. Let $\varepsilon>0$ and let $P=\sum_k a_k x^k$ be the polynomial that uniformly approximates f, i.e., $P(x)-\varepsilon<f(x)<P(x)+\varepsilon$, for all $x\in[0,1]$. It follows that

$$\sqrt{n}\int_0^{\pi/2}\sin^n x(P(\cos x)-\varepsilon)dx<\sqrt{n}\int_0^{\pi/2}\sin^n xf(\cos x)dx$$

$$<\sqrt{n}\int_0^{\pi/2}\sin^n x(P(\cos x)+\varepsilon)dx.\qquad(1.38)$$

On the other hand,

$$\sqrt{n}\int_0^{\pi/2}\sin^n x(P(\cos x)\pm\varepsilon)dx=\sum_k a_k\sqrt{n}\int_0^{\pi/2}\sin^n x\cos^k x\,dx\pm\varepsilon\sqrt{n}\int_0^{\pi/2}\sin^n x\,dx.$$

It follows from Lemma 1.10 that

$$\lim_{n\to\infty}\sqrt{n}\int_0^{\pi/2}\sin^n x(P(\cos x)\pm\varepsilon)dx=\sum_k a_k\left(\lim_{n\to\infty}\sqrt{n}\int_0^{\pi/2}\sin^n x\cos^k x\,dx\right)$$

$$\pm\varepsilon\lim_{n\to\infty}\sqrt{n}\int_0^{\pi/2}\sin^n x\,dx$$

$$=(a_0\pm\varepsilon)\sqrt{\frac{\pi}{2}}.\qquad(1.39)$$

Letting n tend to ∞ in (1.38) and using (1.39), we get that

$$(a_0-\varepsilon)\sqrt{\frac{\pi}{2}}\le\lim_{n\to\infty}\sqrt{n}\int_0^{\pi/2}\sin^n xf(\cos x)dx\le(a_0+\varepsilon)\sqrt{\frac{\pi}{2}}.$$

Letting $\varepsilon\to0^+$ in the preceding inequality and using that $a_0=f(0)$, we get that

$$\lim_{n\to\infty}\sqrt{n}\int_0^{\pi/2}\sin^n xf(\cos x)dx=f(0)\sqrt{\frac{\pi}{2}}.$$

(b) The limit equals $\sqrt{\frac{\pi}{2}}g\left(\frac{\pi}{2}\right)$. Let $f:[0,1]\to\mathbb{R}$ be the function defined by $f(x)=g(\arccos x)$. We have, based on part (a), that the limit equals

$$\sqrt{\frac{\pi}{2}}f(0)=\sqrt{\frac{\pi}{2}}g(\arccos 0)=\sqrt{\frac{\pi}{2}}g\left(\frac{\pi}{2}\right).$$

For an alternative solution, see [137].

Remark. More generally, one can prove that if $f : [0, \pi/2] \to \mathbb{R}$ is a bounded function such that $\lim_{x \to \pi/2} f(x) = L$, then

$$\lim_{n \to \infty} \sqrt{n} \int_0^{\pi/2} \sin^n x f(x) \mathrm{d}x = L \cdot \sqrt{\pi/2}.$$

1.47. The limit equals $M = \sup_{x \in [a,b]} f(x)$. We have $\int_a^b f^{n+1}(x) \mathrm{d}x \leq M \int_a^b f^n(x) \mathrm{d}x$, and hence

$$\frac{\int_a^b f^{n+1}(x) \mathrm{d}x}{\int_a^b f^n(x) \mathrm{d}x} \leq M. \tag{1.40}$$

On the other hand, we have, based on Hölder's inequality,[6] with $p = (n+1)/n$ and $q = n+1$, that

$$\int_a^b f^n(x) \mathrm{d}x \leq \left(\int_a^b f^{n+1}(x) \mathrm{d}x \right)^{n/(n+1)} \cdot \left(\int_a^b \mathrm{d}x \right)^{1/(n+1)},$$

and hence,

$$\frac{1}{\sqrt[n+1]{b-a}} \sqrt[n+1]{\int_a^b f^{n+1}(x) \mathrm{d}x} \leq \frac{\int_a^b f^{n+1}(x) \mathrm{d}x}{\int_a^b f^n(x) \mathrm{d}x}. \tag{1.41}$$

Letting n tend to ∞ in (1.41) and using that $\lim_{n \to \infty} \sqrt[n]{\int_a^b f^n(x) \mathrm{d}x} = M$[7] we get that

$$M \leq \underline{\lim} \frac{\int_a^b f^{n+1}(x) \mathrm{d}x}{\int_a^b f^n(x) \mathrm{d}x},$$

which combined with (1.40) solves the problem.

---

[6]Hölder's inequality states that if $p > 1$ and q are positive real numbers such that $1/p + 1/q = 1$ and f and g are functions defined on $[a,b]$, then

$$\int_a^b |f(x)g(x)| \mathrm{d}x \leq \left(\int_a^b |f(x)|^p \mathrm{d}x \right)^{1/p} \left(\int_a^b |g(x)|^q \mathrm{d}x \right)^{1/q}.$$

Equality holds if and only if $A|f(x)|^p = B|g(x)|^q$ for all $x \in [a,b]$.

[7]This is also the topic of a classical problem of Pólya and Szegő [104, Problem 198, p. 78]. It is worth mentioning that another closely related problem states that if $f : [a,b] \to \mathbb{R}$ is integrable, then the function $g : [1, \infty) \to \mathbb{R}$ defined by

$$g(p) = \left(\frac{1}{b-a} \int_a^b |f(x)|^p \mathrm{d}x \right)^{1/p},$$

is monotone increasing.

Remark. It is worth mentioning that this problem is a particular case of a problem of Pólya and Szegö [104, Problem 199, p. 78]. However, the solution, which is based on an application of Hölder's inequality, is different than the solution from Pólya and Szegö. For a more general problem, see also [112, Problem 23, p. 74].

1.48. The limit equals $||f||_\infty \cdot \ln ||f||_\infty$. We have

$$n\left(\sqrt[n]{\int_a^b f^{n+1}(x)dx} - \sqrt[n]{\int_a^b f^n(x)dx}\right)$$

$$= n\left(\exp\left(\frac{1}{n}\ln\int_a^b f^{n+1}(x)dx\right) - \exp\left(\frac{1}{n}\ln\int_a^b f^n(x)dx\right)\right)$$

$$= \ln\left(\frac{\int_a^b f^{n+1}(x)dx}{\int_a^b f^n(x)dx}\right)\exp(\theta_n),$$

where the last equality follows based on Lagrange's Mean Value Theorem applied to the exponential function and θ_n is between $\ln\sqrt[n]{\int_a^b f^{n+1}(x)dx}$ and $\ln\sqrt[n]{\int_a^b f^n(x)dx}$. This implies that $\lim_{n\to\infty}\theta_n = \ln||f||_\infty$ and the solution is completed based on Problem **1.47**.

Remark. Similarly, one can prove that if $f : [a,b] \to [0,\infty)$ is a continuous function, then

$$\lim_{n\to\infty} n\left(\sqrt[n+1]{\int_a^b f^{n+1}(x)dx} - \sqrt[n]{\int_a^b f^n(x)dx}\right) = 0.$$

1.49. and **1.50.** The solutions of these two problems are given in [1].

1.51. Let $M = ||f||_\infty$. If $M = 0$ the equality to prove follows by triviality, so we consider the case when $M > 0$. We have $\sqrt[n]{\int_0^1 f(x)f(x^2)\cdots f(x^n)dx} \le M$. Thus, it suffices to prove that

$$\varliminf \sqrt[n]{\int_0^1 f(x)f(x^2)\cdots f(x^n)dx} \ge M.$$

First we consider the case when f attains its maximum at 1, i.e., $M = f(1)$. Let $0 < \varepsilon < M$. Using the continuity of f at 1, we get that there is $\delta = \delta(\varepsilon) > 0$ such that $M - \varepsilon \le f(x) \le M$ for $\delta \le x \le 1$. Since the functions $x \to f(x^k)$, $k = 1,\dots,n$, also attain their maximum at 1, we have that $M - \varepsilon \le f(x^k) \le M$ for $\sqrt[k]{\delta} \le x \le 1$.

On the other hand, since $\delta < 1$ and $\delta < \sqrt{\delta} < \cdots < \sqrt[n]{\delta}$, we have that for all $k = 1,\ldots,n$, one has $M - \varepsilon \leq f(x^k) \leq M$, for $\sqrt[n]{\delta} \leq x \leq 1$. Thus,

$$\int_0^1 f(x)f(x^2)\cdots f(x^n)\mathrm{d}x \geq \int_{\sqrt[n]{\delta}}^1 f(x)f(x^2)\cdots f(x^n)\mathrm{d}x \geq (M-\varepsilon)^n \left(1 - \sqrt[n]{\delta}\right),$$

and it follows that

$$\sqrt[n]{\int_0^1 f(x)f(x^2)\cdots f(x^n)\mathrm{d}x} \geq (M-\varepsilon)\sqrt[n]{1 - \sqrt[n]{\delta}}.$$

Using that $\lim_{n\to\infty} \sqrt[n]{1 - \sqrt[n]{\delta}} = 1$, we get that $\underline{\lim} \sqrt[n]{\int_0^1 f(x)f(x^2)\cdots f(x^n)\mathrm{d}x} \geq M - \varepsilon$, and since ε was arbitrary taken the result follows.

Now we consider the case when f attains its maximum at 0, i.e., $M = f(0)$. Let $0 < \varepsilon < f(0)$ be fixed. Using the continuity of f at 0 we get that there is $\delta > 0$ such that $0 < f(0) - \varepsilon < f(x) < f(0)$ for all $0 < x < \delta$. Since $x^k < x$ for $k \in \mathbb{N}$ and $x \in (0,\delta)$, one has that $f(x^k) > f(0) - \varepsilon > 0$. We have

$$\sqrt[n]{\int_0^1 f(x)f(x^2)\cdots f(x^n)\mathrm{d}x} \geq \sqrt[n]{\int_0^\delta f(x)f(x^2)\cdots f(x^n)\mathrm{d}x}. \tag{1.42}$$

It follows, based on Bernoulli's Integral Inequality, that

$$\sqrt[n]{\int_0^\delta f(x)f(x^2)\cdots f(x^n)\mathrm{d}x} \geq \delta^{\frac{1}{n}-1} \int_0^\delta \sqrt[n]{f(x)f(x^2)\cdots f(x^n)}\mathrm{d}x. \tag{1.43}$$

Combining (1.42) and (1.43), we obtain that

$$\sqrt[n]{\int_0^1 f(x)f(x^2)\cdots f(x^n)\mathrm{d}x} \geq \delta^{\frac{1}{n}-1} \cdot \int_0^\delta \sqrt[n]{f(x)f(x^2)\cdots f(x^n)}\mathrm{d}x. \tag{1.44}$$

We prove that

$$\lim_{n\to\infty} \int_0^\delta \sqrt[n]{f(x)f(x^2)\cdots f(x^n)}\mathrm{d}x = \delta \cdot f(0). \tag{1.45}$$

Let

$$h_n(x) = \sqrt[n]{f(x)f(x^2)\cdots f(x^n)}, \quad x \in (0,\delta),$$

and let v be the constant function $v(x) = M = f(0)$. Then $h_n(x) \leq v(x)$ for all $x \in (0,\delta)$. On the other hand, $\ln h_n(x) = \frac{1}{n}\sum_{k=1}^n \ln f(x^k)$, and note that $\ln$ is well defined since $f(x^k) > 0$ for $x \in (0,\delta)$. It follows, based on Stolz–Cesàro lemma (the ∞/∞ case), that

$$\lim_{n\to\infty} \ln h_n(x) = \lim_{n\to\infty} \ln f(x^{n+1}) = \ln f(0).$$

Thus, $\lim_{n\to\infty} h_n(x) = f(0)$, and equality (1.45) follows based on Lebesgue Convergence Theorem. Combining (1.44) and (1.45), we obtain that

$$\underline{\lim} \sqrt[n]{\int_0^1 f(x)f(x^2)\cdots f(x^n)dx} \geq \underline{\lim}\left(\delta^{\frac{1}{n}-1}\cdot\int_0^\delta \sqrt[n]{f(x)f(x^2)\cdots f(x^n)}dx\right) = f(0).$$

(b) The second part of the problem can be solved by using a similar technique.

Remark. The problem was motivated by an exercise in classical analysis which states that if f is continuous on $[a,b]$, then $\lim_{n\to\infty}\left(\int_a^b |f(x)|^n dx\right)^{1/n} = ||f||_\infty$ (see [104, Problem 198, p. 78]).

1.52. The limit equals $L/f'(b)$. Without losing the generality we assume that $L > 0$. Let $\varepsilon > 0$ be such that $0 < \varepsilon < \min\{L, f'(b)\}$. Since $\lim_{x\to b^-} g(x) = L$, we have that there exists $\delta_1 > 0$ such that

$$L - \varepsilon < g(x) < L + \varepsilon, \quad b - \delta_1 < x < b. \tag{1.46}$$

On the other hand, we have, based on the differentiability of f at b, that there is $\delta_2 > 0$ such that

$$\left|\frac{f(b) - f(x)}{b - x} - f'(b)\right| < \varepsilon, \quad b - \delta_2 < x < b,$$

from which it follows that

$$1 - (b-x)(\varepsilon + f'(b)) < f(x) < 1 - (b-x)(f'(b) - \varepsilon), \quad b - \delta_2 < x < b.$$

Let $\delta = \min\{\delta_1, \delta_2, 2/(\varepsilon + f'(b)), 2/(f'(b) - \varepsilon)\}$. It follows that for $b - \delta < x < b$, one has

$$[1 - (b-x)(\varepsilon + f'(b))]^n < f^n(x) < [1 - (b-x)(f'(b) - \varepsilon)]^n. \tag{1.47}$$

We have, based on (1.46) and (1.47), that for $b - \delta < x < b$, one has

$$(L - \varepsilon)[1 - (b-x)(\varepsilon + f'(b))]^n < f^n(x)g(x) < (L+\varepsilon)[1 - (b-x)(f'(b) - \varepsilon)]^n. \tag{1.48}$$

We have

$$n\int_a^b f^n(x)g(x)dx = n\int_a^{b-\delta} f^n(x)g(x)dx + n\int_{b-\delta}^b f^n(x)g(x)dx. \tag{1.49}$$

On the other hand,

$$0 \leq \left|n\int_a^{b-\delta} f^n(x)g(x)dx\right| \leq nM\int_a^{b-\delta} f^n(x)dx \leq nMf^n(b-\delta),$$

where $M = \sup_{x \in [a,b]} |g(x)|$. It follows, since $0 < f(b-\delta) < 1$, that $\lim_{n \to \infty} n f^n(b - \delta) = 0$, and hence

$$\lim_{n \to \infty} n \int_a^{b-\delta} f^n(x)g(x)dx = 0. \tag{1.50}$$

We have, based on (1.48), that

$$(L-\varepsilon)n \int_{b-\delta}^b \left[1-(b-x)(\varepsilon+f'(b))\right]^n dx < n \int_{b-\delta}^b f^n(x)g(x)dx$$

$$< (L+\varepsilon)n \int_{b-\delta}^b \left[1-(b-x)(f'(b)-\varepsilon)\right]^n dx.$$

Thus,

$$(L-\varepsilon)\frac{n}{n+1} \cdot \frac{[1-(b-x)(\varepsilon+f'(b))]^{n+1}}{\varepsilon+f'(b)} \bigg|_{b-\delta}^b$$

$$< n \int_{b-\delta}^b f^n(x)g(x)dx < (L+\varepsilon)\frac{n}{n+1} \cdot \frac{[1-(b-x)(f'(b)-\varepsilon)]^{n+1}}{f'(b)-\varepsilon} \bigg|_{b-\delta}^b,$$

and hence,

$$(L-\varepsilon)\frac{n}{(n+1)(\varepsilon+f'(b))}\left\{1 - \left[1-\delta(\varepsilon+f'(b))\right]^{n+1}\right\}$$

$$< n \int_{b-\delta}^b f^n(x)g(x)dx < (L+\varepsilon)\frac{n}{(n+1)(f'(b)-\varepsilon)}\left\{1 - \left[1-\delta(f'(b)-\varepsilon)\right]^{n+1}\right\}.$$

We have based on the definition of δ that $-1 < 1 - \delta(\varepsilon + f'(b)) < 1$ and $-1 < 1 - \delta(f'(b) - \varepsilon) < 1$, from which it follows that $\lim_{n \to \infty} [1 - \delta(\varepsilon + f'(b))]^{n+1} = 0$ and $\lim_{n \to \infty} [1 - \delta(f'(b) - \varepsilon)]^{n+1} = 0$. Letting $n \to \infty$ in the preceding inequalities we get that

$$\frac{L-\varepsilon}{\varepsilon+f'(b)} \le \lim_{n \to \infty} n \int_{b-\delta}^b f^n(x)g(x)dx \le \frac{L+\varepsilon}{f'(b)-\varepsilon}. \tag{1.51}$$

We have, based on (1.49)–(1.51), that

$$\frac{L-\varepsilon}{\varepsilon+f'(b)} \le \lim_{n \to \infty} n \int_a^b f^n(x)g(x)dx \le \frac{L+\varepsilon}{f'(b)-\varepsilon},$$

and since ε was arbitrary taken, the result follows.

Remark. If f and g are both continuous then, the integral $\int_a^b f^n(x)g(x)dx$ exists for each positive integer n. However, we give below a counterexample (see [6, p. 96]) when the integral fails to exist for suitable functions f and g. Let $f(x) = x$ and let

$$g(x) = \begin{cases} -1 & \text{if } x \in [0, 1/2] \cap \mathbb{Q}, \\ 1 & \text{if } x \in [0, 1/2] \setminus \mathbb{Q}, \\ 0 & \text{if } x \in (1/2, 1]. \end{cases}$$

Then, for each n, the integral $\int_0^1 f^n(x)g(x)dx$ is not Riemann integrable.

Next, let $f(x) = x$ and $g(x) = (1 - \chi_A(x))(x - 1)$ where A is a nonmeasurable subset of $[0, 1]$ and χ_A is the characteristic function of A. Then g is nonmeasurable and hence $f^n g$ is nonmeasurable.

1.53. The limit equals

$$\int_0^1 x^{f'(1)-1} g(x)dx.$$

Let $0 < \varepsilon < \min\{1, f'(1)\}$. It follows, based on the differentiability of f at 1, that there is $\delta > 0$ such that $1 - (1-x)(f'(1)+\varepsilon) \leq f(x) \leq 1 - (1-x)(f'(1)-\varepsilon)$ for $1 - \delta < x < 1$. We observe that this condition implies that $f'(1) > 0$ and $f(x) < 1$ for all $x < 1$. We have, based on the substitution $x^n = y$, that

$$n \int_0^1 f^n(x)g(x^n)dx = \int_0^1 f^n(\sqrt[n]{y})y^{\frac{1}{n}-1}g(y)dy.$$

Let $M = \sup_{x \in [0,1]} |g(x)|$. We have

$$n \int_0^1 f^n(x)g(x^n)dx = \int_0^1 f^n(\sqrt[n]{y})y^{\frac{1}{n}-1}g(y)dy$$

$$= \int_0^\varepsilon f^n(\sqrt[n]{y})y^{\frac{1}{n}-1}g(y)dy + \int_\varepsilon^1 f^n(\sqrt[n]{y})y^{\frac{1}{n}-1}g(y)dy$$

$$= I_n + J_n.$$

On the other hand,

$$I_n = \int_0^\varepsilon f^n(\sqrt[n]{y})y^{\frac{1}{n}-1}g(y)dy \overset{x^n=y}{=} n \int_0^{\varepsilon^n} f^n(x)g(x^n)dx.$$

It follows, since $\varepsilon^n < \varepsilon$, that

$$|I_n| = \left| n \int_0^{\varepsilon^n} f^n(x)g(x^n)dx \right| \leq n \cdot M \int_0^{\varepsilon^n} f^n(x)dx \leq n \cdot M \int_0^\varepsilon f^n(x)dx \leq M \cdot nf^n(\varepsilon).$$

We get, since $f(\varepsilon) < 1$, that $\lim_{n \to \infty} nf^n(\varepsilon) = 0$, and hence $\lim_{n \to \infty} I_n = 0$.

Let $h_n : [\varepsilon, 1] \to \mathbb{R}$ be the sequence of functions defined by

$$h_n(y) = f^n(\sqrt[n]{y})y^{\frac{1}{n}-1}g(y).$$

We have, since $0 \le f(x) \le 1$, that $|h_n(y)| \le M/\varepsilon$ for $y \in [\varepsilon, 1]$ and $n \in \mathbb{N}$. We calculate the limit function $h(y) = \lim_{n \to \infty} h_n(y)$. We note that $h(1) = \lim_{n \to \infty} h_n(1) = g(1)$. For $y \in [\varepsilon, 1)$, we have that

$$h_n(y) = (1 + f(\sqrt[n]{y}) - 1)^n y^{\frac{1}{n} - 1} g(y) = e^{\frac{\ln(1 + f(\sqrt[n]{y}) - 1)}{f(\sqrt[n]{y}) - 1} \cdot \frac{f(\sqrt[n]{y}) - 1}{\sqrt[n]{y} - 1} \cdot \frac{\sqrt[n]{y} - 1}{\frac{1}{n}}} \frac{1}{n} g(y) y^{\frac{1}{n} - 1}.$$

It follows, since $\lim_{x \to 0} \ln(1 + x)/x = 1$, that

$$\lim_{n \to \infty} \frac{\ln(1 + f(\sqrt[n]{y}) - 1)}{f(\sqrt[n]{y}) - 1} = 1.$$

Also, $\lim_{n \to \infty} n(\sqrt[n]{y} - 1) = \ln y$, for $y \in [\varepsilon, 1)$. On the other hand, we have, based on the differentiability of f at 1, that

$$\lim_{n \to \infty} \frac{f(\sqrt[n]{y}) - 1}{\sqrt[n]{y} - 1} = f'(1).$$

Combining the preceding limits, we get that

$$h(y) = \lim_{n \to \infty} h_n(y) = \lim_{n \to \infty} e^{\frac{\ln(1 + f(\sqrt[n]{y}) - 1)}{f(\sqrt[n]{y}) - 1} \cdot \frac{f(\sqrt[n]{y}) - 1}{\sqrt[n]{y} - 1} \cdot \frac{\sqrt[n]{y} - 1}{\frac{1}{n}}} g(y) y^{\frac{1}{n} - 1}$$

$$= e^{1 \cdot f'(1) \cdot \ln y} y^{-1} g(y) = y^{f'(1) - 1} g(y).$$

It follows from the Bounded Convergence Theorem that

$$\lim_{n \to \infty} J_n = \lim_{n \to \infty} \int_\varepsilon^1 h_n(y) dy = \int_\varepsilon^1 \left(\lim_{n \to \infty} h_n(y) \right) dy = \int_\varepsilon^1 x^{f'(1) - 1} g(x) dx.$$

Thus,

$$\lim_{n \to \infty} n \int_0^1 f^n(x) g(x^n) dx = \lim_{n \to \infty} (I_n + J_n) = \int_\varepsilon^1 x^{f'(1) - 1} g(x) dx,$$

and since ε is arbitrary the result follows.

Remark 1. If the condition $f'(1) \ne 0$ is not satisfied, then $\lim_{n \to \infty} n \int_0^1 f^n(x) g(x^n) dx$ need not be finite. To see this, let $f(x) = 2x - x^2$ and $g(x) = 1$. The function f is increasing, with $f(1) = 1$ and $f'(1) = 0$. However, in this case one can prove that $\lim_{n \to \infty} n \int_0^1 (2x - x^2)^n dx = \infty$.

Remark 2. When $g(x) = 1$, one has that

$$\lim_{n \to \infty} n \int_0^1 f^n(x) dx = \int_0^1 x^{f'(1) - 1} dx = \frac{1}{f'(1)}.$$

1.54. The limit equals $f(0)\Gamma(1 + 1/k)$. We record below the solution due to Chip Curtis (see [30]). Using the substitution $t = y/\sqrt[k]{n}$, we obtain that

$$\sqrt[k]{n}\int_0^x \frac{f(t)}{(1+t^k)^n}\,dt = \int_0^{x\sqrt[k]{n}} \frac{f(y/\sqrt[k]{n})}{(1+y^k/n)^n}\,dy = \int_0^\infty \frac{f(y/\sqrt[k]{n})}{(1+y^k/n)^n}\chi_n(y)\,dy,$$

where

$$\chi_n(y) = \begin{cases} 1 & \text{if } y \le x\sqrt[k]{n}, \\ 0 & \text{if } y > x\sqrt[k]{n}. \end{cases}$$

Let $f_n : [0,\infty) \to \mathbb{R}$ be the sequence of functions defined by

$$f_n(y) = \frac{f(y/\sqrt[k]{n})}{(1+y^k/n)^n}\chi_n(y).$$

We have, based on the continuity of f at 0, that $\lim_{n\to\infty} f_n(y) = f(0)e^{-y^k}$. Also, f_n is bounded by an integrable function, namely,

$$\left| \frac{f(y/\sqrt[k]{n})}{(1+y^k/n)^n}\chi_n(y) \right| \le \frac{M}{(1+y^k/n)^n} \le \frac{M}{1+y^k},$$

where $M = \sup_{x\ge 0} |f(x)|$. Now the result follows from Lebesgue Convergence Theorem, since

$$\lim_{n\to\infty} \int_0^\infty \frac{f(y/\sqrt[k]{n})}{(1+y^k/n)^n}\chi_n(y)\,dy = \lim_{n\to\infty} \int_0^\infty f_n(y)\,dy$$

$$= \int_0^\infty f(0)e^{-y^k}\,dy$$

$$= f(0)\Gamma\left(1 + \frac{1}{k}\right).$$

1.55. The solution is given in [49].

1.56. The limit equals $2\Gamma(1+1/k)h(x)$. We have, by making the substitution $t = -z$, in the first integral, that

$$I_n = \sqrt[k]{n}\int_{-\infty}^\infty h(x+t)e^{-nt^k}\,dt$$

$$= \sqrt[k]{n}\int_{-\infty}^0 h(x+t)e^{-nt^k}\,dt + \sqrt[k]{n}\int_0^\infty h(x+t)e^{-nt^k}\,dt$$

$$= \sqrt[k]{n}\int_0^\infty h(x-z)e^{-nz^k}\,dz + \sqrt[k]{n}\int_0^\infty h(x+t)e^{-nt^k}\,dt$$

$$= \frac{1}{k}\int_0^\infty \left(h\left(x - \sqrt[k]{\frac{u}{n}}\right) + h\left(x + \sqrt[k]{\frac{u}{n}}\right) \right) e^{-u}u^{\frac{1}{k}-1}\,du.$$

Let $f_n(u) = \left(h\left(x + \sqrt[k]{\frac{u}{n}}\right) + h\left(x - \sqrt[k]{\frac{u}{n}}\right)\right) e^{-u} u^{\frac{1}{k}-1}$. We have that $\lim_{n\to\infty} f_n(u) = 2h(x) e^{-u} u^{\frac{1}{k}-1}$. Also, $|f_n(u)| \le 2\sup_{x\in\mathbb{R}} |h(x)| e^{-u} u^{\frac{1}{k}-1}$, which is integrable over $(0, \infty)$. It follows from Lebesgue Convergence Theorem that

$$\lim_{n\to\infty} I_n = \frac{1}{k} \int_0^\infty 2h(x) e^{-u} u^{\frac{1}{k}-1} du = 2h(x) \frac{1}{k} \Gamma\left(\frac{1}{k}\right) = 2h(x) \Gamma\left(1 + \frac{1}{k}\right).$$

1.57. The limit equals $\Gamma(1 + 1/k)$. Using the substitution $x^k = y$ we have that

$$I_n = \int_0^{\sqrt[k]{n}} \left(1 - \frac{x^k}{n}\right)^n dx = \frac{1}{k} \int_0^n \left(1 - \frac{y}{n}\right)^n y^{1/k-1} dy.$$

One can check (see [133, p. 242]) that $0 \le e^{-x} - (1 - x/n)^n \le x^2 e^{-x}/n$, for $0 \le x \le n$. Thus,

$$\frac{1}{k} \int_0^n e^{-y} y^{1/k-1} dy - \frac{1}{kn} \int_0^n e^{-y} y^{1+1/k} dy \le I_n \le \frac{1}{k} \int_0^n e^{-y} y^{1/k-1} dy.$$

Passing to the limit, as $n \to \infty$, in the preceding inequalities, the result follows.

1.58. The limit equals $k!/2^{k+1}$. We have, by the substitution $(1-x)/(1+x) = y$, that

$$n^{k+1} \int_0^1 \left(\frac{1-x}{1+x}\right)^n x^k dx = 2n^{k+1} \int_0^1 y^n (1-y)^k \cdot \frac{dy}{(1+y)^{k+2}} = 2n^{k+1} \int_0^1 y^n f(y) dy,$$

where

$$f(y) = \frac{(1-y)^k}{(1+y)^{k+2}}.$$

We observe that

$$f(1) = f'(1) = \cdots = f^{(k-1)}(1) = 0. \tag{1.52}$$

We calculate the integral $\int_0^1 y^n f(y) dy$ by parts, k times, and we get, based on (1.52), that

$$\int_0^1 y^n f(y) dy = \frac{(-1)^k}{(n+1)(n+2)\cdots(n+k)} \int_0^1 y^{n+k} f^{(k)}(y) dy,$$

and one more integration implies that $\int_0^1 y^n f(y) dy$ equals

$$\frac{(-1)^k}{(n+1)(n+2)\cdots(n+k)} \left(\frac{f^{(k)}(y)}{n+k+1}\bigg|_0^1 - \frac{1}{n+k+1} \int_0^1 y^{n+k+1} f^{(k+1)}(y) dy\right)$$

$$= \frac{(-1)^k f^{(k)}(1)}{(n+1)(n+2)\cdots(n+k+1)}$$

$$+ \frac{(-1)^{k+1}}{(n+1)(n+2)\cdots(n+k+1)} \int_0^1 y^{n+k+1} f^{(k+1)}(y) dy.$$

It follows that

$$\lim_{n\to\infty} 2n^{k+1} \int_0^1 y^n f(y)dy = (-1)^k \cdot 2 \cdot f^{(k)}(1),$$

since

$$\lim_{n\to\infty} \int_0^1 y^{n+k+1} f^{(k+1)}(y)dy = 0,$$

$f^{(k+1)}$ being continuous and hence bounded. Using Leibniz's formula we get that $f^{(k)}(1) = (-1)^k \cdot k!/2^{k+2}$, and the problem is solved.

Remark. This problem, proposed by the author of this book, was given as Problem 4 to SEEMOUS 2012, Blagoevgrad, Bulgaria.

1.59. We prove that $l = \int_1^\infty f(x)/x\,dx$. Making the substitution $x^n = y$, we get that

$$I_n = n \int_1^a f(x^n)dx = \int_1^{a^n} \frac{f(y)}{y} y^{\frac{1}{n}}dy = \int_1^\infty \frac{f(y)}{y} y^{\frac{1}{n}}dy - \int_{a^n}^\infty \frac{f(y)}{y} y^{\frac{1}{n}}dy.$$

Let $L = \lim_{x\to\infty} x^\alpha f(x)$. For $\varepsilon > 0$, there exists $\delta = \delta(\varepsilon) > 1$ such that for $y > \delta$ one has $|f(y)y^\alpha - L| < \varepsilon$. It follows that $|f(y)| \le (\varepsilon + |L|)/y^\alpha$ for $y > \delta$. Since for large n, $a^n > \delta$, we get that

$$\left| \int_{a^n}^\infty \frac{f(y)}{y} y^{\frac{1}{n}}dy \right| \le (\varepsilon + |L|) \int_{a^n}^\infty \frac{dy}{y^{1+\alpha-1/n}} = \frac{\varepsilon + |L|}{(\alpha - 1/n)a^{n\alpha-1}},$$

and hence,

$$\lim_{n\to\infty} \int_{a^n}^\infty \frac{f(y)}{y} y^{\frac{1}{n}}dy = 0 \quad \text{and} \quad \lim_{n\to\infty} n \int_{a^n}^\infty \frac{f(y)}{y} y^{\frac{1}{n}}dy = 0. \tag{1.53}$$

On the other hand,

$$\int_1^\infty \frac{f(y)}{y} y^{\frac{1}{n}}dy = \int_1^\delta \frac{f(y)}{y} y^{\frac{1}{n}}dy + \int_\delta^\infty \frac{f(y)}{y} y^{\frac{1}{n}}dy.$$

Since f is continuous and $f_n(y) = \frac{f(y)}{y} y^{\frac{1}{n}} \to \frac{f(y)}{y}$ we get, by the Bounded Convergence Theorem, that

$$\lim_{n\to\infty} \int_1^\delta \frac{f(y)}{y} y^{\frac{1}{n}}dy = \int_1^\delta \frac{f(y)}{y}dy. \tag{1.54}$$

Let $f_n(y) = \frac{f(y)}{y} y^{\frac{1}{n}}$, for $y \ge \delta$, and we note that for large n (i.e., $n > 2/\alpha$), we have

$$|f_n(y)| \le (\varepsilon + |L|) \cdot \frac{y^{1/n}}{y^{\alpha+1}} = (\varepsilon + |L|) \cdot \frac{1}{y^{\alpha/2+1}} \cdot \frac{1}{y^{\alpha/2-1/n}} \le \frac{\varepsilon + |L|}{y^{\alpha/2+1}}.$$

Since the function $y \to 1/y^{\alpha/2+1}$ is integrable over (δ, ∞), we get, based on the Lebesgue Convergence Theorem, that

$$\lim_{n \to \infty} \int_\delta^\infty \frac{f(y)}{y} y^{\frac{1}{n}} \, dy = \int_\delta^\infty \frac{f(y)}{y} \, dy. \tag{1.55}$$

Combining (1.54) and (1.55), we get that

$$\lim_{n \to \infty} \int_1^\infty \frac{f(y)}{y} y^{\frac{1}{n}} \, dy = \int_1^\infty \frac{f(y)}{y} \, dy. \tag{1.56}$$

From (1.53) and (1.56), we get that the first part of the problem is solved.

The second limit equals $\int_1^\infty (f(y) \ln y)/y \, dy$. A calculation shows that

$$n \left(n \int_1^a f(x^n) \, dx - l \right) = n \int_1^\infty \frac{f(y)}{y} \left(y^{\frac{1}{n}} - 1 \right) dy - n \int_{a^n}^\infty \frac{f(y)}{y} y^{\frac{1}{n}} \, dy.$$

We have

$$n \int_1^\infty \frac{f(y)}{y} \left(y^{\frac{1}{n}} - 1 \right) dy = n \int_1^\delta \frac{f(y)}{y} \left(y^{\frac{1}{n}} - 1 \right) dy + n \int_\delta^\infty \frac{f(y)}{y} \left(y^{\frac{1}{n}} - 1 \right) dy.$$

We note that, for $y > 1$ fixed, the function $x \to (y^x - 1)/x$ increases on $(0, \infty)$ and this implies that the sequence $(y^{1/n} - 1)/1/n$ decreases as n increases.

Let $u_n(y) = \frac{f(y)}{y} \cdot \frac{y^{1/n} - 1}{1/n}$, for $y \in (1, \delta)$ and note that $\lim_{n \to \infty} u_n(y) = f(y) \ln y / y$. The continuity of f implies that

$$|u_n(y)| \le \left| \frac{f(y)}{y} \right| \cdot \frac{y^{1/n} - 1}{1/n} \le |f(y)|(y - 1) \le (\delta - 1) \sup_{y \in [1, \delta]} |f(y)|,$$

and we get, based on the Bounded Convergence Theorem, that

$$\lim_{n \to \infty} n \int_1^\delta \frac{f(y)}{y} \left(y^{\frac{1}{n}} - 1 \right) dy = \int_1^\delta \frac{f(y)}{y} \ln y \, dy. \tag{1.57}$$

Now we consider the sequence $(u_n(y))$ when $y > \delta$. Let n_0 be such that $n_0 > 1/\alpha$, and we note that for $n \ge n_0$, one has $(y^{1/n} - 1)/1/n \le (y^{1/n_0} - 1)/1/n_0$. This implies that

$$|u_n(y)| = \left| \frac{f(y)}{y} \cdot \frac{y^{1/n} - 1}{1/n} \right| \le \frac{\varepsilon + |L|}{y^{\alpha+1}} \cdot \frac{y^{1/n_0} - 1}{1/n_0} = n_0 (\varepsilon + |L|) \left(\frac{1}{y^{1+\alpha - 1/n_0}} - \frac{1}{y^{\alpha+1}} \right).$$

Since the function $y \to 1/y^{1+\alpha - 1/n_0} - 1/y^{\alpha+1}$ is integrable over (δ, ∞), we get, based on the Lebesgue Convergence Theorem, that

$$\lim_{n \to \infty} n \int_\delta^\infty \frac{f(y)}{y} \left(y^{\frac{1}{n}} - 1 \right) dy = \int_\delta^\infty \frac{f(y)}{y} \ln y \, dy. \tag{1.58}$$

Combining (1.57) and (1.58), we get that

$$\lim_{n\to\infty} n \int_1^\infty \frac{f(y)}{y}\left(y^{\frac{1}{n}}-1\right)dy = \int_1^\infty \frac{f(y)}{y}\ln y\,dy.$$

Remark. The existence of $\lim_{x\to\infty} x^\alpha f(x)$ implies the convergence of the two improper integrals

$$\int_1^\infty \frac{f(x)}{x}dx \quad \text{and} \quad \int_1^\infty \frac{f(x)}{x}\ln x\,dx.$$

1.60. First we note that $f(0)=1$. Since any integrable function is bounded, we get that there are two real numbers, m and M, such that $m \le (\ln f(x))/x \le M$, for all $x \in [0,1]$. It follows that $\exp(mx) \le f(x) \le \exp(Mx)$, and this implies that $f(0)=1$.

On the other hand, we note that we have an indeterminate form of type $0\cdot\infty$. We have

$$0 < a = \inf_{x\in[0,1]} \exp(mx) \le f(x) \le \exp(Mx) \le \sup_{x\in[0,1]} \exp(Mx) = b,$$

which implies that $\sqrt[n]{a} \le \sqrt[n]{f(x^n)} \le \sqrt[n]{b}$. This shows that

$$\lim_{n\to\infty} \int_0^1 \sqrt[n]{f(x^n)}g(x)dx = \int_0^1 g(x)dx.$$

We have, based on the substitution $x^n = y$, that

$$n^2\left(\int_0^1 \sqrt[n]{f(x^n)}g(x)dx - \int_0^1 g(x)dx\right) = n^2 \int_0^1 \left(\sqrt[n]{f(x^n)} - \sqrt[n]{f(0)}\right)g(x)dx$$

$$= n\int_0^1 \left(\sqrt[n]{f(y)} - \sqrt[n]{f(0)}\right)g\left(y^{\frac{1}{n}}\right)y^{\frac{1}{n}-1}dy$$

$$= n\int_0^1 \left(\exp\left(\frac{1}{n}\ln f(y)\right) - \exp\left(\frac{1}{n}\ln f(0)\right)\right)g\left(y^{\frac{1}{n}}\right)y^{\frac{1}{n}-1}dy$$

$$= \int_0^1 (\ln f(y) - \ln f(0))\exp(\theta_n(y))g\left(y^{\frac{1}{n}}\right)y^{\frac{1}{n}-1}dy$$

$$= \int_0^1 \frac{\ln f(y)}{y}\exp(\theta_n(y))g\left(y^{\frac{1}{n}}\right)y^{\frac{1}{n}}dy,$$

where the last equality follows based on Lagrange's Mean Value Theorem applied to the function $x \to \exp(x)$ and $\theta_n(y)$ is between $\frac{1}{n}\ln f(y)$ and $\frac{1}{n}\ln f(0) = 0$. We note that for all $y \in [0,1]$ one has that $\lim_{n\to\infty}\theta_n(y)=0$.

Let $f_n:[0,1]\to\mathbb{R}$ be the sequence of functions defined by

$$f_n(y) = \frac{\ln f(y)}{y}\exp(\theta_n(y))g\left(y^{\frac{1}{n}}\right)y^{\frac{1}{n}},$$

and we note that the limit function equals

$$f(y) = \lim_{n\to\infty} f_n(y) = g(1)\frac{\ln f(y)}{y}, \quad y \in (0,1].$$

On the other hand,

$$|f_n(y)| \le \left|\frac{\ln f(y)}{y}\right| \cdot |\exp(\theta_n(y))| \cdot |g(y^{\frac{1}{n}})| \le \exp(1) \cdot \sup_{x\in[0,1]} |g(x)| \cdot \left|\frac{\ln f(y)}{y}\right|.$$

Here, we used the fact that, for large n

$$|\theta_n(y)| \le \frac{1}{n}|\ln f(y)| \le \frac{1}{n} \cdot \sup_{x\in[0,1]} |\ln f(y)| \le 1.$$

It follows, based on Lebesgue Convergence Theorem, that

$$\lim_{n\to\infty} \int_0^1 \frac{\ln f(y)}{y} \exp(\theta_n(y)) g\left(y^{\frac{1}{n}}\right) y^{\frac{1}{n}} dy = g(1) \int_0^1 \frac{\ln f(y)}{y} dy,$$

and the problem is solved.

Remark. It is worth mentioning that if f is a function such that $(\ln f(x))/x \in L^1[0,1]$, then

$$\lim_{n\to\infty} n^2 \left(\int_0^1 \frac{dx}{\sqrt[n]{f(x^n)}} - 1\right) = -\int_0^1 \frac{\ln f(x)}{x} dx$$

and

$$\lim_{n\to\infty} \left(\int_0^1 \sqrt[n]{f(x^n)} dx\right)^{n^2} = \exp\left(\int_0^1 \frac{\ln f(x)}{x} dx\right).$$

1.61. Follow the technique from the solution of Problem **1.60.**

1.63. (a) We have that

$$n\int_0^1 f(x^n) g(x) dx = \int_0^1 \frac{f(y)}{y} g(y^{1/n}) y^{1/n} dy.$$

Let $l = \lim_{x\to 0^+} f(x)/x$ and let $\varepsilon > 0$. Then $|f(x)/x| \le \varepsilon + |l|$, for $x \in [0,\delta)$. On the other hand, $|f(x)/x| \le \|f\|_\infty/\delta$, for $x \in [\delta,1]$. This implies that $|f(x)/x| \le \max\{|l| + \varepsilon, \|f\|_\infty/\delta\} = M_1$. Let

$$h_n(y) = \begin{cases} \frac{f(y)}{y} g(y^{1/n}) y^{1/n} & \text{if } y \in (0,1], \\ 0 & \text{if } y = 0. \end{cases}$$

We used the fact that

$$h_n(0) = \lim_{y \to 0^+} h_n(y) = \lim_{y \to 0^+} \frac{f(y)}{y} g(y^{1/n}) y^{1/n} = l \cdot 0 \cdot g(0) = 0.$$

We calculate the limit function and we get that

$$h(y) = \begin{cases} \frac{f(y)}{y} g(1) & \text{if } y \in (0,1], \\ 0 & \text{if } y = 0. \end{cases}$$

Since $|h_n(y)| \le M_1 \cdot \|g\|_\infty$, the result follows in view of the Bounded Convergence Theorem.

(b) The limit equals $f(1)/g(1)$. Use part (a) of the problem.

1.64. Making the substitution $x^n = y$, we get that

$$n \int_0^1 x^n f(x^n) g(x) dx = \int_0^1 y^{1/n} f(y) g(y^{1/n}) dy.$$

Let $h_n : [0,1] \to \mathbb{R}$ be the sequence of functions defined by $h_n(x) = x^{1/n} f(x) g(x^{1/n})$. We have, since f and g are continuous functions, that the sequence $(h_n)_{n \in \mathbb{N}}$ is uniformly bounded, i.e., there is $M > 0$ such that $|h_n(x)| \le M$ for all $n \in \mathbb{N}$ and all $x \in [0,1]$. On the other hand, a calculation shows that the limit function is given by

$$h(x) = \lim_{n \to \infty} h_n(x) = \begin{cases} f(x)g(1) & \text{if } x \in (0,1), \\ 0 & \text{if } x = 0, \\ f(1)g(1) & \text{if } x = 1. \end{cases}$$

It follows from the Bounded Convergence Theorem that

$$\lim_{n \to \infty} n \int_0^1 x^n f(x^n) g(x) dx = \lim_{n \to \infty} \int_0^1 h_n(x) dx = \int_0^1 \left(\lim_{n \to \infty} h_n(x) \right) dx = g(1) \int_0^1 f(x) dx.$$

Let

$$x_n = n \left(n \int_0^1 x^n f(x^n) g(x) dx - g(1) \int_0^1 f(x) dx \right)$$

$$= n \int_0^1 f(x) \left(x^{1/n} g(x^{1/n}) - g(1) \right) dx.$$

We integrate by parts, with

$$u(x) = x^{1/n} g(x^{1/n}) - g(1), \quad u'(x) = (1/n)x^{1/n-1} g(x^{1/n}) + (1/n)x^{2/n-1} g'(x^{1/n}),$$

$$v'(x) = f(x), \quad v(x) = \int_0^x f(t) dt,$$

and we get that

$$x_n = n\left[x^{1/n}g(x^{1/n}) - g(1)\right] \int_0^x f(t)dt \Big|_{x=0}^{x=1}$$

$$- \int_0^1 \frac{\int_0^x f(t)dt}{x}\left[x^{1/n}g(x^{1/n}) + x^{2/n}g'(x^{1/n})\right]dx$$

$$= -\int_0^1 \frac{\int_0^x f(t)dt}{x}\left[x^{1/n}g(x^{1/n}) + x^{2/n}g'(x^{1/n})\right]dx.$$

Let $v_n : [0,1] \to \mathbb{R}$ be the sequence of functions defined by

$$v_n(x) = \frac{\int_0^x f(t)dt}{x}\left[x^{1/n}g(x^{1/n}) + x^{2/n}g'(x^{1/n})\right].$$

We note that $|\int_0^x f(t)dt| \le \int_0^x |f(t)|dt \le x\sup_{x\in[0,1]}|f(x)|$, and it follows that (v_n) is uniformly bounded, i.e., there is $A > 0$ such that $|v_n(x)| \le A$ for all $n \in \mathbb{N}$ and all $x \in [0,1]$. A calculation shows that the limit function equals

$$v(x) = \lim_{n\to\infty} v_n(x) = \begin{cases} (g(1) + g'(1))\frac{\int_0^x f(t)dt}{x} & \text{if } x \in (0,1), \\ 0 & \text{if } x = 0, \\ (g(1) + g'(1))\int_0^1 f(x)dx & \text{if } x = 1. \end{cases}$$

It follows from the Bounded Convergence Theorem that

$$\lim_{n\to\infty} n\left(n\int_0^1 x^n f(x^n)g(x)dx - g(1)\int_0^1 f(x)dx\right) = \lim_{n\to\infty} -\int_0^1 v_n(x)dx$$

$$= -\int_0^1 \left(\lim_{n\to\infty} v_n(x)\right)dx = -(g(1) + g'(1))\int_0^1 \frac{\int_0^x f(t)dt}{x}dx.$$

1.65. (a) Use the $\varepsilon - \delta$ definition of $\lim_{x\to\infty} f(x)$ and the fact that $\sigma\int_0^\infty e^{-\sigma t}dt = 1$.
(b) We need the following lemma.

Lemma 1.11. *Let α be a positive real number and let $f : (0,\infty) \to \mathbb{R}$ be a locally integrable function such that $\lim_{t\to\infty} f(t) = L$. Then,*

$$\lim_{p\to 0^+} p\int_0^\infty f(t)\alpha t^{\alpha-1}e^{-pt^\alpha}dt = L.$$

Proof. First we note that $\int_0^\infty p\alpha t^{\alpha-1}e^{-pt^\alpha}dt = 1$. Let $\varepsilon > 0$ be fixed. We have, since $\lim_{t\to\infty} f(t) = L$, that there is $\delta = \delta(\varepsilon, t) > 0$ such that $|f(t) - L| < \varepsilon$ for all $t > \delta$. Thus,

$$\left| \int_0^\infty f(t) p\alpha t^{\alpha-1} e^{-pt^\alpha} dt - L \right| = \left| \int_0^\infty (f(t) - L) p\alpha t^{\alpha-1} e^{-pt^\alpha} dt \right|$$

$$\leq \int_0^\infty |f(t) - L| p\alpha t^{\alpha-1} e^{-pt^\alpha} dt$$

$$= \int_0^\delta |f(t) - L| p\alpha t^{\alpha-1} e^{-pt^\alpha} dt$$

$$+ \int_\delta^\infty |f(t) - L| p\alpha t^{\alpha-1} e^{-pt^\alpha} dt$$

$$\leq p\alpha \int_0^\delta |f(t) - L| t^{\alpha-1} dt + \varepsilon \int_\delta^\infty p\alpha t^{\alpha-1} e^{-pt^\alpha} dt$$

$$\leq p\alpha \int_0^\delta |f(t) - L| t^{\alpha-1} dt + \varepsilon.$$

On the other hand, f is a locally integrable function, so we get, since any integrable function is also bounded, that there is $M = M(\delta)$ such that $|f(t)| \leq M$ for all $t \in [0, \delta]$. It follows that

$$\left| \int_0^\infty f(t) p\alpha t^{\alpha-1} e^{-pt^\alpha} dt - L \right| \leq p\alpha(M + |L|) \int_0^\delta t^{\alpha-1} dt + \varepsilon = p(M + |L|)\delta^\alpha + \varepsilon.$$

Letting $p \to 0^+$ in the preceding inequality, we obtain that

$$\lim_{p \to 0^+} \left| \int_0^\infty f(t) p\alpha t^{\alpha-1} e^{-pt^\alpha} dt - L \right| \leq \varepsilon.$$

Since ε is arbitrary the result follows and the lemma is proved.

Now we are ready to solve the problem. Integrating by parts we obtain that

$$\int_0^\infty g(t) e^{-pt^\alpha} dt = e^{-pt^\alpha} \int_0^t g(s) ds \Big|_0^\infty + p\alpha \int_0^\infty \left(\int_0^t g(s) ds \right) e^{-pt^\alpha} t^{\alpha-1} dt$$

$$= p\alpha \int_0^\infty \left(\int_0^t g(s) ds \right) e^{-pt^\alpha} t^{\alpha-1} dt.$$

We have, based on Lemma 1.11, with $f(t) = \int_0^t g(s) ds$, that

$$\lim_{p \to 0^+} \int_0^\infty g(t) e^{-pt^\alpha} dt = \lim_{p \to 0^+} \int_0^\infty \left(\int_0^t g(s) ds \right) e^{-pt^\alpha} p\alpha t^{\alpha-1} dt$$

$$= \lim_{t \to \infty} \int_0^t g(s) ds$$

$$= \int_0^\infty g(s) ds.$$

Remark. We mention that this problem, which may be of independent interest for the reader, is used for studying the behavior of special integrals that are connected to various topics of Fourier analysis and wavelets such as the *method of stationary phase* (see [102, Proposition 2.7.4, p. 167]).

1.66. We need the following lemma.

Lemma 1.12. *The following equality holds*

$$\int_0^1 (1-x^n)^n dx = \frac{1}{n} B\left(n+1, \frac{1}{n}\right) \quad and \quad \lim_{n \to \infty} \int_0^1 (1-x^n)^n dx = 1,$$

where B *denotes the Beta function.*

Proof. The lemma can be proved by using the substitution $x^n = y$ and the definition of the Beta function.

Let $M = \sup_{x \in [0,1]} |f(x)|$, which exists since any integrable function is bounded. We have

$$\left| \int_0^1 (1-x^n)^n f(x) dx - \int_0^1 f(x) dx \right| \leq M \int_0^1 (1-(1-x^n)^n) dx. \tag{1.59}$$

Letting n tend to ∞ in the preceding inequality, we get, based on the lemma, that the first part of the problem is solved. To show that the second limit equals 0, it suffices to prove, based on (1.59), that

$$\lim_{n \to \infty} n^\alpha \int_0^1 (1-(1-x^n)^n) dx = 0.$$

We have

$$I_n = n^\alpha \int_0^1 (1-(1-x^n)^n) dx = n^\alpha \left(1 - \frac{\Gamma(n+1)\Gamma(1+1/n)}{\Gamma(n+1+1/n)}\right)$$

$$\sim n^\alpha \left(1 - \frac{\Gamma(1+1/n)}{\sqrt[n]{n}}\right) = \frac{1}{\sqrt[n]{n}} \cdot \frac{\sqrt[n]{n} - \Gamma(1+1/n)}{n^{-\alpha}},$$

where the approximation holds based on Stirling's formula, $\Gamma(n+1)/\Gamma(n+1+1/n) \sim n^{-1/n}$. A calculation, based on l'Hôpital's rule, shows that when $\alpha \in (0,1)$, one has that $\lim_{n \to \infty} I_n = 0$.

1.67. Let $h_n : [0,1] \to R$ be given by $h_n(x) = f(x^n)g(x)$. Since f and g are continuous functions, we get that h_n is bounded, i.e., $|h_n(x)| \leq M$ for all n and $x \in [0,1]$. Let $h(x) = \lim_{n \to \infty} h_n(x)$ be the limit function. A calculation shows that $h(x) = f(0)g(x)$ for $x \in [0,1)$ and $h(1) = f(1)g(1)$. It follows from the Bounded Convergence Theorem that

$$\lim_{n \to \infty} \int_0^1 f(x^n)g(x) dx = \lim_{n \to \infty} \int_0^1 h_n(x) dx = \int_0^1 h(x) dx = f(0) \int_0^1 g(x) dx.$$

1.68. The limit equals 0. Let $h_n : [0,1] \times [0,1] \to \mathbb{R}$ be the sequence of functions defined by $h_n(x,y) = (x-y)^n f(x,y)$. A calculation shows that the limit function equals

$$h(x,y) = \begin{cases} f(1,0) & \text{if } (x,y) = (1,0), \\ 0 & \text{if } x,y \in (0,1), \\ \text{undefined} & \text{if } (x,y) = (0,1), \end{cases}$$

i.e., $h_n \to 0$ a.e. Also, we have, since $|x-y| \le 1$, that

$$|h_n(x,y)| \le |x-y|^n |f(x,y)| \le |f(x,y)|.$$

Using Lebesgue Convergence Theorem one has that

$$\lim_{n\to\infty} \int_0^1 \int_0^1 (x-y)^n f(x,y)\,dxdy = \lim_{n\to\infty} \int_0^1 \int_0^1 h_n(x,y)\,dxdy = \int_0^1 \int_0^1 h(x,y)\,dxdy = 0.$$

1.6.3 Non-standard Limits

> *On earth there is nothing great but man; in man there is nothing great but mind.*
>
> Sir William Rowan Hamilton (1805–1865)

1.69. See [48].

1.70. We need the following lemma.

Lemma 1.13. *Let k and l be two nonnegative integers. Then,*

$$\lim_{n\to\infty} \int_0^1 \left\{ \frac{n}{x} \right\}^k x^l\,dx = \frac{1}{(k+1)(l+1)}.$$

Proof. Using the substitution $n/x = y$ we get that

$$I_n = \int_0^1 \left\{ \frac{n}{x} \right\}^k x^l\,dx = n^{l+1} \int_n^\infty \frac{\{y\}^k}{y^{l+2}}\,dy = n^{l+1} \sum_{p=n}^\infty \int_p^{p+1} \frac{\{y\}^k}{y^{l+2}}\,dy$$

$$= n^{l+1} \sum_{p=n}^\infty \int_p^{p+1} \frac{(y-p)^k}{y^{l+2}}\,dy = n^{l+1} \sum_{p=n}^\infty \int_0^1 \frac{t^k}{(p+t)^{l+2}}\,dt$$

$$= n^{l+1} \int_0^1 t^k \left(\sum_{p=n}^\infty \frac{1}{(p+t)^{l+2}} \right)\,dt.$$

We have, based on Stolz–Cesàro lemma (the $0/0$ case), that for $t \in [0,1]$, one has

$$\lim_{n\to\infty} n^{l+1} \sum_{p=n}^{\infty} \frac{1}{(p+t)^{l+2}} = \lim_{n\to\infty} \frac{\sum_{p=n}^{\infty} \frac{1}{(p+t)^{l+2}}}{\frac{1}{n^{l+1}}}$$

$$= \lim_{n\to\infty} \frac{\sum_{p=n+1}^{\infty} \frac{1}{(p+t)^{l+2}} - \sum_{p=n}^{\infty} \frac{1}{(p+t)^{l+2}}}{\frac{1}{(n+1)^{l+1}} - \frac{1}{n^{l+1}}}$$

$$= \lim_{n\to\infty} \frac{-\frac{1}{(t+n)^{l+2}}}{\frac{1}{(n+1)^{l+1}} - \frac{1}{n^{l+1}}}$$

$$= \frac{1}{l+1}.$$

Let $h_n : [0,1] \to \mathbb{R}$ be the sequence of functions defined by

$$h_n(t) = n^{l+1} \sum_{p=n}^{\infty} \frac{1}{(p+t)^{l+2}},$$

and we note that the limit function equals $h(t) = \lim_{n\to\infty} h_n(t) = 1/(l+1)$ and $h_n(t) \leq n^{l+1} \sum_{p=n}^{\infty} 1/p^{l+2}$. Thus, the sequence $(h_n(t))$ is bounded for all $n \in \mathbb{N}$ and all $t \in [0,1]$ since $\lim_{n\to\infty} n^{l+1} \sum_{p=n}^{\infty} 1/p^{l+2} = 1/(1+l)$. It follows, in view of the Bounded Convergence Theorem, that

$$\lim_{n\to\infty} I_n = \int_0^1 t^k \lim_{n\to\infty} \left(n^{l+1} \sum_{p=n}^{\infty} \frac{1}{(p+t)^{l+2}} \right) dt = \int_0^1 \frac{t^k}{l+1} dt = \frac{1}{(k+1)(l+1)},$$

and the lemma is proved.

Now we are ready to solve part (a) of the problem. Let $\varepsilon > 0$ and let P and Q be the polynomials that uniformly approximate f and g, i.e., $|f(x) - P(x)| < \varepsilon$ and $|g(x) - Q(x)| < \varepsilon$ for all $x \in [0,1]$. We have that

$$\int_0^1 f\left(\left\{\frac{n}{x}\right\}\right) g(x) dx = \int_0^1 \left(f\left(\left\{\frac{n}{x}\right\}\right) - P\left(\left\{\frac{n}{x}\right\}\right) \right) g(x) dx$$

$$+ \int_0^1 P\left(\left\{\frac{n}{x}\right\}\right) (g(x) - Q(x)) dx + \int_0^1 P\left(\left\{\frac{n}{x}\right\}\right) Q(x) dx.$$

It follows that

$$\left| \int_0^1 f\left(\left\{\frac{n}{x}\right\}\right) g(x) dx - \int_0^1 P\left(\left\{\frac{n}{x}\right\}\right) Q(x) dx \right| \leq \varepsilon \int_0^1 |g(x)| dx$$

$$+ \varepsilon \int_0^1 \left| P\left(\left\{\frac{n}{x}\right\}\right) \right| dx \leq \varepsilon M_g + \varepsilon(\varepsilon + M_f), \tag{1.60}$$

where the last inequality follows from the fact that $|P| \leq |P - f| + |f|$, and $M_f = \sup_{x \in [0,1]} |f(x)|$ and $M_g = \sup_{x \in [0,1]} |g(x)|$. Let $P(x) = \sum_{k=0}^{m} a_k x^k$ and let $Q(x) = \sum_{l=0}^{r} b_l x^l$. We have, based on the lemma, that

$$
\lim_{n \to \infty} \int_0^1 P\left(\left\{\frac{n}{x}\right\}\right) Q(x) dx = \sum_{k=0}^{m} \sum_{l=0}^{r} a_k b_l \lim_{n \to \infty} \int_0^1 \left\{\frac{n}{x}\right\}^k x^l dx
$$

$$
= \sum_{k=0}^{m} \sum_{l=0}^{r} \frac{a_k b_l}{(k+1)(l+1)}
$$

$$
= \int_0^1 P(x) dx \int_0^1 Q(x) dx. \tag{1.61}
$$

Letting $n \to \infty$ in (1.60), we get, based on (1.61), that

$$
\left| \lim_{n \to \infty} \int_0^1 f\left(\left\{\frac{n}{x}\right\}\right) g(x) dx - \int_0^1 P(x) dx \int_0^1 Q(x) dx \right| \leq \varepsilon M_g + \varepsilon (\varepsilon + M_f). \tag{1.62}
$$

On the other hand, when $\varepsilon \to 0$, one has

$$
\int_0^1 P(x) dx \to \int_0^1 f(x) dx \quad \text{and} \quad \int_0^1 Q(x) dx \to \int_0^1 g(x) dx. \tag{1.63}
$$

Letting ε converge to 0 in (1.62), we get, based on (1.63), that

$$
\lim_{n \to \infty} \int_0^1 f\left(\left\{\frac{n}{x}\right\}\right) g(x) dx = \int_0^1 f(x) dx \int_0^1 g(x) dx,
$$

and part (a) of the problem is solved.

(b) To solve this part of the problem we use a different technique. We have, with the substitution $n/x = y$, that

$$
\int_0^1 f\left(\left\{\frac{n}{x}\right\}\right) dx = n \int_n^\infty f(\{y\}) \frac{dy}{y^2} = \int_0^1 f(t) \left(\sum_{k=n}^\infty \frac{n}{(k+t)^2} \right) dt.
$$

Thus,

$$
n \left(\int_0^1 f\left(\left\{\frac{n}{x}\right\}\right) dx - \int_0^1 f(x) dx \right) = n \int_0^1 f(t) \left(\sum_{k=n}^\infty \frac{n}{(k+t)^2} - 1 \right) dt.
$$

Let $t \in [0, 1]$ and let

$$
g_n(t) = \sum_{k=n}^\infty \frac{1}{(k+t)^2} \quad \text{and} \quad h_n(t) = n(ng_n(t) - 1).
$$

It is easy to prove that for all $t \in [0,1]$, we have

$$n\left(\frac{\pi^2}{6} - \sum_{k=1}^{n} \frac{1}{k^2}\right) \le n g_n(t) \le n\left(\frac{\pi^2}{6} - \sum_{k=1}^{n-1} \frac{1}{k^2}\right)$$

and

$$n\left(n\left(\frac{\pi^2}{6} - \sum_{k=1}^{n} \frac{1}{k^2}\right) - 1\right) \le h_n(t) \le n\left(n\left(\frac{\pi^2}{6} - \sum_{k=1}^{n-1} \frac{1}{k^2}\right) - 1\right).$$

It follows that $\lim_{n\to\infty} n g_n(t) = 1$ and the sequence $(h_n(t))_{n\in\mathbb{N}}$ is bounded (why?).

We prove that, for $t \in [0,1]$, we have $\lim_{n\to\infty} h_n(t) = (1-2t)/2$. Using Stolz–Cesàro lemma (the $0/0$ case) we have that

$$
\begin{aligned}
l = \lim_{n\to\infty} h_n(t) &= \lim_{n\to\infty} \frac{(n+1)g_{n+1}(t) - n g_n(t)}{1/(n+1) - 1/n} \\
&= -\lim_{n\to\infty} n(n+1)(n(g_{n+1}(t) - g_n(t)) + g_{n+1}(t)) \\
&= -\lim_{n\to\infty} n^2\left(\frac{-n}{(n+t)^2} + g_{n+1}(t)\right) \\
&= -\lim_{n\to\infty} n\left(n g_{n+1}(t) - \frac{n^2}{(n+t)^2}\right) \\
&= -\lim_{n\to\infty} n\left(((n+1)g_{n+1}(t) - 1) - g_{n+1}(t) + \frac{2nt + t^2}{(n+t)^2}\right) \\
&= -\lim_{n\to\infty} h_{n+1}(t) + \lim_{n\to\infty} n g_{n+1}(t) - 2t \\
&= -l + 1 - 2t.
\end{aligned}
$$

Now, the result follows from the Lebesgue Convergence Theorem.

For an alternative solution of part (a), see [129].

Remark. It is worth mentioning that this result holds in greater generality (see [67]). More precisely, if $\beta : \mathbb{R} \to \mathbb{R}$ is a bounded measurable function that is periodic with period 1, so that β satisfies $\beta(z+1) = \beta(z)$, and if $g \in L^1([0,1])$, then

$$\lim_{n\to\infty} \int_0^1 \beta\left(\frac{n}{x}\right) g(x)dx = \int_0^1 \beta(z)dz \int_0^1 g(x)dx.$$

1.71. Before we give the solution to this problem, we need some auxiliary results.

Lemma 1.14. *Let n be an integer and let k and m be two nonnegative integers. Then,*

$$\int_n^{n+1} y^k(\{y\})^m dy = \sum_{p=0}^{k} \binom{k}{p} \frac{n^p}{k+m-p+1}.$$

Proof. We have

$$\int_n^{n+1} y^k(\{y\})^m \mathrm{d}y \stackrel{y-n=t}{=} \int_0^1 (n+t)^k t^m \mathrm{d}t = \int_0^1 \left(\sum_{p=0}^k \binom{k}{p} n^p t^{k-p} \right) t^m \mathrm{d}t$$

$$= \sum_{p=0}^k \binom{k}{p} n^p \int_0^1 t^{k+m-p} \mathrm{d}t = \sum_{p=0}^k \binom{k}{p} \frac{n^p}{k+m-p+1}.$$

Lemma 1.15. *Let k, m be two nonnegative integers. Then,*

$$\lim_{n\to\infty} \int_0^1 x^k(\{nx\})^m \mathrm{d}x = \frac{1}{(k+1)(m+1)} = \int_0^1 x^k \mathrm{d}x \int_0^1 x^m \mathrm{d}x.$$

Proof. Using the substitution $nx = y$ we get that

$$\int_0^1 x^k(\{nx\})^m \mathrm{d}x = \frac{1}{n^{k+1}} \int_0^n y^k(\{y\})^m \mathrm{d}y.$$

We have, based on Stolz–Cesàro lemma (the ∞/∞ case) combined with Lemma 1.14, that

$$\lim_{n\to\infty} \int_0^1 x^k(\{nx\})^m \mathrm{d}x = \lim_{n\to\infty} \frac{1}{n^{k+1}} \int_0^n y^k(\{y\})^m \mathrm{d}y = \lim_{n\to\infty} \frac{\int_n^{n+1} y^k(\{y\})^m \mathrm{d}y}{(n+1)^{k+1} - n^{k+1}}$$

$$= \lim_{n\to\infty} \frac{\sum_{p=0}^k \binom{k}{p} \frac{n^p}{k+m-p+1}}{(n+1)^{k+1} - n^{k+1}} = \frac{1}{(k+1)(m+1)}.$$

Lemma 1.16. *Let $f, g : [0,1] \to \mathbb{R}$ be two continuous functions. Then,*

$$\lim_{n\to\infty} \int_0^1 f(x)g(\{nx\})\mathrm{d}x = \int_0^1 f(x)\mathrm{d}x \int_0^1 g(x)\mathrm{d}x.$$

Proof. Let $\varepsilon > 0$ and let P and Q be the polynomials that uniformly approximate f and g, i.e., $|f(x) - P(x)| < \varepsilon$ and $|g(x) - Q(x)| < \varepsilon$ for all $x \in [0,1]$. We have

$$\int_0^1 f(x)g(\{nx\})\mathrm{d}x = \int_0^1 (f(x) - P(x))g(\{nx\})\mathrm{d}x$$

$$+ \int_0^1 P(x)(g(\{nx\}) - Q(\{nx\}))\mathrm{d}x + \int_0^1 P(x)Q(\{nx\})\mathrm{d}x.$$

$$(1.64)$$

On the other hand,

$$\left| \int_0^1 (f(x) - P(x))g(\{nx\})\mathrm{d}x \right| \le \int_0^1 |f(x) - P(x)||g(\{nx\})|\mathrm{d}x \le \varepsilon M_g,$$

where $M_g = \sup_{x \in [0,1]} |g(x)|$. It follows

$$- \varepsilon M_g \leq \int_0^1 (f(x) - P(x))g(\{nx\})dx \leq \varepsilon M_g. \tag{1.65}$$

Also,

$$\left| \int_0^1 P(x)(g(\{nx\}) - Q(\{nx\}))dx \right| \leq \varepsilon \int_0^1 |P(x)|dx$$

$$= \varepsilon \int_0^1 |P(x) - f(x) + f(x)|dx$$

$$\leq \varepsilon(\varepsilon + M_f),$$

where $M_f = \sup_{x \in [0,1]} |f(x)|$, and it follows that

$$- \varepsilon(\varepsilon + M_f) \leq \int_0^1 P(x)(g(\{nx\}) - Q(\{nx\}))dx \leq \varepsilon(\varepsilon + M_f). \tag{1.66}$$

Combining (1.64)–(1.66), we get that

$$-\varepsilon(\varepsilon + M_f) - \varepsilon M_g + \int_0^1 P(x)Q(\{nx\})dx \leq \int_0^1 f(x)g(\{nx\})dx$$

$$\leq \varepsilon(\varepsilon + M_f) + \varepsilon M_g$$

$$+ \int_0^1 P(x)Q(\{nx\})dx.$$

Letting n tend to infinity in the preceding inequalities, we get that

$$\left| \lim_{n \to \infty} \int_0^1 f(x)g(\{nx\})dx - \lim_{n \to \infty} \int_0^1 P(x)Q(\{nx\})dx \right| \leq \varepsilon(\varepsilon + M_f) + \varepsilon M_g. \tag{1.67}$$

Let $P(x) = \sum_k a_k x^k$ and $Q(x) = \sum_m b_m x^m$. We have, based on Lemma 1.15, that

$$\lim_{n \to \infty} \int_0^1 P(x)Q(\{nx\})dx = \lim_{n \to \infty} \int_0^1 \sum_k \sum_m a_k b_m x^k (\{nx\})^m dx$$

$$= \sum_k \sum_m a_k b_m \left(\lim_{n \to \infty} \int_0^1 x^k (\{nx\})^m dx \right)$$

$$= \sum_k \sum_m \frac{a_k b_m}{(k+1)(m+1)}$$

$$= \int_0^1 P(x)dx \int_0^1 Q(x)dx. \tag{1.68}$$

Combining (1.67) and (1.68), we obtain that

$$\left| \lim_{n \to \infty} \int_0^1 f(x)g(\{nx\})dx - \int_0^1 P(x)dx \int_0^1 Q(x)dx \right| \le \varepsilon(\varepsilon + M_f) + \varepsilon M_g. \quad (1.69)$$

It is easy to see that

$$\lim_{\varepsilon \to 0} \left(\int_0^1 P(x)dx \int_0^1 Q(x)dx \right) = \int_0^1 f(x)dx \int_0^1 g(x)dx. \quad (1.70)$$

Letting ε tend to zero in (1.69) and using (1.70), we get that the lemma is proved.

Now we are ready to solve the problem. Let $\varepsilon > 0$ and recall that any L^1 function is approximated in the L^1 norm by a continuous function (see [112, Theorem 3.14]). Thus, since $f \in L^1[0,1]$, there is $h : [0,1] \to \mathbb{R}$, a continuous function, such that $||f - h||_{L^1[0,1]} < \varepsilon$, i.e., $\int_0^1 |f(x) - h(x)|dx \le \varepsilon$. We have

$$\left| \int_0^1 f(x)g(\{nx\})dx - \int_0^1 h(x)g(\{nx\})dx \right| \le \int_0^1 |f(x) - h(x)||g(\{nx\})|dx$$

$$\le M_g ||f - h||_{L^1[0,1]}$$

$$\le M_g \varepsilon.$$

Letting n tend to infinity in the preceding inequality, we get that

$$\left| \lim_{n \to \infty} \int_0^1 f(x)g(\{nx\})dx - \lim_{n \to \infty} \int_0^1 h(x)g(\{nx\})dx \right| \le M_g \varepsilon.$$

This implies, based on Lemma 1.16, that

$$\left| \lim_{n \to \infty} \int_0^1 f(x)g(\{nx\})dx - \int_0^1 h(x)dx \int_0^1 g(x)dx \right| \le M_g \varepsilon.$$

Now,

$$\left| \lim_{n \to \infty} \int_0^1 f(x)g(\{nx\})dx - \int_0^1 f(x)dx \int_0^1 g(x)dx \right|$$

$$\le \left| \lim_{n \to \infty} \int_0^1 f(x)g(\{nx\})dx - \int_0^1 h(x)dx \int_0^1 g(x)dx \right|$$

$$+ \left| \int_0^1 (h(x) - f(x))dx \int_0^1 g(x)dx \right|$$

$$\le M_g \varepsilon + ||f - h||_{L^1[0,1]} \int_0^1 |g(x)|dx$$

$$\le M_g \varepsilon + \varepsilon \int_0^1 |g(x)|dx.$$

Since ε is arbitrary taken the result follows.

Remark. The problem is motivated by the following theorem of Fejer: If f and g are continuous functions on $\mathbb{R}$ of period 1, then

$$\lim_{n\to\infty}\int_0^1 f(x)g(nx)\,dx = \int_0^1 f(x)\,dx\int_0^1 g(x)\,dx.$$

1.72. The solution to this problem is similar to the solution of the previous problem. Use that a continuous function on a compact interval can be uniformly approximated by polynomials.

1.73. We need the following lemma.

Lemma 1.17. *Let m,n be two nonnegative integers. Then,*

$$\lim_{t\to\infty}\int_{-1}^1 \sin^n(tx)\cos^m(tx)\,dx = \frac{1}{\pi}\int_0^{2\pi}\sin^n x\cos^m x\,dx.$$

Proof. We have, based on Euler's formulae,

$$\sin(tx) = \frac{e^{itx}-e^{-itx}}{2i} \quad\text{and}\quad \cos(tx) = \frac{e^{itx}+e^{-itx}}{2},$$

that

$$\int_{-1}^1 \sin^n(tx)\cos^m(tx)\,dx$$

$$= \int_{-1}^1 \frac{1}{2^{n+m}i^n}\left(e^{itx}-e^{-itx}\right)^n\left(e^{itx}+e^{-itx}\right)^m dx$$

$$= \frac{1}{2^{n+m}i^n}\int_{-1}^1 e^{-itx(m+n)}(e^{2itx}-1)^n(e^{2itx}+1)^m dx$$

$$= \frac{1}{2^{n+m}i^n}\int_{-1}^1 e^{-itx(m+n)}\left(\sum_{k=0}^n \binom{n}{k}e^{2itxk}(-1)^{n-k}\cdot\sum_{p=0}^m \binom{m}{p}e^{2itxp}\right)dx$$

$$= \frac{1}{2^{n+m}i^n}\sum_{k=0}^n\sum_{p=0}^m \binom{n}{k}\binom{m}{p}(-1)^{n-k}\int_{-1}^1 e^{txi(2k+2p-m-n)}dx. \qquad (1.71)$$

We have that

$$\int_{-1}^1 e^{txi(2k+2p-m-n)}\,dx = \begin{cases} 2 & \text{if } 2k+2p-m-n=0, \\ \frac{2\sin t(2k+2p-n-m)}{t(2k+2p-n-m)} & \text{if } 2k+2p-n-m\neq 0. \end{cases} \qquad (1.72)$$

Combining (1.71) and (1.72), we get that

$$\int_{-1}^{1} \sin^n(tx)\cos^m(tx)dx = \frac{2}{2^{n+m}\,i^n} \left(\sum_{2k+2p-n-m=0} \binom{n}{k}\binom{m}{p}(-1)^{n-k} \right)$$

$$+ \frac{1}{t} \cdot \frac{2}{2^{n+m}\,i^n} \left(\sum_{2k+2p-n-m\neq 0} \binom{n}{k}\binom{m}{p}(-1)^{n-k}\frac{\sin t(2k+2p-n-m)}{2k+2p-n-m} \right),$$

and in the limit,

$$\lim_{t\to\infty} \int_{-1}^{1} \sin^n(tx)\cos^m(tx)dx = \frac{2}{2^{n+m}\,i^n} \left(\sum_{2k+2p-n-m=0} \binom{n}{k}\binom{m}{p}(-1)^{n-k} \right).$$

$$(1.73)$$

Now we calculate

$$\int_{0}^{2\pi} \sin^n x\cos^m x\,dx = \int_{0}^{2\pi} \left(\frac{e^{ix}-e^{-ix}}{2i} \right)^n \cdot \left(\frac{e^{ix}+e^{-ix}}{2} \right)^m dx$$

$$= \frac{1}{2^{m+n}\,i^n} \int_{0}^{2\pi} e^{-ix(m+n)}(e^{2ix}-1)^n(e^{2ix}+1)^m dx$$

$$= \frac{1}{2^{m+n}\,i^n} \int_{0}^{2\pi} e^{-ix(m+n)} \left(\sum_{k=0}^{n}\binom{n}{k}e^{2ixk}(-1)^{n-k} \right) \left(\sum_{p=0}^{m}\binom{m}{p}e^{2ixp} \right) dx$$

$$= \frac{1}{2^{m+n}\,i^n} \sum_{k=0}^{n}\sum_{p=0}^{m}\binom{n}{k}\binom{m}{p}(-1)^{n-k}\int_{0}^{2\pi} e^{i(2k+2p-m-n)x}dx.$$

On the other hand,

$$\int_{0}^{2\pi} e^{i(2k+2p-m-n)x}dx = \begin{cases} 2\pi & \text{if } 2k+2p-n-m=0, \\ 0 & \text{if } 2k+2p-n-m\neq 0. \end{cases}$$

It follows that

$$\int_{0}^{2\pi} \sin^n x\cos^m x\,dx = \frac{2\pi}{2^{m+n}\,i^n} \left(\sum_{2k+2p-n-m=0} \binom{n}{k}\binom{m}{p}(-1)^{n-k} \right). \qquad (1.74)$$

Combining (1.73) and (1.74), we get that the lemma is proved.

Now we are ready to solve the problem. Let $\varepsilon > 0$ and let $P(x,y) = \sum_{k,p} a_{k,p}x^k y^p$ be the polynomial that uniformly approximates f. We have that $-\varepsilon \leq f(x,y) - P(x,y) \leq \varepsilon$ for all $x,y \in [-1,1]$, and it follows that

$$-\varepsilon \leq f(\sin(xt),\cos(xt)) - P(\sin(xt),\cos(xt)) \leq \varepsilon. \qquad (1.75)$$

We have, based on Lemma 1.17, that

$$\lim_{t\to\infty}\int_{-1}^{1}P(\sin(xt),\cos(xt))dx=\sum_{k,p}a_{k,p}\left(\lim_{t\to\infty}\int_{-1}^{1}\sin^k(xt)\cos^p(xt)dx\right)$$

$$=\sum_{k,p}a_{k,p}\frac{1}{\pi}\int_0^{2\pi}\sin^k x\cos^p x\,dx$$

$$=\frac{1}{\pi}\int_0^{2\pi}\left(\sum_{k,p}a_{k,p}\sin^k x\cos^p x\right)$$

$$=\frac{1}{\pi}\int_0^{2\pi}P(\sin x,\cos x)dx. \tag{1.76}$$

Integrating (1.75), we get that

$$-2\varepsilon+\int_{-1}^{1}P(\sin(xt),\cos(xt))dx\le\int_{-1}^{1}f(\sin(xt),\cos(xt))dx$$

$$\le 2\varepsilon+\int_{-1}^{1}P(\sin(xt),\cos(xt))dx,$$

and this implies

$$-2\varepsilon+\lim_{t\to\infty}\int_{-1}^{1}P(\sin(xt),\cos(xt))dx\le\lim_{t\to\infty}\int_{-1}^{1}f(\sin(xt),\cos(xt))dx$$

$$\le 2\varepsilon+\lim_{t\to\infty}\int_{-1}^{1}P(\sin(xt),\cos(xt))dx,$$

It follows from (1.76) that

$$-2\varepsilon+\frac{1}{\pi}\int_0^{2\pi}P(\sin x,\cos x)dx\le\lim_{t\to\infty}\int_{-1}^{1}f(\sin(xt),\cos(xt))dx$$

$$\le 2\varepsilon+\frac{1}{\pi}\int_0^{2\pi}P(\sin x,\cos x)dx. \tag{1.77}$$

On the other hand, since $-\varepsilon+f(\sin x,\cos x)\le P(\sin x,\cos x)\le\varepsilon+f(\sin x,\cos x)$, we get, by integration, that

$$-2\varepsilon+\frac{1}{\pi}\int_0^{2\pi}f(\sin x,\cos x)dx\le\frac{1}{\pi}\int_0^{2\pi}P(\sin x,\cos x)dx$$

$$\le 2\varepsilon+\frac{1}{\pi}\int_0^{2\pi}f(\sin x,\cos x)dx. \tag{1.78}$$

It follows, based on (1.77) and (1.78), that

$$-4\varepsilon + \frac{1}{\pi}\int_0^{2\pi} f(\sin x, \cos x)dx \le \lim_{t\to\infty}\int_{-1}^1 f(\sin(xt), \cos(xt))dx$$

$$\le 4\varepsilon + \frac{1}{\pi}\int_0^{2\pi} f(\sin x, \cos x)dx.$$

This implies that

$$\left| \lim_{t\to\infty}\int_{-1}^1 f(\sin(xt), \cos(xt))dx - \frac{1}{\pi}\int_0^{2\pi} f(\sin x, \cos x)dx \right| \le 4\varepsilon,$$

and since ε is arbitrary taken, the result follows. The second limit can be proved by using a similar technique.

Remark. The problem was motivated by the following problem in classical analysis (see [14, Problem 5.75]). Determine

$$\lim_{t\to\infty}\int_a^b f(x, \sin(tx))dx, \quad \text{for} \quad f \in C([a,b] \times [-1,1]).$$

We have, as a consequence of this problem, that if $f : [-1,1] \to \mathbb{R}$ is a continuous function, then

$$\lim_{t\to\infty}\int_{-1}^1 f(\sin(xt))dx = \frac{1}{\pi}\int_0^{2\pi} f(\sin x)dx.$$

1.74. The limit equals $f(0)$. See [53] or the solution of Problem **1.76**.

1.75. The limit equals $f(1/e)$. See the solution of the next problem.

1.76. The limit equals $f(0)g(1/e)$. We need the following lemma.

Lemma 1.18. *(a) Let $k \ge 0$ be an integer and let l be a positive integer. Then,*

$$\lim_{n\to\infty}\int_0^1 \cdots \int_0^1 (\sqrt[n]{x_1 x_2 \cdots x_n})^k \left(\frac{n}{\frac{1}{x_1} + \cdots + \frac{1}{x_n}} \right)^l dx_1 \cdots dx_n = 0.$$

(b) Let $k \ge 0$ be an integer. Then,

$$\lim_{n\to\infty}\int_0^1 \cdots \int_0^1 (\sqrt[n]{x_1 x_2 \cdots x_n})^k dx_1 \cdots dx_n = \left(\frac{1}{e} \right)^k.$$

Proof. (a) Let I_n be the integral in question. Using the substitutions $\frac{1}{x_i} = y_i$, $i = 1, 2, \ldots, n$, we obtain that

$$I_n = n^l \int_1^\infty \cdots \int_1^\infty \frac{dy_1 \cdots dy_n}{(y_1 + \cdots + y_n)^l y_1^{2+k/n} \cdots y_n^{2+k/n}}.$$

Since

$$\frac{1}{(y_1 + \cdots + y_n)^l} = \frac{1}{\Gamma(l)}\int_0^\infty e^{-t(y_1 + \cdots + y_n)} t^{l-1}dt,$$

we obtain that

$$I_n = n^l \int_1^\infty \cdots \int_1^\infty \left(\frac{1}{\Gamma(l)} \int_0^\infty e^{-t(y_1+\cdots+y_n)} t^{l-1} dt \right) \frac{dy_1 \cdots dy_n}{y_1^{2+k/n} \cdots y_n^{2+k/n}}$$

$$= \frac{n^l}{\Gamma(l)} \int_0^\infty t^{l-1} \left(\int_1^\infty \frac{e^{-ty_1}}{y_1^{2+k/n}} dy_1 \right) \cdots \left(\int_1^\infty \frac{e^{-ty_n}}{y_n^{2+k/n}} dy_n \right) dt$$

$$= \frac{n^l}{\Gamma(l)} \int_0^\infty t^{l-1} \left(\int_1^\infty \frac{e^{-ty}}{y^{2+k/n}} dy \right)^n dt$$

$$= \frac{1}{\Gamma(l)} \int_0^\infty s^{l-1} \left(\int_1^\infty \frac{e^{-sy/n}}{y^{2+k/n}} dy \right)^n ds,$$

where the last equality follows by using the substitution $t = \frac{s}{n}$ in the outer integral.

Let

$$f_n(s) = s^{l-1} \left(\int_1^\infty \frac{e^{-sy/n}}{y^{2+k/n}} dy \right)^n, \quad s > 0.$$

We have

$$f_n(s) \le s^{l-1} \left(e^{-s/n} \int_1^\infty \frac{dy}{y^2} \right)^n = s^{l-1} e^{-s}, \tag{1.79}$$

and we note that the function $s \to s^{l-1} e^{-s}$ is integrable over $(0, \infty)$.

Now, we prove that

$$\lim_{n \to \infty} f_n(s) = 0. \tag{1.80}$$

We have that

$$0 < \left(\int_1^\infty \frac{e^{-sy/n}}{y^{2+k/n}} dy \right)^n < \left(\int_1^\infty \frac{e^{-sy/n}}{y^2} dy \right)^n.$$

We note that we have an indeterminate form of type 1^∞ and this implies that

$$\lim_{n \to \infty} \left(\int_1^\infty \frac{e^{-sy/n}}{y^2} dy \right)^n = e^{\lim_{n\to\infty} n \left(\int_1^\infty \frac{e^{-sy/n}}{y^2} dy - 1 \right)} = e^{\lim_{n\to\infty} n \int_1^\infty \frac{e^{-sy/n}-1}{y^2} dy} = e^{-\infty} = 0.$$

The last equality can be proved as follows. Integrating by parts we obtain that

$$\lim_{n \to \infty} n \int_1^\infty \frac{e^{-sy/n} - 1}{y^2} dy = \lim_{n \to \infty} n \left(\left. \frac{1 - e^{-sy/n}}{y} \right|_1^\infty - \frac{s}{n} \int_1^\infty \frac{e^{-sy/n}}{y} dy \right)$$

$$= -s - s \lim_{n \to \infty} \int_1^\infty \frac{e^{-sy/n}}{y} dy$$

$$= -\infty,$$

and the limit, $\lim_{n\to\infty} \int_1^\infty e^{-sy/n}/y\,dy = \infty$, follows from the Monotone Convergence Theorem. Combining (1.79) and (1.80) and using Lebesgue Convergence Theorem, we obtain that

$$\lim_{n\to\infty} I_n = \frac{1}{\Gamma(l)} \cdot \lim_{n\to\infty} \int_0^\infty f_n(s)\,ds = 0,$$

and the first part of the lemma is proved.

(b) We have

$$\lim_{n\to\infty} \int_0^1 \cdots \int_0^1 (\sqrt[n]{x_1 x_2 \cdots x_n})^k \, dx_1 \cdots dx_n = \lim_{n\to\infty} \left(\frac{n}{n+k}\right)^n = e^{-k},$$

and the second part of the lemma is proved.

Let $\varepsilon > 0$ and let P and Q be the polynomials that uniformly approximate f and g, i.e., $|f(x) - P(x)| < \varepsilon$ and $|g(x) - Q(x)| < \varepsilon$, for all $x \in [0,1]$. Also, let $M_g = \sup_{x \in [0,1]} |g(x)|$. We note that

$$f \cdot g = (f - P) \cdot g + P(g - Q) + P \cdot Q. \tag{1.81}$$

We have

$$\left| \int_0^1 \cdots \int_0^1 \left(f\left(\frac{n}{\frac{1}{x_1} + \cdots + \frac{1}{x_n}}\right) - P\left(\frac{n}{\frac{1}{x_1} + \cdots + \frac{1}{x_n}}\right) \right) g(\sqrt[n]{x_1 \cdots x_n}) dx_1 \cdots dx_n \right|$$

is less than or equal to $\varepsilon \cdot M_g$. Since ε was arbitrary we obtain that

$$\lim_{n\to\infty} \int_0^1 \cdots \int_0^1 (f - P) \left(\frac{n}{\frac{1}{x_1} + \cdots + \frac{1}{x_n}}\right) g(\sqrt[n]{x_1 \cdots x_n}) dx_1 \cdots dx_n = 0. \tag{1.82}$$

Similarly,

$$0 < \left| \int_0^1 \cdots \int_0^1 (g(\sqrt[n]{x_1 \cdots x_n}) - Q(\sqrt[n]{x_1 \cdots x_n})) P\left(\frac{n}{\frac{1}{x_1} + \cdots + \frac{1}{x_n}}\right) dx_1 \cdots dx_n \right|$$

$$\leq \varepsilon \int_0^1 \cdots \int_0^1 \left| P\left(\frac{n}{\frac{1}{x_1} + \cdots + \frac{1}{x_n}}\right) \right| dx_1 \cdots dx_n. \tag{1.83}$$

Let $P(x) = a_m x^m + \cdots + a_0$ and let $Q(x) = b_p x^p + \cdots + b_0$. An application of part (a) of Lemma 1.18, with $k = 0$, shows that

$$\lim_{n\to\infty} \int_0^1 \left| P\left(\frac{n}{\frac{1}{x_1} + \cdots + \frac{1}{x_n}}\right) \right| dx_1 \cdots dx_n$$

$$\leq \lim_{n\to\infty} \sum_{j=0}^m |a_j| \int_0^1 \cdots \int_0^1 \left(\frac{n}{\frac{1}{x_1} + \cdots + \frac{1}{x_n}}\right)^j dx_1 \cdots dx_n = |a_0|. \tag{1.84}$$

Combining (1.83) and (1.84), we obtain, since ε was arbitrary, that

$$\lim_{n\to\infty} \int_0^1 \cdots \int_0^1 (g(\sqrt[n]{x_1\cdots x_n}) - Q(\sqrt[n]{x_1\cdots x_n}))P\left(\frac{n}{\frac{1}{x_1}+\cdots+\frac{1}{x_n}}\right) dx_1\cdots dx_n = 0.$$

$$(1.85)$$

On the other hand,

$$P\left(\frac{n}{\frac{1}{x_1}+\cdots+\frac{1}{x_n}}\right) \cdot Q(\sqrt[n]{x_1\cdots x_n}) = \sum_{i=0}^m \sum_{j=0}^p a_i b_j \left(\frac{n}{\frac{1}{x_1}+\cdots+\frac{1}{x_n}}\right)^i (\sqrt[n]{x_1\cdots x_n})^j.$$

Integrating the preceding equality and taking the limit, we obtain, as a consequence of Lemma 1.18, that

$$\lim_{n\to\infty} \int_0^1 \cdots \int_0^1 P\left(\frac{n}{\frac{1}{x_1}+\cdots+\frac{1}{x_n}}\right) \cdot Q(\sqrt[n]{x_1\cdots x_n}) dx_1\cdots dx_n$$

$$= a_0 Q\left(\frac{1}{e}\right) = P(0)Q\left(\frac{1}{e}\right).$$

$$(1.86)$$

Combining (1.82), (1.85), and (1.86), we obtain, based on (1.81), that

$$\lim_{n\to\infty} \int_0^1 \cdots \int_0^1 f\left(\frac{n}{\frac{1}{x_1}+\cdots+\frac{1}{x_n}}\right) \cdot g(\sqrt[n]{x_1 x_2\cdots x_n}) dx_1\cdots dx_n = P(0)Q\left(\frac{1}{e}\right).$$

Since $P(0)Q(1/e) \to f(0)g(1/e)$, as $\varepsilon \to 0$, we obtain that the desired limit holds and the problem is solved.

Remark. It is worth mentioning that the limit

$$\lim_{n\to\infty} \int_0^1 \cdots \int_0^1 f\left(\frac{n}{\frac{1}{x_1}+\cdots+\frac{1}{x_n}}\right) \cdot g(\sqrt[n]{x_1 x_2\cdots x_n}) dx_1\cdots dx_n = f(0)g\left(\frac{1}{e}\right)$$

equals the product of the two limits

$$\lim_{n\to\infty} \int_0^1 \cdots \int_0^1 f\left(\frac{n}{\frac{1}{x_1}+\cdots+\frac{1}{x_n}}\right) dx_1\cdots dx_n = f(0)$$

and

$$\lim_{n\to\infty} \int_0^1 \cdots \int_0^1 g(\sqrt[n]{x_1 x_2\cdots x_n}) dx_1\cdots dx_n = g\left(\frac{1}{e}\right).$$

1.77. The limit equals $f(1/e)$. Approximate f by a polynomial and use Lemma 1.18. See the solution of Problem **1.76**.

1.78. The limit equals $f(0)$. Approximate f by a polynomial and use Lemma 1.18. See the solution of Problem **1.76**.

1.6.4 Comments on Two Open Problems

An expert is someone who knows some of the worst mistakes that can be made in his subject, and how to avoid them.

Werner Heisenberg (1901–1976)

1.80. The motivation behind this problem is a standard exercise from the theory of infinite series (see [14, Problem 3.80, p. 26]) or Problem **3.3** which is about studying the convergence of the infinite series $\sum_{n=2}^{\infty}((2 - \sqrt{e})(2 - \sqrt[3]{e})\cdots(2 - \sqrt[n]{e}))^{\alpha}$. One possible approach of this problem is based on proving that the general term of the series behaves like $1/n$. To see this, one can prove that if $x > 1$, then $1 - e^{1/x} + 1/(x-1) > 0$. Let $x_n = (2 - \sqrt{e})(2 - \sqrt[3]{e})\cdots(2 - \sqrt[n]{e})$. It follows that for all $k \geq 3$, one has $2 - \sqrt[k]{e} > 1 - \frac{1}{k-1}$. Thus,

$$x_n = (2 - \sqrt{e})\prod_{k=3}^{n}(2 - \sqrt[k]{e}) > (2 - \sqrt{e})\prod_{k=3}^{n}\left(1 - \frac{1}{k-1}\right) = \frac{2 - \sqrt{e}}{n-1}. \qquad (1.87)$$

On the other hand, since for all $x \geq 0$ one has that $e^x \geq 1 + x$, we obtain that $2 - \sqrt[k]{e} \leq 1 - 1/k$, which implies that

$$x_n = \prod_{k=2}^{n}(2 - \sqrt[k]{e}) < \prod_{k=2}^{n}\left(1 - \frac{1}{k}\right) = \frac{1}{n}. \qquad (1.88)$$

It follows from (1.87) and (1.88) that $(2 - \sqrt{e})/(n-1) < x_n < 1/n$, and our goal is achieved. Now, one can prove, and this is left as an exercise to the reader, that $\lim_{n \to \infty} n x_n$ exists, so the question of determining whether this limit can be calculated in terms of well-known constants is natural and appealing.

1.81. This problem is motivated by Problem **1.12**. A heuristic motivation of the problem is as follows. One can show (prove it!) that $\lim_{n \to \infty}(n - \Gamma(1/n)) = \gamma$. This means that when n is large, one has that $\Gamma(1/n) \sim n$. Replacing n by n/k in the preceding estimation, and bear with me even if this is not correct, we have that $\Gamma(k/n) \sim n/k$. This in turn implies that $(\Gamma(k/n))^{-k} \sim (k/n)^k$. Hence, $\sum_{k=1}^{n}(\Gamma(k/n))^{-k} \sim \sum_{k=1}^{n}(k/n)^k$ and since the limit of the right-hand side sum exists and can be calculated (this is Problem **1.12**), one may wonder what would be the limit of the left-hand side sum. Numerical calculations show that the limit of the sum involving the Gamma function would be $e^{\gamma}/(e^{\gamma} - 1)$.

Chapter 2
Fractional Part Integrals

If I had the theorems! Then I should find the proofs easily enough.

Bernhard Riemann (1826–1866)

2.1 Single Integrals

I have had my results for a long time: but I do not yet know how I am to arrive at them.

Carl Friedrich Gauss (1777–1855)

2.1. A de la Vallée Poussin integral. Calculate

$$\int_0^1 \left\{ \frac{1}{x} \right\} dx.$$

2.2. Prove that

$$\int_0^1 \left\{ \frac{1}{x} \right\}^2 dx = \ln(2\pi) - \gamma - 1.$$

Let A denote the Glaisher–Kinkelin constant defined by the limit

$$A = \lim_{n \to \infty} n^{-n^2/2 - n/2 - 1/12} e^{n^2/4} \prod_{k=1}^{n} k^k = 1.28242\,71291\,00622\,63687\dots.$$

(continued)

O. Furdui, *Limits, Series, and Fractional Part Integrals: Problems in Mathematical Analysis*, Problem Books in Mathematics, DOI 10.1007/978-1-4614-6762-5_2, © Springer Science+Business Media New York 2013

(continued)

2.3. A cubic integral. Prove that

$$\int_0^1 \left\{\frac{1}{x}\right\}^3 dx = -\frac{1}{2} - \gamma + \frac{3}{2}\ln(2\pi) - 6\ln A.$$

2.4. Let $k \geq 0$ be an integer. Prove that

$$\int_0^1 \left\{\frac{1}{x}\right\}^k dx = \sum_{p=1}^{\infty} \frac{\zeta(p+1)-1}{\binom{k+p}{p}}.$$

2.5. (a) Let $k \geq 1$ be an integer. Prove that

$$\int_0^1 \left\{\frac{k}{x}\right\} dx = k\left(1 + \frac{1}{2} + \cdots + \frac{1}{k} - \ln k - \gamma\right),$$

where γ denotes the Euler–Mascheroni constant.

(b) More generally, if q is a positive real number, then

$$\int_0^1 \left\{\frac{q}{x}\right\} dx = \begin{cases} q(1 - \gamma - \ln q) & \text{if } q \leq 1, \\ q\left(1 + \frac{1}{2} + \cdots + \frac{1}{1+\lfloor q\rfloor} - \gamma - \ln q + \frac{\lfloor q\rfloor(\{q\}-1)}{q(1+\lfloor q\rfloor)}\right) & \text{if } q > 1. \end{cases}$$

2.6. Let $k \geq 1$ be an integer. Prove that

$$\int_0^1 \left\{\frac{k}{x}\right\}^2 dx = k\left(\ln(2\pi) - \gamma + 1 + \frac{1}{2} + \cdots + \frac{1}{k} + 2k\ln k - 2k - 2\ln k!\right).$$

2.7. Let $k \geq 2$ be an integer. Prove that

$$\int_0^1 \left\{\frac{1}{\sqrt[k]{x}}\right\} dx = \frac{k}{k-1} - \zeta(k),$$

where ζ denotes the Riemann zeta function.

2.8. Let $k \geq 2$ be an integer. Prove that

$$\int_0^1 \left\{ \frac{k}{\sqrt[k]{x}} \right\} dx = \frac{k}{k-1} - k^k \left(\zeta(k) - \frac{1}{1^k} - \frac{1}{2^k} - \cdots - \frac{1}{k^k} \right).$$

2.9. Let $k \geq 2$ be an integer. Prove that

$$\int_0^1 \left\{ \frac{1}{k\sqrt[k]{x}} \right\} dx = \frac{1}{k-1} - \frac{\zeta(k)}{k^k}.$$

2.10. A Euler's constant integral. Prove that

$$\int_0^1 \left\{ \frac{1}{x} \right\} \left\{ \frac{1}{1-x} \right\} dx = 2\gamma - 1.$$

2.11. (a) A quadratic integral and Euler's constant. Prove that

$$\int_0^1 \left\{ \frac{1}{x} \right\}^2 \left\{ \frac{1}{1-x} \right\}^2 dx = 4\ln(2\pi) - 4\gamma - 5.$$

(b) A class of fractional part integrals. Let $n \geq 2$ be an integer. Then

$$\int_0^1 \left\{ \frac{1}{x} \right\}^n \left\{ \frac{1}{1-x} \right\}^n dx = 2\sum_{j=2}^{n-1} (-1)^{n+j-1}(\zeta(j)-1) + (-1)^n$$

$$- 2n \sum_{m=0}^{\infty} \frac{\zeta(2m+n) - \zeta(2m+n+1)}{n+m},$$

where ζ is the Riemann zeta function and the first sum is missing when $n = 2$.

2.12. Prove that

$$\int_0^1 \left\{ \frac{1}{x} \right\}^2 \left\{ \frac{1}{1-x} \right\} dx = \frac{5}{2} - \gamma - \ln(2\pi).$$

2.13. Prove that

$$\int_0^1 \left\{ (-1)^{\lfloor \frac{1}{x} \rfloor} \frac{1}{x} \right\} dx = 1 + \ln\frac{2}{\pi},$$

where $\lfloor a \rfloor$ denotes the greatest integer not exceeding a.

2.14. (a) Prove that

$$\int_0^1 x \left\{ \frac{1}{x} \right\} \left\lfloor \frac{1}{x} \right\rfloor dx = \frac{\pi^2}{12} - \frac{1}{2}.$$

(b) More generally, let α, β, and γ be positive real numbers. Study the convergence of

$$\int_0^1 x^\alpha \left\{ \frac{1}{x} \right\}^\beta \left\lfloor \frac{1}{x} \right\rfloor^\gamma dx,$$

where $\lfloor a \rfloor$ denotes the greatest integer not exceeding a.

2.15. Prove that

$$\int_0^1 \left\{ \frac{1}{x} \right\} \frac{x}{1-x} dx = \gamma.$$

2.16. Let $m > -1$ be a real number. Prove that

$$\int_0^1 \{\ln x\} x^m dx = \frac{e^{m+1}}{(m+1)(e^{m+1} - 1)} - \frac{1}{(1+m)^2}.$$

2.17. The first Stieltjes constant, γ_1, is the special constant defined by

$$\gamma_1 = \lim_{n \to \infty} \left(\sum_{k=1}^n \frac{\ln k}{k} - \frac{\ln^2 n}{2} \right).$$

Prove that

$$\int_0^1 \left\{ \frac{1}{x} \right\} \ln x \, dx = \gamma + \gamma_1 - 1.$$

2.18. Let k be a positive real number. Find the value of

$$\lim_{n \to \infty} \int_0^1 \left\{ \frac{n}{x} \right\}^k dx.$$

2.19. Calculate

$$L = \lim_{n \to \infty} \int_0^1 \left\{ \frac{n}{x} \right\}^n dx \quad \text{and} \quad \lim_{n \to \infty} n \left(\int_0^1 \left\{ \frac{n}{x} \right\}^n dx - L \right).$$

2.20. Let $m > 0$ be a real number. Calculate

$$\int_0^1 t^m \left\{ \frac{1}{t} \right\} dt.$$

2.21. Let $m \geq 1$ be an integer. Prove that

$$\int_0^1 \left\{ \frac{1}{x} \right\}^m x^m dx = 1 - \frac{\zeta(2) + \zeta(3) + \cdots + \zeta(m+1)}{m+1}.$$

2.22. (a) Let m and k be positive real numbers. Prove that

$$\int_0^1 \left\{ \frac{1}{x} \right\}^k x^m dx = \frac{k!}{(m+1)!} \sum_{j=1}^{\infty} \frac{(m+j)!}{(k+j)!} (\zeta(m+j+1) - 1).$$

(b) **A new integral formula for Euler's constant.** If $m \geq 1$ is an integer, then

$$\int_0^1 x^m \left\{ \frac{1}{x} \right\}^{m+1} dx = H_{m+1} - \gamma - \sum_{j=2}^{m+1} \frac{\zeta(j)}{j},$$

where H_{m+1} denotes the $(m+1)$th harmonic number.

2.2 Double Integrals

Nature laughs at the difficulties of integration.

Pierre-Simon de Laplace (1749–1827)

2.23. (a) Calculate

$$\int_0^1 \int_0^1 x \left\{ \frac{\bullet 1}{1 - xy} \right\} dxdy.$$

(b) More generally, if $k \geq 1$ is an integer, calculate

$$\int_0^1 \int_0^1 x^k \left\{ \frac{1}{1 - xy} \right\} dxdy.$$

2.24. Let $m \geq 1$ be an integer. Calculate

$$\int_0^1 \int_0^1 \left\{ \frac{1}{x+y} \right\}^m dxdy.$$

2.25. Let $k \geq 1$ be an integer. Calculate

$$\int_0^1 \int_0^{1-x} \frac{dxdy}{\left(\left\lfloor \frac{x}{y} \right\rfloor + 1 \right)^k},$$

where $\lfloor a \rfloor$ denotes the greatest integer not exceeding a.

2.26. Let $k \geq 1$ be an integer. Calculate

$$\int_0^1 \int_0^1 \frac{dxdy}{\left(\left\lfloor \frac{x}{y} \right\rfloor + 1 \right)^k},$$

where $\lfloor a \rfloor$ denotes the greatest integer not exceeding a.

2.27. Calculate

$$\int_0^1 \int_0^1 \left\{ \frac{x}{y} \right\} dxdy.$$

2.28. Let $k \geq 1$ be an integer. Prove that

$$\int_0^1 \int_0^1 \left\{ k\frac{x}{y} \right\} dxdy = \frac{k}{2} \left(1 + \frac{1}{2} + \cdots + \frac{1}{k} - \ln k - \gamma \right) + \frac{1}{4}.$$

2.29. Let m and n be positive integers such that $m \leq n$. Prove that

$$\int_0^1 \int_0^1 \left\{ \frac{mx}{ny} \right\} dxdy = \frac{m}{2n} \left(\ln \frac{n}{m} + \frac{3}{2} - \gamma \right).$$

2.30. Let $k \geq 0$ be a real number. Prove that

$$\int_0^1 \int_0^1 \left\{ \frac{x^k}{y} \right\} dxdy = \frac{2k+1}{(k+1)^2} - \frac{\gamma}{k+1}.$$

2.31. Prove that

$$\int_0^1 \int_0^1 \left\{ \frac{x}{y} \right\}^2 dxdy = -\frac{1}{3} - \frac{\gamma}{2} + \frac{\ln(2\pi)}{2}.$$

2.32. Let $k \geq 0$ be an integer. Prove that

$$\int_0^1 \int_0^1 \left\{ \frac{x}{y} \right\}^k dxdy = \frac{1}{2} \int_0^1 \left\{ \frac{1}{x} \right\}^k dx + \frac{1}{2(k+1)} = \frac{1}{2} \sum_{p=1}^{\infty} \frac{\zeta(p+1)-1}{\binom{k+p}{p}} + \frac{1}{2(k+1)}.$$

2.33. Let $k \geq 1$ be an integer. Prove that

$$\int_0^1 \int_0^1 \left\{ \frac{x}{y} \right\}^k \left(\frac{y}{x} \right)^k dxdy = 1 - \frac{\zeta(2) + \zeta(3) + \cdots + \zeta(k+1)}{2(k+1)}.$$

2.34. Let $k \geq 1$ be an integer and let p be a real number such that $k - p > -1$. Prove that

$$\int_0^1 \int_0^1 \left\{ \frac{x}{y} \right\}^k \frac{y^k}{x^p} dxdy = \frac{1}{k-p+1} - \frac{\zeta(2) + \zeta(3) + \cdots + \zeta(k+1)}{(k+2-p)(k+1)}.$$

2.35. Let $k \geq 1$ and let m, p be nonnegative integers such that $m - p > -2$ and $k - p > -1$. Prove that

$$\int_0^1 \int_0^1 \left\{ \frac{x}{y} \right\}^k \frac{y^m}{x^p} dxdy$$

$$= \frac{1}{m-p+2} \left(\frac{k!}{(m+1)!} \sum_{j=1}^{\infty} \frac{(m+j)!}{(k+j)!} (\zeta(m+j+1) - 1) + \frac{1}{k-p+1} \right).$$

2.36. A surprising appearance of $\zeta(2)$. Prove that

$$\int_0^1 \int_0^1 \left\{ \frac{x}{y} \right\} \left\{ \frac{y}{x} \right\} dxdy = 1 - \frac{\pi^2}{12}.$$

2.37. Let $n, m > -1$ be real numbers. Prove that

$$\int_0^1 \int_0^1 x^m y^n \left\{ \frac{x}{y} \right\} \left\{ \frac{y}{x} \right\} dxdy$$

$$= \frac{1}{m+n+2} \left(\frac{1}{n+1} + \frac{1}{m+1} - \frac{\zeta(n+2)}{n+2} - \frac{\zeta(m+2)}{m+2} \right).$$

2.38. Let $n > -1$ be a real number. Prove that

$$\int_0^1 \int_0^1 x^n y^n \left\{ \frac{x}{y} \right\} \left\{ \frac{y}{x} \right\} dxdy = \frac{1}{(n+1)^2} - \frac{\zeta(n+2)}{(n+1)(n+2)}.$$

2.39. Let $k \geq 1$ be an integer and let

$$I_k = \int_0^1 \int_0^1 \left\{ k\frac{x}{y} \right\} \left\{ k\frac{y}{x} \right\} dxdy.$$

Prove that

(a) $I_k = \int_0^1 \{kx\}\{k/x\}\,dx$.

(b) $I_2 = 49/6 - 2\pi^2/3 - 2\ln 2$.

(c) **Open problem.** Find an explicit formula for I_k when $k \geq 3$.

2.40. Let $m \geq 1$ be an integer. Prove that

$$\int_0^1 \int_0^1 \left\{\frac{x}{y}\right\}^m \left\{\frac{y}{x}\right\}^m dxdy = 1 - \frac{\zeta(2) + \zeta(3) + \cdots + \zeta(m+1)}{m+1}.$$

2.41. Let $m, k \geq 1$ be two integers. Prove that

$$\int_0^1 \int_0^1 \left\{\frac{x}{y}\right\}^m \left\{\frac{y}{x}\right\}^k dxdy = \frac{k!}{2(m+1)!} \sum_{p=1}^{\infty} \frac{(m+p)!}{(k+p)!}(\zeta(m+p+1) - 1)$$

$$+ \frac{m!}{2(k+1)!} \sum_{p=1}^{\infty} \frac{(k+p)!}{(m+p)!}(\zeta(k+p+1) - 1).$$

2.42. 1. Let $a > 0$ be a real number and let k be a positive integer. Prove that

$$\int_a^{a+k} \{x\}\,dx = \frac{k}{2}.$$

2. Let $n \geq 1$ be an integer and let $a_1, \ldots, a_n$ be positive integers. Calculate

$$\int_0^{a_1} \cdots \int_0^{a_n} \{k(x_1 + x_2 + \cdots + x_n)\}\,dx_1 dx_2 \cdots dx_n.$$

The Stieltjes constants, γ_m, are the special constants defined by

$$\gamma_m = \lim_{n \to \infty} \left(\sum_{k=1}^{n} \frac{(\ln k)^m}{k} - \frac{(\ln n)^{m+1}}{m+1} \right).$$

2.43. A multiple integral in terms of Stieltjes constants. Let $m \geq 1$ be an integer. Prove that

$$\int_0^1 \cdots \int_0^1 \left\{\frac{1}{x_1 x_2 \cdots x_m}\right\} dx_1 dx_2 \cdots dx_m = 1 - \sum_{k=0}^{m-1} \frac{\gamma_k}{k!}.$$

2.3 Quickies

In my opinion, a mathematician, in so far as he is a mathematician, need not preoccupy himself with philosophy, an opinion, moreover, which has been expressed by many philosophers.

Henri Lebesgue (1875–1941)

2.44. Let $k > -1$ be a real number and let $n \geq 1$ be an integer. Prove that

$$\int_0^1 \{nx\}^k \, dx = \frac{1}{k+1}.$$

2.45. Calculate

$$\int_0^1 \left\{ \frac{1}{x} - \frac{1}{1-x} \right\} dx.$$

2.46. Calculate

$$\int_0^1 \left\{ \frac{1}{x} - \frac{1}{1-x} \right\} \left\{ \frac{1}{x} \right\} \left\{ \frac{1}{1-x} \right\} dx.$$

2.47. Let $k > 0$ and let $m \geq 0$ be real numbers. Calculate

$$\int_0^1 \left\{ \left(\frac{1}{x} \right)^k - \left(\frac{1}{1-x} \right)^k \right\} x^m (1-x)^m dx.$$

2.48. Let n be a positive integer and let k be a natural number. Prove that

$$\int_0^1 (x - x^2)^k \{nx\} \, dx = \frac{(k!)^2}{2(2k+1)!}.$$

2.49. Let f and g be functions on $[0,1]$ with g integrable and $g(x) = g(1-x)$. Prove that

$$\int_0^1 \{f(x) - f(1-x)\} g(x) dx = \frac{1}{2} \int_0^1 g(x) dx,$$

where $\{a\}$ denotes the fractional part of a.

2.50. Calculate the double integral

$$\int_0^1 \int_0^1 \{x - y\} \, dx dy.$$

2.51. Prove that

$$\int_0^1 \int_0^1 \left\{ \frac{x-y}{x+y} \right\} dx dy = \int_0^1 \int_0^1 \left\{ \frac{x+y}{x-y} \right\} dx dy = \frac{1}{2}.$$

2.52. Let $k > 0$ be a real number. Prove that

$$\int_0^1 \int_0^1 \left\{ \frac{k}{x-y} \right\} \left\{ \frac{1}{x} \right\} \left\{ \frac{1}{y} \right\} dxdy = \frac{1}{2}(1-\gamma)^2.$$

2.53. Let $k > 0$ be a real number. Find

$$\int_0^1 \int_0^1 \left\{ \left(\frac{1}{x} \right)^k - \left(\frac{1}{y} \right)^k \right\} dxdy.$$

2.54. Let k and m be positive real numbers. Calculate

$$\int_0^1 \int_0^1 \left\{ \left(\frac{x}{y} \right)^k - \left(\frac{y}{x} \right)^k \right\} x^m y^m dxdy.$$

2.55. Let $n \geq 1$ be an integer and let $m > -1$ be a real number. Calculate

$$\int_0^1 \int_0^1 \left\{ \frac{nx}{x+y} \right\} x^m y^m dxdy.$$

2.56. Let $n \geq 1$ be an integer and let $m > -1$ be a real number. Calculate

$$\int_0^1 \int_0^1 \left\{ \frac{nx}{x-y} \right\} x^m y^m dxdy.$$

2.57. Let $a > 0$, let $k \neq 0$ be a real number, and let $g : [0,a] \times [0,a] \to \mathbb{R}$ be an integrable and symmetric function in x and y. Prove that

$$\int_0^a \int_0^a \left\{ x^k - y^k \right\} g(x,y)dxdy = \frac{1}{2} \int_0^a \int_0^a g(x,y)dxdy,$$

where $\{x\}$ denotes the fractional part of x.

2.4 Open Problems

I could never resist an integral.

G. H. Hardy (1877–1947)

2.58. Integrating a product of fractional parts. Let $n \geq 3$ be an integer. Calculate

$$\int_0^1 \cdots \int_0^1 \left\{ \frac{x_1}{x_2} \right\} \left\{ \frac{x_2}{x_3} \right\} \cdots \left\{ \frac{x_n}{x_1} \right\} dx_1 dx_2 \cdots dx_n.$$

2.59. A power integral. Let $n \geq 3$ and $m \geq 1$ be integers. Calculate, in closed form, the integral

$$\int_0^1 \cdots \int_0^1 \left\{ \frac{1}{x_1 + x_2 + \cdots + x_n} \right\}^m dx_1 \cdots dx_n.$$

2.5 Hints

> *To myself I am only a child playing on the beach, while vast oceans of truth lie undiscovered before me.*
>
> Sir Isaac Newton (1642–1727)

2.5.1 Single Integrals

> *It is not certain that everything is uncertain.*
>
> Blaise Pascal (1642–1662)

2.1. Make the substitution $1/x = t$.

2.3. Make the substitution $1/x = t$ and the integral reduces to the calculation of the series $\sum_{k=1}^{\infty} \left(3k^2 \ln(k+1)/k + 3/2 - 3k - 1/(k+1) \right)$. Then, calculate the nth partial sum of the series by using the definition of Glaisher–Kinkelin constant.

2.5. and **2.6.** Make the substitution $k/x = t$.

2.7. Make the substitution $x = 1/y^k$.

2.8. Calculate the integral by using the substitution $x = k^k/y^k$.

2.9. Make the substitution $x = 1/(k^k y^k)$.

2.10. Make the substitution $x = 1/t$ and show the integral reduces to the calculation of the series $2\sum_{k=1}^{\infty} \left(k \ln \frac{k}{k+1} + k \ln \frac{k+2}{k+1} + \frac{1}{k+2} \right)$.

2.13. Make the substitution $1/x = y$ and observe that if y is a positive real number, which is not an integer, one has $\{-y\} = 1 - \{y\}$.

2.14., 2.15., and **2.17.** Make the substitution $x = 1/y$.

2.18. Substitute $n/x = t$ and apply Stolz–Cesàro lemma (the $0/0$ case).

2.19. Make the substitution $n/x = t$ and prove that $x_n = \int_0^1 \{n/x\}^n dx$ verifies the inequalities

$$\frac{n}{n+1} \left(\frac{\pi^2}{6} - \sum_{k=1}^{n} \frac{1}{k^2} \right) < x_n < \frac{n}{n+1} \left(\frac{\pi^2}{6} - \sum_{k=1}^{n-1} \frac{1}{k^2} \right)$$

and

$$\frac{n^2}{n+1}\left(\frac{\pi^2}{6}-\sum_{k=1}^{n}\frac{1}{k^2}\right)<nx_n<\frac{n^2}{n+1}\left(\frac{\pi^2}{6}-\sum_{k=1}^{n-1}\frac{1}{k^2}\right).$$

2.21. and **2.22.** Use the substitution $1/x = y$ and the improper integral formula $1/a^n = 1/\Gamma(n)\int_0^\infty e^{-ax}x^{n-1}dx$ where $a > 0$.

2.5.2 Double Integrals

As for everything else, so for a mathematical theory; beauty can be perceived but not explained.

Arthur Cayley (1821–1895)

In general, these double integrals can be calculated by writing the double integral $\int_0^1\int_0^1 f(x,y)dxdy$ as an iterated integral $\int_0^1(\int_0^1 f(x,y)dy)dx$, making a particular change of variables in the inner integral, and then integrating by parts.

2.23. Write the integral in the form $\int_0^1 x\left(\int_0^1\left\{\frac{1}{1-xy}\right\}dy\right)dx$; make the substitution $xy = t$, in the inner integral; and integrate by parts.

2.24. Write the integral as $\int_0^1\left(\int_x^{x+1}\left\{\frac{1}{t}\right\}^m dt\right)dx$ and integrate by parts.

2.26. Use that $\int_0^1\left(\int_0^1\frac{dy}{(\lfloor x/y\rfloor+1)^k}\right)dx = \int_0^1 x\left(\int_0^{1/x}\frac{dt}{(\lfloor 1/t\rfloor+1)^k}\right)dx$ and integrate by parts.

2.28. Substitute $kx/y = t$, in the inner integral, to get that $\int_0^1\left(\int_0^1\{kx/y\}dy\right)dx = k\int_0^1 x\left(\int_{kx}^\infty\{t\}/t^2dt\right)dx$ and integrate by parts.

2.29. Write $\int_0^1\left(\int_0^1\left\{\frac{mx}{ny}\right\}dy\right)dx \overset{t=\frac{mx}{ny}}{=} \frac{m}{n}\int_0^1 x\left(\int_{\frac{mx}{n}}^\infty\frac{\{t\}}{t^2}dt\right)dx$ and integrate by parts.

2.30. $\int_0^1\left(\int_0^1\left\{\frac{x^k}{y}\right\}dy\right)dx \overset{t=x^k/y}{=} \int_0^1 x^k\left(\int_{x^k}^\infty\frac{\{t\}}{t^2}dt\right)dx$ and integrate by parts.

2.31.–2.35. Use the substitution $x/y = t$ and integrate by parts.

2.38. Write $\int_0^1 x^n\left(\int_0^1 y^n\left\{\frac{x}{y}\right\}\left\{\frac{y}{x}\right\}dy\right)dx = \int_0^1 x^{2n+1}\left(\int_0^{1/x}t^n\{t\}\left\{\frac{1}{t}\right\}dt\right)dx$ and integrate by parts.

2.40. and **2.41.** Use the substitution $x/y = t$, in the inner integral, to get that $\int_0^1\left(\int_0^1\{x/y\}^m\{y/x\}^k dy\right)dx = \int_0^1 x\left(\int_x^\infty\{t\}^m\left\{\frac{1}{t}\right\}^k\frac{dt}{t^2}\right)dx$ and integrate by parts.

2.42. 1. Let $m = \lfloor a\rfloor$ be the floor of a and calculate the integral on intervals of the form $[j, j+1]$.
2. Write the multiple integral as an iterated integral and use part 1 of the problem.

2.5.3 Quickies

> *Nature not only suggests to us problems, she suggests their solution.*
>
> Henri Poincaré (1854–1912)

These integrals are solved by using either symmetry or the following identity involving the fractional part function.

A fractional part identity. If $x \in \mathbb{R} \setminus \mathbb{Z}$, then $\{x\} + \{-x\} = 1$.

2.44. Make the substitution $nx = y$.

2.45.–2.47. and **2.49.** Make the substitution $x = 1 - y$ and use the fractional part identity.

2.48. Observe that if n is a positive integer, then $\{n(1-y)\} = 1 - \{ny\}$, for all $y \in [0,1]$ except for $0, 1/n, 2/n, \ldots, (n-1)/n, 1$. Then, use the substitution $x = 1 - y$ and the definition of the Beta function.

2.50.–2.54. and **2.57.** Use symmetry and the fractional part identity.

2.55. and **2.56.** Observe that if n is a positive integer, then $\{n(1-y)\} = 1 - \{ny\}$, for all $y \in [0,1]$ except for $0, 1/n, 2/n, \ldots, (n-1)/n, 1$. Then, use symmetry.

2.6 Solutions

> *Everything you add to the truth subtracts from the truth.*
>
> Alexander Solzhenitsyn (1918–2008)

This section contains the solutions to the problems from the first three sections of the chapter.

2.6.1 Single Integrals

> *Sir, I have found you an argument. I am not obliged to find you an understanding.*
>
> Samuel Johnson (1709–1784)

2.1. The integral equals $1 - \gamma$. We have

$$\int_0^1 \left\{\frac{1}{x}\right\} dx = \int_1^\infty \frac{\{t\}}{t^2} dt = \sum_{k=1}^\infty \int_k^{k+1} \frac{t-k}{t^2} dt = \sum_{k=1}^\infty \left(\ln \frac{k+1}{k} - \frac{1}{k+1} \right) = 1 - \gamma.$$

Remark. This integral, which is well known in the mathematical literature ([43, p. 32], [65, pp. 109–111]), is related to a surprising result, due to de la Vallée Poussin [105], which states that

$$\lim_{n \to \infty} \frac{1}{n} \sum_{k=1}^n \left\{ \frac{n}{k} \right\} = 1 - \gamma.$$

In words, if a large integer n is divided by each integer $1 \le k \le n$, then the average fraction by which the quotient n/k falls short of the next integer is not $1/2$, but γ!

2.2. See the solution of Problem **2.6**.

2.3. Using the substitution $1/x = t$ the integral becomes

$$\int_1^\infty \frac{\{t\}^3}{t^2} dt = \sum_{k=1}^\infty \int_k^{k+1} \frac{(t-k)^3}{t^2} dt = \sum_{k=1}^\infty \left(3k^2 \ln \frac{k+1}{k} + \frac{3}{2} - 3k - \frac{1}{k+1} \right).$$

Let $S_n = \sum_{k=1}^n \left(3k^2 \ln(k+1)/k + 3/2 - 3k - 1/(k+1) \right)$ be the nth partial sum of the series. A calculation shows that

$$S_n = 1 - \left(1 + \frac{1}{2} + \cdots + \frac{1}{n+1} - \ln n \right) - \frac{3}{2} n^2 - \ln n + 3 \sum_{k=1}^n k^2 \ln \frac{k+1}{k}.$$

On the other hand,

$$\sum_{k=1}^n k^2 \ln \frac{k+1}{k} = \ln \prod_{k=1}^n \left(\frac{k+1}{k} \right)^{k^2} = \ln \left(\frac{(n+1)^{n^2} n!}{(2^2 3^3 \cdots n^n)^2} \right)$$

and

$$-\frac{3}{2} n^2 - \ln n + 3 \sum_{k=1}^n k^2 \ln \frac{k+1}{k} = \ln \left(\frac{(n+1)^{3n^2} (n!)^3}{(2^2 3^3 \cdots n^n)^6 e^{\frac{3n^2}{2}} n} \right).$$

Let

$$x_n = \frac{(n+1)^{3n^2} (n!)^3}{(2^2 3^3 \cdots n^n)^6 e^{\frac{3n^2}{2}} n} = \frac{n^{3n^2+3n+\frac{1}{2}} e^{\frac{-3n^2}{2}}}{(2^2 3^3 \cdots n^n)^6} \cdot \frac{(n+1)^{3n^2} (n!)^3}{n^{3n^2+3n+\frac{3}{2}}}.$$

The first fraction converges to $1/A^6$, and for calculating the limit of the second fraction, we have, based on Stirling's formula, $n! \sim \sqrt{2\pi n}\,(n/e)^n$, that

$$\frac{(n+1)^{3n^2}(n!)^3}{n^{3n^2+3n+\frac{3}{2}}} \sim (2\pi)^{\frac{3}{2}}\left(\left(\frac{n+1}{n}\right)^n \frac{1}{e}\right)^{3n} \to (2\pi)^{\frac{3}{2}}e^{-\frac{3}{2}}.$$

Thus, $x_n \to (2\pi)^{\frac{3}{2}}e^{-\frac{3}{2}}/A^6$, which implies that $\lim\limits_{n\to\infty} S_n = -\frac{1}{2}-\gamma+\frac{3}{2}\ln(2\pi)-6\ln A$.

2.4. See the solution of Problem **2.22**.

2.5. (a) Using the substitution $k/x=t$, we get that

$$\int_0^1 \left\{\frac{k}{x}\right\} dx = k\int_k^\infty \frac{\{t\}}{t^2}\,dt = k\sum_{l=k}^\infty \int_l^{l+1} \frac{t-l}{t^2}\,dt = k\sum_{l=k}^\infty \left(\ln\frac{l+1}{l} - \frac{1}{l+1}\right).$$

A calculation shows that

$$S_n = \sum_{l=k}^n \left(\ln\frac{l+1}{l} - \frac{1}{l+1}\right) = -\ln k - \left(\frac{1}{k+1}+\frac{1}{k+2}+\cdots+\frac{1}{1+n} - \ln(n+1)\right),$$

and this implies that $\lim_{n\to\infty} S_n = 1+1/2+\cdots+1/k - \ln k - \gamma$.
(b) Using the substitution $x=q/y$, we obtain that

$$I = \int_0^1 \left\{\frac{q}{x}\right\} dx = q\int_q^\infty \frac{\{y\}}{y^2}\,dy.$$

We distinguish here two cases.

Case 1. $q \le 1$. We have

$$I = q\left(\int_q^1 \frac{1}{y}\,dy + \sum_{k=1}^\infty \int_k^{k+1} \frac{y-k}{y^2}\,dy\right)$$

$$= q\left(\ln\frac{1}{q} + \sum_{k=1}^\infty \left(\ln\frac{k+1}{k} - \frac{1}{k+1}\right)\right)$$

$$= q(1-\gamma-\ln q).$$

Case 2. $q > 1$. A calculation shows that

$$I = q\left(\int_q^{\lfloor q\rfloor+1} \frac{y-\lfloor q\rfloor}{y^2}\,dy + \int_{\lfloor q\rfloor+1}^\infty \frac{\{y\}}{y^2}\,dy\right)$$

$$= q\left(1+\frac{1}{2}+\cdots+\frac{1}{1+\lfloor q\rfloor} - \gamma - \ln q + \frac{\lfloor q\rfloor(\{q\}-1)}{q(1+\lfloor q\rfloor)}\right).$$

Remark. If $k \geq 2$ is an integer, then

$$\int_0^1 \left(\left\lfloor \frac{k}{x} \right\rfloor - k \cdot \left\lfloor \frac{1}{x} \right\rfloor \right) dx = k \ln k - k \left(\frac{1}{2} + \frac{1}{3} + \cdots + \frac{1}{k} \right),$$

and this generalizes an integral from Pólya and Szegö (see [104, p. 43]).

2.6. Using the substitution $k/x = t$, the integral becomes

$$\int_0^1 \left\{ \frac{k}{x} \right\}^2 dx = k \int_k^\infty \frac{\{t\}^2}{t^2} dt = k \sum_{l=k}^\infty \left(2 - 2l \ln \frac{l+1}{l} - \frac{1}{l+1} \right).$$

Let S_n be the nth partial sum of the series. We have

$$\sum_{l=k}^n \left(2 - 2l \ln \frac{l+1}{l} - \frac{1}{l+1} \right) = 2(n - k + 1) - \left(\frac{1}{k+1} + \frac{1}{k+2} + \cdots + \frac{1}{n+1} \right)$$

$$- 2n \ln(n+1) + 2k \ln k + 2 \ln n! - 2 \ln k!.$$

Since $2 \ln n! \sim \ln(2\pi) + (2n + 1) \ln n - 2n$, it follows that S_n is approximated by

$$2(1 - k) + \ln(2\pi) + 2k \ln k - 2 \ln k! - 2n \ln \frac{n+1}{n} - \left(\frac{1}{k+1} + \cdots + \frac{1}{n+1} - \ln n \right)$$

and hence, $\lim_{n \to \infty} S_n = \ln(2\pi) - \gamma + 1 + \frac{1}{2} + \cdots + \frac{1}{k} + 2k \ln k - 2k - 2 \ln k!$.

2.7. We have, based on the substitution $x = 1/y^k$, that

$$\int_0^1 \left\{ \frac{1}{\sqrt[k]{x}} \right\} dx = k \int_1^\infty \frac{\{y\}}{y^{k+1}} dy = k \sum_{m=1}^\infty \int_m^{m+1} \frac{\{y\}}{y^{k+1}} dy = k \sum_{m=1}^\infty \int_m^{m+1} \frac{y - m}{y^{k+1}} dy$$

$$= k \sum_{m=1}^\infty \left(\frac{y^{-k+1}}{-k+1} \Big|_m^{m+1} + m \frac{y^{-k}}{k} \Big|_m^{m+1} \right)$$

$$= \frac{k}{k-1} \sum_{m=1}^\infty \left(\frac{1}{m^{k-1}} - \frac{1}{(m+1)^{k-1}} \right) + \sum_{m=1}^\infty m \left(\frac{1}{(m+1)^k} - \frac{1}{m^k} \right)$$

$$= \frac{k}{k-1} + \sum_{m=1}^\infty \left(\frac{1}{(m+1)^{k-1}} - \frac{1}{m^{k-1}} \right) - \sum_{m=1}^\infty \frac{1}{(m+1)^k}$$

$$= \frac{k}{k-1} - \zeta(k).$$

2.8. Using the substitution $x = k^k/y^k$ we obtain that

$$\int_0^1 \left\{\frac{k}{\sqrt[k]{x}}\right\} dx = k^{k+1} \int_k^\infty \frac{\{y\}}{y^{k+1}} dy = k^{k+1} \sum_{m=k}^\infty \left(\int_m^{m+1} \frac{y-m}{y^{k+1}} dy\right)$$

$$= k^{k+1} \sum_{m=k}^\infty \left(\frac{y^{-k+1}}{-k+1}\Big|_m^{m+1} + m\frac{y^{-k}}{k}\Big|_m^{m+1}\right)$$

$$= \frac{k^{k+1}}{k-1} \sum_{m=k}^\infty \left(\frac{1}{m^{k-1}} - \frac{1}{(m+1)^{k-1}}\right) + k^k \sum_{m=k}^\infty m\left(\frac{1}{(m+1)^k} - \frac{1}{m^k}\right)$$

$$= \frac{k^2}{k-1} + k^k \sum_{m=k}^\infty \left(\frac{1}{(m+1)^{k-1}} - \frac{1}{m^{k-1}} - \frac{1}{(m+1)^k}\right)$$

$$= \frac{k^2}{k-1} - k - k^k \left(\zeta(k) - \frac{1}{1^k} - \frac{1}{2^k} - \cdots - \frac{1}{k^k}\right)$$

$$= \frac{k}{k-1} - k^k \left(\zeta(k) - \frac{1}{1^k} - \frac{1}{2^k} - \cdots - \frac{1}{k^k}\right).$$

2.9. The substitution $x = 1/(k^k y^k)$ implies that

$$\int_0^1 \left\{\frac{1}{k\sqrt[k]{x}}\right\} dx = \frac{1}{k^{k-1}} \int_{\frac{1}{k}}^\infty \frac{\{y\}}{y^{k+1}} dy = \frac{1}{k^{k-1}} \left(\int_{\frac{1}{k}}^1 \frac{y}{y^{k+1}} dy + \int_1^\infty \frac{\{y\}}{y^{k+1}} dy\right)$$

$$= \frac{1}{k^{k-1}} \left(\frac{y^{1-k}}{1-k}\Big|_{\frac{1}{k}}^1 + \frac{1}{k-1} - \frac{\zeta(k)}{k}\right)$$

$$= \frac{1}{k^{k-1}} \left(\frac{1}{1-k} - \frac{1}{1-k}\left(\frac{1}{k}\right)^{1-k} + \frac{1}{k-1} - \frac{\zeta(k)}{k}\right)$$

$$= \frac{1}{k-1} - \frac{\zeta(k)}{k^k}.$$

We used that (see the solution of Problem **2.7**)

$$\int_1^\infty \frac{\{y\}}{y^{k+1}} dy = \frac{1}{k-1} - \frac{\zeta(k)}{k}.$$

2.10. The substitution $x = 1/t$ implies that

$$I = \int_0^1 \left\{\frac{1}{x}\right\}\left\{\frac{1}{1-x}\right\} dx = \int_1^\infty \frac{\{t\}}{t^2}\left\{\frac{t}{t-1}\right\} dt$$

$$= \int_1^2 \frac{\{t\}}{t^2}\left\{\frac{t}{t-1}\right\} dt + \sum_{k=2}^\infty \int_k^{k+1} \frac{t-k}{t^2}\left\{\frac{t}{t-1}\right\} dt.$$

We calculate the integral and the sum separately. We have that

$$\sum_{k=2}^{\infty} \int_{k}^{k+1} \frac{t-k}{t^2} \left\{ \frac{t}{t-1} \right\} dt = \sum_{k=2}^{\infty} \int_{k}^{k+1} \frac{t-k}{t^2} \left(\frac{t}{t-1} - 1 \right) dt$$

$$= \sum_{k=2}^{\infty} \int_{k}^{k+1} \frac{t-k}{t^2} \cdot \frac{dt}{t-1}$$

$$= \sum_{k=1}^{\infty} \left(k \ln \frac{k}{k+1} + k \ln \frac{k+2}{k+1} + \frac{1}{k+2} \right).$$

On the other hand, the substitution $t - 1 = u$ implies that

$$\int_{1}^{2} \frac{\{t\}}{t^2} \left\{ \frac{t}{t-1} \right\} dt = \int_{0}^{1} \frac{u}{(u+1)^2} \left\{ \frac{u+1}{u} \right\} du = \int_{0}^{1} \frac{u}{(u+1)^2} \left\{ \frac{1}{u} \right\} du$$

$$= \int_{1}^{\infty} \frac{\{t\}}{t(1+t)^2} dt = \sum_{k=1}^{\infty} \int_{k}^{k+1} \frac{t-k}{t(1+t)^2} dt$$

$$= \sum_{k=1}^{\infty} \left(k \ln \frac{k}{k+1} + k \ln \frac{k+2}{k+1} + \frac{1}{k+2} \right).$$

Putting all these together we get that

$$I = 2 \sum_{k=1}^{\infty} \left(k \ln \frac{k}{k+1} + k \ln \frac{k+2}{k+1} + \frac{1}{k+2} \right).$$

Let S_n be the nth partial sum of the series. A calculation shows that

$$S_n = 2 \sum_{k=1}^{n} \left(k \ln \frac{k}{k+1} + k \ln \frac{k+2}{k+1} + \frac{1}{k+2} \right)$$

$$= 2 \left(n \ln \frac{n+2}{n+1} - \ln(n+1) + \frac{1}{3} + \frac{1}{4} + \cdots + \frac{1}{n+2} \right),$$

and this implies $\lim_{n \to \infty} S_n = 2\gamma - 1$.

Remark. It is worth mentioning that the following integral formulae hold

$$\int_{0}^{\frac{1}{2}} \left\{ \frac{1}{x} \right\} \left\{ \frac{1}{1-x} \right\} dx = \int_{\frac{1}{2}}^{1} \left\{ \frac{1}{x} \right\} \left\{ \frac{1}{1-x} \right\} dx = \gamma - \frac{1}{2}.$$

2.11. For a solution of this problem, see [45].

2.12. We need the following lemma:

Lemma 2.1. *The following formulae hold:*

1. $\lim_{n \to \infty} n \left(1 - (n+2) \ln \frac{n+2}{n+1} \right) = -\frac{1}{2}.$

2. $\int_k^{k+1} \frac{x-k}{(x+1)^2 x^2} dx = -\frac{1}{k+1} - \frac{1}{k+2} + (2k+1)\ln\frac{k+1}{k} - (2k+1)\ln\frac{k+2}{k+1}$, $k \geq 1$.

3. $\int_k^{k+1} \frac{(x-k)^2}{x^2(x-1)} dx = -\frac{k}{k+1} + (k-1)^2\ln\frac{k}{k-1} - (k-2)k\ln\frac{k+1}{k}$, $k > 1$.

The lemma can be proved by elementary calculations.

Using the substitution $x = 1/t$, the integral becomes

$$I = \int_1^\infty \frac{\{t\}^2}{t^2} \left\{ \frac{t}{t-1} \right\} dt$$

$$= \int_1^2 \frac{\{t\}^2}{t^2} \left\{ \frac{t}{t-1} \right\} dt + \sum_{k=2}^\infty \int_k^{k+1} \frac{(t-k)^2}{t^2} \left\{ \frac{t}{t-1} \right\} dt. \qquad (2.1)$$

We calculate the integral and the sum separately. We have

$$S = \sum_{k=2}^\infty \int_k^{k+1} \frac{(t-k)^2}{t^2} \left\{ \frac{t}{t-1} \right\} dt$$

$$= \sum_{k=2}^\infty \int_k^{k+1} \frac{(t-k)^2}{t^2} \left(\frac{t}{t-1} - 1 \right) dt$$

$$= \sum_{k=2}^\infty \int_k^{k+1} \frac{(t-k)^2}{t^2} \cdot \frac{dt}{t-1}.$$

Using part (3) of Lemma 2.1 we get that

$$S = \sum_{k=2}^\infty \left((k-1)^2\ln\frac{k}{k-1} + (2k-k^2)\ln\frac{k+1}{k} - \frac{k}{k+1} \right)$$

$$= \sum_{k=1}^\infty \left(k^2\ln\frac{k+1}{k} + (1-k^2)\ln\frac{k+2}{k+1} - \frac{k+1}{k+2} \right). \qquad (2.2)$$

On the other hand, we have based on the substitution $t - 1 = u$ that

$$J = \int_1^2 \frac{\{t\}^2}{t^2} \left\{ \frac{t}{t-1} \right\} dt = \int_0^1 \frac{\{u+1\}^2}{(u+1)^2} \left\{ \frac{u+1}{u} \right\} du = \int_0^1 \frac{u^2}{(u+1)^2} \left\{ \frac{1}{u} \right\} du.$$

The substitution $1/u = t$, combined with part (2) of Lemma 2.1, shows that

$$J = \int_1^\infty \frac{\{t\}}{t^2(1+t)^2} dt = \sum_{k=1}^\infty \int_k^{k+1} \frac{t-k}{t^2(1+t)^2} dt$$

$$= \sum_{k=1}^\infty \left(-\frac{1}{k+1} - \frac{1}{k+2} + (2k+1)\ln\frac{k+1}{k} - (2k+1)\ln\frac{k+2}{k+1} \right). \qquad (2.3)$$

Combining (2.1)–(2.3), we get that

$$\int_0^1 \left\{\frac{1}{x}\right\}^2 \left\{\frac{1}{1-x}\right\} dx = \sum_{k=1}^{\infty} \left((k+1)^2 \ln\frac{k+1}{k} - (k^2+2k)\ln\frac{k+2}{k+1} - 1 - \frac{1}{k+1} \right).$$

Let S_n be the nth partial sum of the series, i.e.,

$$S_n = \sum_{k=1}^{n} \left((k+1)^2 \ln\frac{k+1}{k} - (k^2+2k)\ln\frac{k+2}{k+1} - 1 - \frac{1}{k+1} \right).$$

A calculation shows

$$\sum_{k=1}^{n}(k+1)^2\ln\frac{k+1}{k} - \sum_{k=1}^{n}(k^2+2k)\ln\frac{k+2}{k+1}$$

$$= \sum_{k=1}^{n}(k+1)^2\ln\frac{k+1}{k} - \sum_{k=1}^{n}(k^2-1)\ln\frac{k+1}{k} - n(n+2)\ln\frac{n+2}{n+1}$$

$$= \sum_{k=1}^{n}2(k+1)\ln\frac{k+1}{k} - n(n+2)\ln\frac{n+2}{n+1}$$

$$= 2(n+1)\ln(n+1) - 2\ln n! - n(n+2)\ln\frac{n+2}{n+1}.$$

Thus,

$$S_n = 2(n+1)\ln(n+1) - 2\ln n! - (n^2+2n)\ln\frac{n+2}{n+1} - n$$

$$- \left(\frac{1}{2}+\frac{1}{3}+\cdots+\frac{1}{n+2} - \ln(n+1)\right) - \ln(n+1)$$

$$= (2n+1)\ln(n+1) - 2\ln n! - (n^2+2n)\ln\frac{n+2}{n+1} - n$$

$$- \left(\frac{1}{2}+\frac{1}{3}+\cdots+\frac{1}{n+2} - \ln(n+1)\right).$$

Since $\ln n! = (\ln 2\pi)/2 + (n+1/2)\ln n - n + O(1/n)$, we get that

$$S_n = (2n+1)\ln\frac{n+1}{n} - \ln(2\pi) + n\left(1 - (n+2)\ln\frac{n+2}{n+1}\right)$$

$$- \left(\frac{1}{2}+\frac{1}{3}+\cdots+\frac{1}{n+2} - \ln(n+1)\right) + O\left(\frac{1}{n}\right),$$

and this implies that $\lim_{n\to\infty} S_n = 5/2 - \ln(2\pi) - \gamma$.

Remark. We have, due to symmetry reasons, that the following evaluations hold

$$\int_0^1 \left\{\frac{1}{x}\right\}\left\{\frac{1}{1-x}\right\}^2 dx = \frac{5}{2} - \ln(2\pi) - \gamma.$$

$$\int_0^1 \left\{\frac{1}{x}\right\}\left\{\frac{1}{1-x}\right\}\left(\left\{\frac{1}{x}\right\} + \left\{\frac{1}{1-x}\right\}\right) dx = 5 - 2\ln(2\pi) - 2\gamma.$$

$$\int_0^{\frac{1}{2}} \left\{\frac{1}{x}\right\}\left\{\frac{1}{1-x}\right\}\left(\left\{\frac{1}{x}\right\} + \left\{\frac{1}{1-x}\right\}\right) dx = \frac{5}{2} - \ln(2\pi) - \gamma.$$

$$\int_{\frac{1}{2}}^1 \left\{\frac{1}{x}\right\}\left\{\frac{1}{1-x}\right\}\left(\left\{\frac{1}{x}\right\} + \left\{\frac{1}{1-x}\right\}\right) dx = \frac{5}{2} - \ln(2\pi) - \gamma.$$

2.13. Using the substitution $1/x = y$, the integral becomes $I = \int_1^\infty \left\{(-1)^{\lfloor y\rfloor} y\right\}/y^2 dy$. Since for a positive real number y, which is not an integer, one has $\{-y\} = 1 - \{y\}$, we get that

$$I = \int_1^\infty \frac{\left\{(-1)^{\lfloor y\rfloor} y\right\}}{y^2} dy = \sum_{k=1}^\infty \int_k^{k+1} \frac{\left\{(-1)^{\lfloor y\rfloor} y\right\}}{y^2} dy = \sum_{k=1}^\infty \int_k^{k+1} \frac{\left\{(-1)^k y\right\}}{y^2} dy$$

$$= \sum_{p=1}^\infty \int_{2p-1}^{2p} \frac{\{-y\}}{y^2} dy + \sum_{p=1}^\infty \int_{2p}^{2p+1} \frac{\{y\}}{y^2} dy$$

$$= \sum_{p=1}^\infty \int_{2p-1}^{2p} \frac{1-\{y\}}{y^2} dy + \sum_{p=1}^\infty \int_{2p}^{2p+1} \frac{y-2p}{y^2} dy$$

$$= \sum_{p=1}^\infty \int_{2p-1}^{2p} \frac{1-(y-(2p-1))}{y^2} dy + \sum_{p=1}^\infty \left(\ln y + \frac{2p}{y}\right) \Big|_{2p}^{2p+1}$$

$$= \sum_{p=1}^\infty \int_{2p-1}^{2p} \frac{2p-y}{y^2} dy + \sum_{p=1}^\infty \left(\ln \frac{2p+1}{2p} - \frac{1}{2p+1}\right)$$

$$= \sum_{p=1}^\infty \left(\frac{1}{2p-1} - \ln \frac{2p}{2p-1}\right) + \sum_{p=1}^\infty \left(\ln \frac{2p+1}{2p} - \frac{1}{2p+1}\right)$$

$$= \sum_{p=1}^\infty \left(\frac{1}{2p-1} - \frac{1}{2p+1} + \ln \frac{(2p+1)(2p-1)}{(2p)^2}\right)$$

$$= 1 + \ln\left(\prod_{p=1}^\infty \frac{(2p+1)(2p-1)}{(2p)^2}\right)$$

$$= 1 + \ln \frac{2}{\pi},$$

where the last equality follows from the Wallis product formula.

2.14. (a) Using the substitution $1/x = t$ the integral becomes

$$I = \int_0^1 x \left\{ \frac{1}{x} \right\} \left\lfloor \frac{1}{x} \right\rfloor dx = \int_1^\infty \frac{\{t\} \cdot \lfloor t \rfloor}{t^3} dt = \sum_{k=1}^\infty \int_k^{k+1} \frac{(t-k) \cdot k}{t^3} dt$$

$$= \sum_{k=1}^\infty \int_k^{k+1} \left(\frac{k}{t^2} - \frac{k^2}{t^3} \right) dt = \frac{1}{2} \sum_{k=1}^\infty \frac{1}{(k+1)^2} = \frac{1}{2} \left(\frac{\pi^2}{6} - 1 \right).$$

(b) The integral converges if and only if $\alpha + 1 > \gamma$. Making the substitution $1/x = t$, the integral equals

$$I(\alpha, \beta, \gamma) = \int_0^1 x^\alpha \left\{ \frac{1}{x} \right\}^\beta \left\lfloor \frac{1}{x} \right\rfloor^\gamma dx = \int_1^\infty \frac{\{t\}^\beta \cdot \lfloor t \rfloor^\gamma}{t^{\alpha+2}} dt$$

$$= \sum_{k=1}^\infty \int_k^{k+1} \frac{(t-k)^\beta k^\gamma}{t^{\alpha+2}} dt \overset{t-k=u}{=} \sum_{k=1}^\infty k^\gamma \int_0^1 \frac{u^\beta}{(k+u)^{\alpha+2}} du.$$

We have, since $k^{\alpha+2} < (k+u)^{\alpha+2} < (k+1)^{\alpha+2}$, that

$$\sum_{k=1}^\infty \frac{k^\gamma}{(\beta+1)(k+1)^{\alpha+2}} < I(\alpha, \beta, \gamma) < \sum_{k=1}^\infty \frac{k^\gamma}{(\beta+1)k^{\alpha+2}}.$$

Thus, the integral converges if and only if $\alpha + 1 > \gamma$.

Remark. The convergence of the integral is independent of the parameter β, which can be chosen to be any real number strictly greater than -1.

2.15. We have

$$I = \int_0^1 \left\{ \frac{1}{x} \right\} \frac{x}{1-x} dx = \int_0^1 \left\{ \frac{1}{x} \right\} \left(\sum_{k=1}^\infty x^k \right) dx = \sum_{k=1}^\infty \int_0^1 \left\{ \frac{1}{x} \right\} x^k dx.$$

Let $J_k = \int_0^1 \left\{ \frac{1}{x} \right\} x^k dx$. Using the substitution $1/x = y$ we obtain that

$$J_k = \int_1^\infty \frac{\{y\}}{y^{k+2}} dy = \sum_{m=1}^\infty \int_m^{m+1} \frac{\{y\}}{y^{k+2}} dy = \sum_{m=1}^\infty \int_m^{m+1} \frac{y-m}{y^{k+2}} dy$$

$$= \sum_{m=1}^\infty \int_m^{m+1} \left(\frac{1}{y^{k+1}} - \frac{m}{y^{k+2}} \right) dy = \sum_{m=1}^\infty \left(\frac{1}{-ky^k} + \frac{m}{(k+1)y^{k+1}} \right) \Big|_m^{m+1}$$

$$= \sum_{m=1}^\infty \left(\frac{1}{km^k} - \frac{1}{k(m+1)^k} \right) + \sum_{m=1}^\infty \left(\frac{m}{(k+1)(m+1)^{k+1}} - \frac{1}{(k+1)m^k} \right)$$

$$= \frac{1}{k} + \frac{1}{k+1} \sum_{m=1}^\infty \left[\left(\frac{1}{(m+1)^k} - \frac{1}{m^k} \right) - \frac{1}{(m+1)^{k+1}} \right]$$

$$= \frac{1}{k} + \frac{1}{k+1}(-1-(\zeta(k+1)-1))$$

$$= \frac{1}{k} - \frac{\zeta(k+1)}{k+1}.$$

Thus,

$$I = \sum_{k=1}^{\infty} \left(\frac{1}{k} - \frac{\zeta(k+1)}{k+1} \right) = \sum_{k=1}^{\infty} \left(\frac{1}{k} - \frac{1}{k+1} \right) - \sum_{k=1}^{\infty} \frac{\zeta(k+1)-1}{k+1} = \gamma,$$

since $\sum_{k=1}^{\infty}(\zeta(k+1)-1)/(k+1) = 1 - \gamma$ [122, Entry 135, p. 173].
For an alternative solution, see [31].

2.16. The solution is given in [51].

2.17. For a solution see [50].

2.18. The limit equals $1/(k+1)$. Let $a_n = \int_0^1 \{n/x\}^k dx$. Using the substitution $n/x = t$, we get that

$$a_n = n \int_n^{\infty} \frac{\{t\}^k}{t^2} dt = \frac{\int_n^{\infty} \frac{\{t\}^k}{t^2} dt}{\frac{1}{n}} = \frac{b_n}{c_n},$$

where $b_n = \int_n^{\infty} \{t\}^k / t^2 dt$ and $c_n = 1/n$, and we note that b_n converges to 0 since $b_n = \int_n^{\infty} \{t\}^k / t^2 dt < \int_n^{\infty} 1/t^2 dt = 1/n$. For calculating $\lim_{n\to\infty} a_n$, we use Stolz–Cesàro lemma (the $0/0$ case) and we get that

$$\lim_{n\to\infty} a_n = \lim_{n\to\infty} \frac{b_{n+1} - b_n}{c_{n+1} - c_n} = \lim_{n\to\infty} \frac{-\int_n^{n+1} \frac{\{t\}^k}{t^2} dt}{-\frac{1}{n(n+1)}} = \lim_{n\to\infty} n(n+1) \int_n^{n+1} \frac{\{t\}^k}{t^2} dt$$

$$= \lim_{n\to\infty} n(n+1) \int_n^{n+1} \frac{(t-n)^k}{t^2} dt = \lim_{n\to\infty} n(n+1) \int_0^1 \frac{y^k}{(n+y)^2} dy$$

$$= \lim_{n\to\infty} \int_0^1 y^k \frac{n(n+1)}{(n+y)^2} dy = \int_0^1 y^k dy = \frac{1}{k+1}.$$

For an alternative solution, see Problem **1.70**.

2.19. We prove that $L = 0$. Let $x_n = \int_0^1 \{n/x\}^n dx$. Using the substitution $n/x = t$, we get that

$$x_n = n \int_n^{\infty} \frac{\{t\}^n}{t^2} dt = n \sum_{k=n}^{\infty} \int_k^{k+1} \frac{\{t\}^n}{t^2} dt = n \sum_{k=n}^{\infty} \int_k^{k+1} \frac{(t-k)^n}{t^2} dt$$

$$= n \sum_{k=n}^{\infty} \int_0^1 \frac{y^n}{(k+y)^2} dy = n \int_0^1 y^n \left(\sum_{k=n}^{\infty} \frac{1}{(k+y)^2} \right) dy.$$

Since $1/(k+1)^2 < 1/(k+y)^2 < 1/k^2$, for positive integers k, and $y \in (0,1)$, it follows that

$$n \int_0^1 y^n \left(\sum_{k=n}^\infty \frac{1}{(k+1)^2} \right) dy < x_n < n \int_0^1 y^n \left(\sum_{k=n}^\infty \frac{1}{k^2} \right) dy.$$

Thus,

$$\frac{n}{n+1} \left(\frac{\pi^2}{6} - \sum_{k=1}^n \frac{1}{k^2} \right) < x_n < \frac{n}{n+1} \left(\frac{\pi^2}{6} - \sum_{k=1}^{n-1} \frac{1}{k^2} \right),$$

which implies that $\lim_{n\to\infty} x_n = 0$.

The second limit equals 1. We have, based on the preceding inequalities, that

$$\frac{n^2}{n+1} \left(\frac{\pi^2}{6} - \sum_{k=1}^n \frac{1}{k^2} \right) < nx_n < \frac{n^2}{n+1} \left(\frac{\pi^2}{6} - \sum_{k=1}^{n-1} \frac{1}{k^2} \right),$$

and the result follows since $\lim_{n\to\infty} n \left(\pi^2/6 - \sum_{k=1}^n 1/k^2 \right) = 1$.

2.20. The integral equals $1/m - \zeta(m+1)/(m+1)$. See the calculation of J_k from the solution of Problem **2.15**.

2.21. and 2.22. Let

$$V_{k,m} = \int_0^1 x^m \left\{ \frac{1}{x} \right\}^k dx = \int_1^\infty \frac{\{t\}^k}{t^{m+2}} dt = \sum_{p=1}^\infty \int_p^{p+1} \frac{\{t\}^k}{t^{m+2}} dt = \sum_{p=1}^\infty \int_p^{p+1} \frac{(t-p)^k}{t^{m+2}} dt$$

$$\stackrel{t-p=y}{=} \sum_{p=1}^\infty \int_0^1 \frac{y^k}{(p+y)^{m+2}} dy = \int_0^1 y^k \left(\sum_{p=1}^\infty \frac{1}{(p+y)^{m+2}} \right) dy. \tag{2.4}$$

Since

$$\frac{1}{(p+y)^{2+m}} = \frac{1}{(m+1)!} \int_0^\infty e^{-(p+y)u} u^{m+1} du,$$

we have that

$$\sum_{p=1}^\infty \frac{1}{(p+y)^{m+2}} = \frac{1}{(m+1)!} \sum_{p=1}^\infty \int_0^\infty e^{-(p+y)u} u^{m+1} du$$

$$= \frac{1}{(m+1)!} \int_0^\infty u^{m+1} e^{-yu} \left(\sum_{p=1}^\infty e^{-pu} \right) du$$

$$= \frac{1}{(m+1)!} \int_0^\infty \frac{u^{m+1} e^{-yu}}{e^u - 1} du. \tag{2.5}$$

Combining (2.4) and (2.5), one has that

$$V_{k,m} = \frac{1}{(m+1)!} \int_0^1 y^k \left(\int_0^\infty \frac{u^{m+1}e^{-yu}}{e^u - 1} du \right) dy$$

$$= \frac{1}{(m+1)!} \int_0^\infty \frac{u^{m+1}}{e^u - 1} \left(\int_0^1 y^k e^{-yu} dy \right) du.$$

Let $J_k = \int_0^1 y^k e^{-yu} dy$. Integrating by parts we get the recurrence formula $J_k = -e^{-u}/u + (k/u)J_{k-1}$. Let $a_k = J_k u^k/k!$ and we note that $a_k = -\frac{e^{-u}}{u} \cdot \frac{u^k}{k!} + a_{k-1}$. This implies that

$$a_k = -\frac{e^{-u}}{u} \left(\frac{u^k}{k!} + \frac{u^{k-1}}{(k-1)!} + \cdots + \frac{u}{1!} \right) + \frac{1 - e^{-u}}{u}$$

$$= \frac{e^{-u}}{u} \left(e^u - \left(1 + \frac{u}{1!} + \frac{u^2}{2!} + \cdots + \frac{u^k}{k!} \right) \right)$$

$$= \frac{e^{-u}}{u} \sum_{j=1}^\infty \frac{u^{k+j}}{(k+j)!}.$$

Thus,

$$J_k = k! e^{-u} \sum_{j=1}^\infty \frac{u^{j-1}}{(k+j)!},$$

and it follows that

$$V_{k,m} = \frac{1}{(m+1)!} \sum_{j=1}^\infty \frac{k!}{(k+j)!} \int_0^\infty \frac{u^{m+j}e^{-u}}{e^u - 1} du.$$

On the other hand,

$$\int_0^\infty \frac{u^{m+j}e^{-u}}{e^u - 1} du = \int_0^\infty u^{m+j}e^{-2u} \sum_{p=0}^\infty e^{-pu} du$$

$$= \sum_{p=0}^\infty \int_0^\infty u^{m+j}e^{-(2+p)u} du$$

$$= \sum_{p=0}^\infty \frac{\Gamma(m+j+1)}{(2+p)^{m+j+1}}$$

$$= (m+j)!(\zeta(m+j+1) - 1).$$

Hence,

$$V_{k,m} = \frac{k!}{(m+1)!} \sum_{j=1}^{\infty} \frac{(m+j)!}{(k+j)!} (\zeta(m+j+1) - 1).$$

When $k = m$, one has that

$$V_{m,m} = \frac{1}{m+1} \sum_{j=1}^{\infty} (\zeta(m+j+1) - 1) = 1 - \frac{\zeta(2) + \zeta(3) + \cdots + \zeta(m+1)}{m+1},$$

since $\sum_{j=1}^{\infty} (\zeta(j+1) - 1) = 1$.

Part (b) of Problem **2.22** follows from the calculation of $V_{k,m}$ combined with the formula $\sum_{j=1}^{\infty} (\zeta(j+1) - 1)/(j+1) = 1 - \gamma$.

2.6.2   Double Integrals

> But just as much as it is easy to find the differential of a given
> quantity, so it is difficult to find the integral of a given
> differential. Moreover, sometimes we cannot say with certainty
> whether the integral of a given quantity can be found or not.
>
> Johann Bernoulli (1667–1748)

2.23. (a) The integral equals $1 - \frac{\pi^2}{12}$. We have

$$I = \int_0^1 \int_0^1 x \left\{ \frac{1}{1-xy} \right\} dxdy = \int_0^1 x \left(\int_0^1 \left\{ \frac{1}{1-xy} \right\} dy \right) dx.$$

Using the substitution $xy = t$, in the inner integral, we obtain that

$$I = \int_0^1 \left(\int_0^x \left\{ \frac{1}{1-t} \right\} dt \right) dx.$$

Integrating by parts, with

$$f(x) = \int_0^x \left\{ \frac{1}{1-t} \right\} dt, \quad f'(x) = \left\{ \frac{1}{1-x} \right\}, \quad g'(x) = 1, \quad g(x) = x,$$

we obtain that

$$I = \left(x \int_0^x \left\{ \frac{1}{1-t} \right\} dt \right) \Big|_0^1 - \int_0^1 x \left\{ \frac{1}{1-x} \right\} dx$$

$$= \int_0^1 \left\{ \frac{1}{1-t} \right\} dt - \int_0^1 x \left\{ \frac{1}{1-x} \right\} dx$$

$$= \int_0^1 (1-x) \left\{ \frac{1}{1-x} \right\} dx$$

$$= \int_0^1 x \left\{ \frac{1}{x} \right\} dx$$

$$= 1 - \frac{\zeta(2)}{2},$$

where the last equality follows from Problem **2.20** when $m = 1$.

(b) The integral equals

$$\frac{1}{k} \sum_{j=1}^k \binom{k}{j} (-1)^{j+1} \left(\frac{1}{j} - \frac{\zeta(j+1)}{j+1} \right).$$

Use the same technique as in the solution of part (a) of the problem.

2.24. Let I_m denote the value of the double integral. We have

$$I_m = \int_0^1 \left(\int_0^1 \left\{ \frac{1}{x+y} \right\}^m dy \right) dx \overset{x+y=t}{=} \int_0^1 \left(\int_x^{x+1} \left\{ \frac{1}{t} \right\}^m dt \right) dx.$$

We calculate the integral by parts, with

$$f(x) = \int_x^{x+1} \left\{ \frac{1}{t} \right\}^m dt, \quad f'(x) = \left\{ \frac{1}{x+1} \right\}^m - \left\{ \frac{1}{x} \right\}^m,$$

$g'(x) = 1$ and $g(x) = x$, and we get that

$$I_m = \left(x \int_x^{x+1} \left\{ \frac{1}{t} \right\}^m dt \right) \Big|_{x=0}^{|x=1|} - \int_0^1 x \left(\left\{ \frac{1}{x+1} \right\}^m - \left\{ \frac{1}{x} \right\}^m \right) dx$$

$$= 2 \int_1^2 \left\{ \frac{1}{t} \right\}^m dt - \int_1^2 t \left\{ \frac{1}{t} \right\}^m dt + \int_0^1 x \left\{ \frac{1}{x} \right\}^m dx$$

$$= 2 \int_1^2 t^{-m} dt - \int_1^2 t^{1-m} dt + \int_0^1 x \left\{ \frac{1}{x} \right\}^m dx.$$

When $m = 1$, one has that

$$I_1 = 2\ln 2 - 1 + \int_0^1 x \left\{ \frac{1}{x} \right\} dx = 2\ln 2 - \frac{\pi^2}{12},$$

where the equality follows from Problem **2.20**.
When $m = 2$, one has that

$$I_2 = 1 - \ln 2 + \int_0^1 x \left\{ \frac{1}{x} \right\}^2 dx = \frac{5}{2} - \ln 2 - \gamma - \frac{\pi^2}{12},$$

where the equality follows from part (b) of Problem **2.22**.
When $m \geq 3$, one has that

$$I_m = \frac{m-3}{(m-1)(m-2)} + \frac{2^{2-m}}{(m-1)(m-2)} + \int_0^1 x \left\{ \frac{1}{x} \right\}^m dx$$

$$= \frac{m-3}{(m-1)(m-2)} + \frac{2^{2-m}}{(m-1)(m-2)} + \frac{m!}{2} \sum_{j=1}^{\infty} \frac{(j+1)!}{(m+j)!} (\zeta(j+2) - 1),$$

where the equality follows from part (a) of Problem **2.22**.

It is worth mentioning that the case when $m = 1$ was solved by an alternative method in [106].

2.25. We have, based on the substitution $y = xt$, that

$$I = \int_0^1 \int_0^{1-x} \frac{dxdy}{\left(\left\lfloor \frac{x}{y} \right\rfloor + 1 \right)^k} = \int_0^1 x \left(\int_0^{(1-x)/x} \frac{dt}{\left(\left\lfloor \frac{1}{t} \right\rfloor + 1 \right)^k} \right) dx.$$

We integrate by parts, with

$$f(x) = \int_0^{(1-x)/x} \frac{dt}{\left(\left\lfloor \frac{1}{t} \right\rfloor + 1 \right)^k}, \quad f'(x) = \frac{-1}{x^2 \left(\left\lfloor \frac{x}{1-x} \right\rfloor + 1 \right)^k},$$

$g'(x) = x$ and $g(x) = x^2/2$, and we get that

$$I = \left(\frac{x^2}{2} \int_0^{(1-x)/x} \frac{dt}{\left(\left\lfloor \frac{1}{t} \right\rfloor + 1 \right)^k} \right) \Big|_{x=0}^{x=1} + \frac{1}{2} \int_0^1 \frac{dx}{\left(\left\lfloor \frac{x}{1-x} \right\rfloor + 1 \right)^k}$$

$$= \frac{1}{2} \int_0^1 \frac{dx}{\left(\left\lfloor \frac{x}{1-x} \right\rfloor + 1 \right)^k} = \frac{1}{2} \int_0^1 \frac{dx}{\left(\left\lfloor \frac{1-x}{x} \right\rfloor + 1 \right)^k}$$

$$= \frac{1}{2} \int_0^1 \frac{dx}{\left\lfloor \frac{1}{x} \right\rfloor^k} = \frac{1}{2} \int_1^{\infty} \frac{dt}{t^2 \lfloor t \rfloor^k} = \frac{1}{2} \sum_{m=1}^{\infty} \int_m^{m+1} \frac{dt}{t^2 m^k}$$

$$= \frac{1}{2} \sum_{m=1}^{\infty} \frac{1}{m^k} \left(\frac{1}{m} - \frac{1}{m+1} \right)$$

$$= \frac{1}{2} \zeta(k+1) - \frac{1}{2} \sum_{m=1}^{\infty} \frac{1}{m^k(m+1)}.$$

Let $S_k = \sum_{m=1}^{\infty} 1/(m^k(m+1))$. Since $\frac{1}{m^k(m+1)} = \frac{1}{m^k} - \frac{1}{m^{k-1}(m+1)}$, one has that $S_k = \zeta(k) - S_{k-1}$. This implies that $(-1)^k S_k = (-1)^k \zeta(k) + (-1)^{k-1} S_{k-1}$, and it follows that $S_k = (-1)^{k+1} + \sum_{j=2}^k (-1)^{k+j} \zeta(j)$. Thus,

$$I = \frac{1}{2} \left((-1)^k - \sum_{j=2}^{k+1} (-1)^{k+j} \zeta(j) \right).$$

Remark. We mention that the case when $k = 2$ is due to Paolo Perfetti [99].

2.26. The integral equals $1 + \frac{1}{2}(k - \zeta(2) - \zeta(3) - \cdots - \zeta(k+1))$. We have, based on the substitution $y = xt$, in the inner integral, that

$$I = \int_0^1 \left(\int_0^1 \frac{dy}{(\lfloor x/y \rfloor + 1)^k} \right) dx = \int_0^1 x \left(\int_0^{1/x} \frac{dt}{(\lfloor 1/t \rfloor + 1)^k} \right) dx.$$

We integrate by parts, with

$$f(x) = \int_0^{1/x} \frac{dt}{(\lfloor 1/t \rfloor + 1)^k}, \quad f'(x) = -\frac{1}{x^2 (\lfloor x \rfloor + 1)^k},$$

$g'(x) = x$ and $g(x) = x^2/2$, and we get that

$$I = \left(\frac{x^2}{2} \int_0^{1/x} \frac{dt}{(\lfloor 1/t \rfloor + 1)^k} \right) \Big|_{x=0}^{x=1} + \frac{1}{2} \int_0^1 \frac{dx}{(\lfloor x \rfloor + 1)^k}$$

$$= \frac{1}{2} \int_0^1 \frac{dt}{(\lfloor 1/t \rfloor + 1)^k} + \frac{1}{2} \int_0^1 dx$$

$$= \frac{1}{2} \int_1^{\infty} \frac{du}{u^2(\lfloor u \rfloor + 1)^k} + \frac{1}{2}$$

$$= \frac{1}{2} \sum_{m=1}^{\infty} \int_m^{m+1} \frac{du}{u^2(m+1)^k} + \frac{1}{2}$$

$$= \frac{1}{2} \sum_{m=1}^{\infty} \frac{1}{(m+1)^{k+1}m} + \frac{1}{2},$$

and the result follows from Problem **3.8**.

2.27. See [88] or the solutions of Problems **2.28** or **2.29**.

2.28. We have, based on the substitution $kx/y = t$, that

$$I = \int_0^1 \left(\int_0^1 \left\{ k\frac{x}{y} \right\} dy \right) dx = k \int_0^1 x \left(\int_{kx}^\infty \frac{\{t\}}{t^2} dt \right) dx.$$

Integrating by parts, with

$$f(x) = \int_{kx}^\infty \frac{\{t\}}{t^2} dt, \quad f'(x) = -\frac{\{kx\}}{kx^2},$$

$g'(x) = x$ and $g(x) = x^2/2$, we get that

$$I = k \left(\left(\frac{x^2}{2} \int_{kx}^\infty \frac{\{t\}}{t^2} dt \right) \Big|_0^1 + \frac{1}{2k} \int_0^1 \{kx\} dx \right)$$

$$= k \left(\frac{1}{2} \int_k^\infty \frac{\{t\}}{t^2} dt + \frac{1}{2k} \int_0^1 \{kx\} dx \right). \tag{2.6}$$

A calculation shows that

$$\int_0^1 \{kx\} dx = \frac{1}{k} \int_0^k \{y\} dy = \frac{1}{k} \sum_{j=0}^{k-1} \int_j^{j+1} \{y\} dy = \frac{1}{k} \sum_{j=0}^{k-1} \int_j^{j+1} (y - j) dy = \frac{1}{2}.$$

$$\tag{2.7}$$

On the other hand,

$$\int_k^\infty \frac{\{t\}}{t^2} dt = \sum_{j=k}^\infty \left(\ln \frac{j+1}{j} - \frac{1}{j+1} \right) = 1 + \frac{1}{2} + \cdots + \frac{1}{k} - \ln k - \gamma. \tag{2.8}$$

Combining (2.6)–(2.8), we get that the integral is calculated and the problem is solved.

Remark. We also have, based on symmetry reasons, that

$$\int_0^1 \int_0^1 \left\{ k\frac{y}{x} \right\} dxdy = \frac{k}{2} \left(1 + \frac{1}{2} + \cdots + \frac{1}{k} - \ln k - \gamma \right) + \frac{1}{4}.$$

2.29. We have

$$I = \int_0^1 \int_0^1 \left\{ \frac{mx}{ny} \right\} dxdy = \int_0^1 \left(\int_0^1 \left\{ \frac{mx}{ny} \right\} dy \right) dx \overset{t=\frac{mx}{ny}}{=} \frac{m}{n} \int_0^1 x \left(\int_{\frac{mx}{n}}^\infty \frac{\{t\}}{t^2} dt \right) dx.$$

Integrating by parts, with

$$f(x) = \int_{\frac{mx}{n}}^\infty \frac{\{t\}}{t^2} dt, \quad f'(x) = -\frac{n}{m} \cdot \frac{\left\{ \frac{mx}{n} \right\}}{x^2}, \quad g'(x) = x, \quad g(x) = \frac{x^2}{2},$$

we obtain, since $\left\{\frac{mx}{n}\right\} = \frac{mx}{n}$ for $x \in [0,1)$, that

$$I = \frac{m}{n}\left(\frac{x^2}{2}\int_{\frac{mx}{n}}^{\infty}\frac{\{t\}}{t^2}dt\,\Big|_0^1 + \frac{n}{2m}\int_0^1\left\{\frac{mx}{n}\right\}dx\right)$$

$$= \frac{m}{n}\left(\frac{1}{2}\int_{\frac{m}{n}}^{\infty}\frac{\{t\}}{t^2}dt + \frac{n}{2m}\int_0^1\frac{mx}{n}dx\right)$$

$$= \frac{m}{2n}\left(\int_{\frac{m}{n}}^{\infty}\frac{\{t\}}{t^2}dt + \frac{1}{2}\right)$$

$$= \frac{m}{2n}\left(\int_{\frac{m}{n}}^{1}\frac{\{t\}}{t^2}dt + \int_1^{\infty}\frac{\{t\}}{t^2}dt + \frac{1}{2}\right)$$

$$= \frac{m}{2n}\left(\int_{\frac{m}{n}}^{1}\frac{1}{t}dt + 1 - \gamma + \frac{1}{2}\right)$$

$$= \frac{m}{2n}\left(\ln\frac{n}{m} + \frac{3}{2} - \gamma\right).$$

Remark. It is worth mentioning that, when $m > n$, the integral equals

$$\frac{m}{2n}\left(\ln\frac{n}{m} + 1 + \frac{1}{2} + \cdots + \frac{1}{q} - \gamma + \frac{qn^2 + r^2 + 2mr}{2m^2}\right),$$

where q and r are the integers defined by $m = nq + r$ with $r < n$.

2.30. Using the substitution $t = x^k/y$, in the inner integral, we obtain that

$$I = \int_0^1\int_0^1\left\{\frac{x^k}{y}\right\}dxdy = \int_0^1\left(\int_0^1\left\{\frac{x^k}{y}\right\}dy\right)dx \overset{t=\frac{x^k}{y}}{=} \int_0^1 x^k\left(\int_{x^k}^{\infty}\frac{\{t\}}{t^2}dt\right)dx.$$

Integrating by parts, with

$$f(x) = \int_{x^k}^{\infty}\frac{\{t\}}{t^2}dt, \quad f'(x) = -\frac{k\{x^k\}}{x^{k+1}}, \quad g'(x) = x^k, \quad g(x) = \frac{x^{k+1}}{k+1},$$

we obtain that

$$I = \left(\frac{x^{k+1}}{k+1}\int_{x^k}^{\infty}\frac{\{t\}}{t^2}dt\right)\Big|_0^1 + \frac{k}{k+1}\int_0^1\{x^k\}dx$$

$$= \frac{1}{k+1}\int_1^{\infty}\frac{\{t\}}{t^2}dt + \frac{k}{k+1}\int_0^1 x^k dx$$

$$= \frac{1-\gamma}{k+1} + \frac{k}{(k+1)^2}$$

$$= \frac{2k+1}{(k+1)^2} - \frac{\gamma}{k+1}.$$

2.31. and **2.32.** We have, based on the substitution $x/y = t$, that

$$I_k = \int_0^1 \left(\int_0^1 \left\{ \frac{x}{y} \right\}^k dy \right) dx = \int_0^1 x \left(\int_x^\infty \frac{\{t\}^k}{t^2} dt \right) dx.$$

We integrate by parts, with

$$f(x) = \int_x^\infty \frac{\{t\}^k}{t^2} dt, \quad f'(x) = -\frac{\{x\}^k}{x^2}, \quad g'(x) = x, \quad g(x) = \frac{x^2}{2},$$

and we get that

$$I_k = \left(\frac{x^2}{2} \int_x^\infty \frac{\{t\}^k}{t^2} dt \right) \Bigg|_{x=0}^{x=1} + \frac{1}{2} \int_0^1 \{x\}^k dx = \frac{1}{2} \int_1^\infty \frac{\{t\}^k}{t^2} dt + \frac{1}{2(k+1)}$$

$$= \frac{1}{2} V_{k,0} + \frac{1}{2(k+1)} = \frac{1}{2} \sum_{j=1}^\infty \frac{\zeta(j+1)-1}{\binom{k+j}{j}} + \frac{1}{2(k+1)}, \tag{2.9}$$

where the last equality follows from the calculation of $V_{k,m}$ (see the solutions of Problems **2.21** and **2.22**).

On the other hand, we have, based on the substitution $1/x = t$, that $\int_0^1 \left\{ \frac{1}{x} \right\}^k dx = \int_1^\infty \frac{\{t\}^k}{t^2} dt$, and the equality

$$\int_0^1 \int_0^1 \left\{ \frac{x}{y} \right\}^k dx dy = \frac{1}{2} \int_0^1 \left\{ \frac{1}{x} \right\}^k dx + \frac{1}{2(k+1)},$$

follows from the first line of (2.9).

The case $k = 2$. (this is Problem **2.31**)

$$\int_0^1 \int_0^1 \left\{ \frac{x}{y} \right\}^2 dx dy \stackrel{(2.9)}{=} \sum_{p=1}^\infty \frac{\zeta(p+1)-1}{(p+1)(p+2)} + \frac{1}{6}$$

$$= \sum_{p=1}^\infty \frac{\zeta(p+1)-1}{p+1} - \sum_{p=1}^\infty \frac{\zeta(p+1)-1}{p+2} + \frac{1}{6}$$

$$= -\frac{1}{3} - \frac{\gamma}{2} + \frac{\ln(2\pi)}{2},$$

since $\sum_{p=1}^\infty (\zeta(p+1)-1)/(p+2) = 3/2 - \gamma/2 - \ln(2\pi)/2$ [122, p. 213] and $\sum_{p=2}^\infty (\zeta(p)-1)/p = 1 - \gamma$ [122, p. 173].

2.33.–2.35. We have, based on the substitution $x/y = t$, that

$$\int_0^1 \int_0^1 \left\{\frac{x}{y}\right\}^k \frac{y^m}{x^p} dxdy = \int_0^1 \frac{1}{x^p} \left(\int_0^1 \left\{\frac{x}{y}\right\}^k y^m dy\right) dx$$

$$= \int_0^1 x^{m+1-p} \left(\int_x^\infty \frac{\{t\}^k}{t^{m+2}} dt\right) dx.$$

We integrate by parts, with

$$f(x) = \int_x^\infty \frac{\{t\}^k}{t^{m+2}} dt, \quad f'(x) = -\frac{\{x\}^k}{x^{m+2}}, \quad g'(x) = x^{m+1-p}, \quad g(x) = \frac{x^{m+2-p}}{m+2-p},$$

and we get that

$$\int_0^1 \int_0^1 \left\{\frac{x}{y}\right\}^k \frac{y^m}{x^p} dxdy = \left(\frac{x^{m+2-p}}{m+2-p} \int_x^\infty \frac{\{t\}^k}{t^{m+2}} dt\right)\Bigg|_{x=0}^{x=1} + \frac{1}{m+2-p} \int_0^1 x^{k-p} dx$$

$$= \frac{1}{m+2-p} \int_1^\infty \frac{\{t\}^k}{t^{m+2}} dt + \frac{1}{(m+2-p)(k-p+1)}$$

$$= \frac{V_{k,m}}{m+2-p} + \frac{1}{(m+2-p)(k-p+1)}$$

$$= \frac{k!}{(m+2-p)(m+1)!} \sum_{j=1}^\infty \frac{(m+j)!}{(k+j)!}(\zeta(m+j+1)-1)$$

$$+ \frac{1}{(m-p+2)(k-p+1)}.$$

When $m = k$ (this is Problem **2.34**), we have that

$$\int_0^1 \int_0^1 \left\{\frac{x}{y}\right\}^k \frac{y^k}{x^p} dxdy = \frac{1}{k-p+1} - \frac{\zeta(2)+\zeta(3)+\cdots+\zeta(k+1)}{(k+2-p)(k+1)}.$$

When $m = p = k$ (this is Problem **2.33**), we have that

$$\int_0^1 \int_0^1 \left\{\frac{x}{y}\right\}^k \left(\frac{y}{x}\right)^k dxdy = 1 - \frac{\zeta(2)+\zeta(3)+\cdots+\zeta(k+1)}{2(k+1)}.$$

2.36. See [3].

2.37. See the solution of Problem **2.38**.

2.38. Recall that (see the calculation of J_k in the solution of Problem **2.15**) if $\alpha > 0$ is a real number, then

$$\int_0^1 t^\alpha \left\{ \frac{1}{t} \right\} dt = \frac{1}{\alpha} - \frac{\zeta(\alpha+1)}{\alpha+1}. \tag{2.10}$$

We calculate the double integral by making the substitution $y = xt$, in the inner integral, and we get that

$$I = \int_0^1 x^n \left(\int_0^1 y^n \left\{ \frac{x}{y} \right\} \left\{ \frac{y}{x} \right\} dy \right) dx = \int_0^1 x^{2n+1} \left(\int_0^{1/x} t^n \left\{ t \right\} \left\{ \frac{1}{t} \right\} dt \right) dx.$$

We integrate by parts, with

$$f(x) = \int_0^{1/x} t^n \left\{ t \right\} \left\{ \frac{1}{t} \right\} dt, \quad f'(x) = -\frac{1}{x^{n+2}} \left\{ x \right\} \left\{ \frac{1}{x} \right\},$$

$g'(x) = x^{2n+1}$ and $g(x) = x^{2n+2}/(2n+2)$, and we get, based on (2.10) with $\alpha = n+1$, that

$$I = \left(\frac{x^{2n+2}}{2n+2} \int_0^{1/x} t^n \left\{ t \right\} \left\{ \frac{1}{t} \right\} dt \right) \Bigg|_0^1 + \frac{1}{2n+2} \int_0^1 x^n \left\{ x \right\} \left\{ \frac{1}{x} \right\} dx$$

$$= \frac{1}{2n+2} \int_0^1 t^n \left\{ t \right\} \left\{ \frac{1}{t} \right\} dt + \frac{1}{2n+2} \int_0^1 x^n \left\{ x \right\} \left\{ \frac{1}{x} \right\} dx$$

$$= \frac{1}{n+1} \int_0^1 x^{n+1} \left\{ \frac{1}{x} \right\} dx$$

$$= \frac{1}{n+1} \left(\frac{1}{n+1} - \frac{\zeta(n+2)}{n+2} \right).$$

2.39. See [132].

2.40. and **2.41.** We have, based on the substitution $x/y = t$, that

$$I_{m,k} = \int_0^1 \left(\int_0^1 \left\{ \frac{x}{y} \right\}^m \left\{ \frac{y}{x} \right\}^k dy \right) dx = \int_0^1 x \left(\int_x^\infty \left\{ t \right\}^m \left\{ \frac{1}{t} \right\}^k \frac{dt}{t^2} \right) dx.$$

We integrate by parts, with

$$f(x) = \int_x^\infty \left\{ t \right\}^m \left\{ \frac{1}{t} \right\}^k \frac{dt}{t^2}, \quad f'(x) = -\left\{ x \right\}^m \left\{ \frac{1}{x} \right\}^k \frac{1}{x^2},$$

$g'(x) = x$ and $g(x) = x^2/2$, and we get that

$$I_{m,k} = \left(\frac{x^2}{2}\int_x^\infty \{t\}^m \left\{\frac{1}{t}\right\}^k \frac{dt}{t^2}\right)\Bigg|_{x=0}^{x=1} + \frac{1}{2}\int_0^1 x^m \left\{\frac{1}{x}\right\}^k dx$$

$$= \frac{1}{2}\int_0^1 x^k \left\{\frac{1}{x}\right\}^m dx + \frac{1}{2}\int_0^1 x^m \left\{\frac{1}{x}\right\}^k dx$$

$$= \frac{1}{2}V_{m,k} + \frac{1}{2}V_{k,m}. \tag{2.11}$$

Recall that (see the solutions of Problems **2.21** and **2.22**)

$$V_{k,m} = \int_0^1 x^m \left\{\frac{1}{x}\right\}^k dx = \frac{k!}{(m+1)!}\sum_{j=1}^\infty \frac{(m+j)!}{(k+j)!}(\zeta(m+j+1)-1).$$

Thus,

$$I_{k,m} = \frac{1}{2}V_{m,k} + \frac{1}{2}V_{k,m} = \frac{m!}{2(k+1)!}\sum_{j=1}^\infty \frac{(k+j)!}{(m+j)!}(\zeta(k+j+1)-1)$$

$$+ \frac{k!}{2(m+1)!}\sum_{j=1}^\infty \frac{(m+j)!}{(k+j)!}(\zeta(m+j+1)-1).$$

When $k = m$ (this is Problem **2.40**), one has

$$\int_0^1 \int_0^1 \left\{\frac{x}{y}\right\}^m \left\{\frac{y}{x}\right\}^m dxdy = \frac{1}{m+1}\sum_{j=1}^\infty (\zeta(m+j+1)-1)$$

$$= 1 - \frac{\zeta(2)+\zeta(3)+\cdots+\zeta(m+1)}{m+1},$$

since $\sum_{p=1}^\infty (\zeta(p+1)-1) = 1$ (see [122, p. 178]).

2.42. 1. Let $m = \lfloor a\rfloor$ be the floor of a. We have

$$\int_a^{a+k}\{x\}\,dx = \int_a^{m+1}\{x\}\,dx + \sum_{j=m+1}^{k+m-1}\left(\int_j^{j+1}\{x\}\,dx\right) + \int_{k+m}^{k+a}\{x\}\,dx$$

$$= \int_a^{m+1}(x-m)\,dx + \sum_{j=m+1}^{k+m-1}\left(\int_j^{j+1}(x-j)\,dx\right) + \int_{k+m}^{k+a}(x-k-m)\,dx$$

$$= \frac{k}{2}.$$

2. The integral equals $(1/2)a_1 \cdots a_n$. Let I_n be the value of the integral. We have

$$I_n = \int_0^{a_1} \cdots \int_0^{a_{n-1}} \left(\int_0^{a_n} \{k(x_1 + x_2 + \cdots + x_n)\} \, dx_n \right) dx_1 \cdots dx_{n-1}.$$

Using the substitution $k(x_1 + \cdots + x_n) = y$, in the inner integral, we get that

$$\int_0^{a_n} \{k(x_1 + x_2 + \cdots + x_n)\} \, dx_n = \frac{1}{k} \int_{k(x_1 + \cdots + x_{n-1})}^{k(x_1 + \cdots + x_{n-1}) + ka_n} \{y\} \, dy$$

$$= \frac{a_n}{2},$$

where the last equality follows based on the first part of the problem. Thus,

$$I_n = \int_0^{a_1} \cdots \int_0^{a_{n-1}} \frac{a_n}{2} dx_1 \cdots dx_{n-1} = \frac{1}{2} a_1 \cdots a_n.$$

2.43. See [135].

2.6.3   Quickies

A mathematician's reputation rests on the number of bad proofs he has given.

A.S. Besicovitch

2.44. We have

$$\int_0^1 \{nx\}^k \, dx \overset{nx=y}{=} \frac{1}{n} \int_0^n \{y\}^k \, dy = \frac{1}{n} \sum_{j=0}^{n-1} \int_j^{j+1} (y-j)^k dy = \frac{1}{n} \sum_{j=0}^{n-1} \int_0^1 u^k du = \frac{1}{k+1}.$$

2.45. The integral equals $1/2$. Let I denote the value of the integral. We note that if x is a real number and x is not an integer, then

$$\{x\} + \{-x\} = 1. \qquad (2.12)$$

We have, based on the substitution $x = 1 - y$, that

$$I = \int_0^1 \left\{ \frac{1}{1-y} - \frac{1}{y} \right\} dy = \int_0^1 \left(1 - \left\{ \frac{1}{y} - \frac{1}{1-y} \right\} \right) dy = 1 - I.$$

2.46. The integral equals $\gamma - 1/2$. We have

$$I = \int_0^1 \left\{ \frac{1}{x} - \frac{1}{1-x} \right\} \left\{ \frac{1}{x} \right\} \left\{ \frac{1}{1-x} \right\} dx \overset{x=1-y}{=} \int_0^1 \left\{ \frac{1}{1-y} - \frac{1}{y} \right\} \left\{ \frac{1}{1-y} \right\} \left\{ \frac{1}{y} \right\} dy$$

$$\overset{(2.12)}{=} \int_0^1 \left(1 - \left\{ \frac{1}{y} - \frac{1}{1-y} \right\} \right) \left\{ \frac{1}{1-y} \right\} \left\{ \frac{1}{y} \right\} dy = \int_0^1 \left\{ \frac{1}{1-y} \right\} \left\{ \frac{1}{y} \right\} dy - I,$$

and the result follows from Problem **2.10**.

2.47. The integral equals $(m!)^2/(2(2m+1)!)$. We have

$$I = \int_0^1 \left\{ \left(\frac{1}{x} \right)^k - \left(\frac{1}{1-x} \right)^k \right\} x^m (1-x)^m dx$$

$$\overset{x=1-y}{=} \int_0^1 \left\{ \left(\frac{1}{1-y} \right)^k - \left(\frac{1}{y} \right)^k \right\} y^m (1-y)^m dy$$

$$\overset{(2.12)}{=} \int_0^1 \left(1 - \left\{ \left(\frac{1}{y} \right)^k - \left(\frac{1}{1-y} \right)^k \right\} \right) y^m (1-y)^m dy$$

$$= \int_0^1 y^m (1-y)^m dy - I = B(m+1, m+1) - I,$$

where B denotes the Beta function.

2.48. First, we note that if n is a positive integer, then $\{n(1-y)\} = 1 - \{ny\}$, for all $y \in [0,1]$ except for $0, \frac{1}{n}, \frac{2}{n}, \ldots, \frac{n-1}{n}, 1$. Since the Riemann integral does not depend on sets of Lebesgue measure zero (or sets with a finite number of elements), we get, based on the substitution $x = 1 - y$, that

$$I = \int_0^1 (x - x^2)^k \{nx\} dx = \int_0^1 (y - y^2)^k \{n(1-y)\} dy$$

$$= \int_0^1 (y - y^2)^k (1 - \{ny\}) dy = \int_0^1 (y - y^2)^k dy - I$$

$$= B(k+1, k+1) - I$$

$$= \frac{(k!)^2}{(2k+1)!} - I.$$

Remark. It is worth mentioning that if $f : [0,1] \to \mathbb{R}$ is an integrable function, then

$$\int_0^1 f(x - x^2) \{nx\} dx = \frac{1}{2} \int_0^1 f(x - x^2) dx.$$

2.49. Use the substitution $x = 1 - y$ and identity (2.12).

2.50. The first solution. The integral equals $1/2$. We have

$$I = \int_0^1 \left(\int_0^x \{x - y\} \, dy + \int_x^1 \{x - y\} \, dy \right) dx.$$

We note that when $0 \le y \le x$, then $0 \le x - y \le 1$, and hence $\{x - y\} = x - y$, and when $x \le y \le 1$, then $-1 \le x - y \le 0$, and hence, $\{x - y\} = x - y - \lfloor x - y \rfloor = x - y + 1$. It follows that

$$I = \int_0^1 \left(\int_0^x (x - y) \, dy + \int_x^1 (x - y + 1) \, dy \right) dx = \frac{1}{2}.$$

The second solution. By symmetry,

$$I = \int_0^1 \int_0^1 \{x - y\} \, dx \, dy = \int_0^1 \int_0^1 \{y - x\} \, dx \, dy.$$

Hence,

$$I = \tfrac{1}{2}(I + I) = \tfrac{1}{2} \int_0^1 \int_0^1 (\{x - y\} + \{y - x\}) \, dx \, dy \overset{(2.12)}{=} \tfrac{1}{2} \int_0^1 \int_0^1 dx \, dy = \tfrac{1}{2},$$

because the set on which the integrand is 0 is a set of measure 0.

2.51. We have, based on symmetry, that

$$I = \int_0^1 \int_0^1 \left\{ \frac{x - y}{x + y} \right\} dx \, dy = \int_0^1 \int_0^1 \left\{ \frac{y - x}{x + y} \right\} dx \, dy.$$

Thus,

$$I = \frac{1}{2}(I + I) = \frac{1}{2} \int_0^1 \int_0^1 \left(\left\{ \frac{x - y}{x + y} \right\} + \left\{ \frac{y - x}{x + y} \right\} \right) dx \, dy \overset{(2.12)}{=} \frac{1}{2} \int_0^1 \int_0^1 dx \, dy = \frac{1}{2},$$

because the set on which the integrand is 0 is a set of measure 0.

2.52. Use symmetry and identity (2.12) and Problem **2.1**.

2.53. The integral equals $1/2$. Use symmetry and identity (2.12).

2.54. The integral equals $1/(2(m + 1)^2)$. Use symmetry and identity (2.12).

2.55. The integral equals $1/(2(m + 1)^2)$. First, we note that if n is a fixed positive integer, then $\{n(1 - a)\} = 1 - \{na\}$, for all $a \in [0, 1]$ except for $0, \frac{1}{n}, \frac{2}{n}, \ldots, \frac{n-1}{n}, 1$. We have, by symmetry reasons, that

$$I = \int_0^1 \int_0^1 \left\{ \frac{nx}{x + y} \right\} x^m y^m \, dx \, dy = \int_0^1 \int_0^1 \left\{ \frac{ny}{x + y} \right\} x^m y^m \, dx \, dy,$$

and hence,

$$I = \frac{1}{2}(I+I) = \frac{1}{2}\int_0^1\int_0^1\left(\left\{\frac{nx}{x+y}\right\} + \left\{\frac{ny}{x+y}\right\}\right)x^m y^m dxdy$$

$$= \frac{1}{2}\int_0^1\int_0^1 x^m y^m dxdy = \frac{1}{2}\int_0^1 x^m dx\int_0^1 y^m dy = \frac{1}{2(m+1)^2}.$$

We used that

$$\left\{\frac{nx}{x+y}\right\} + \left\{\frac{ny}{x+y}\right\} = 1,$$

for all $(x,y) \in [0,1]^2$, except for the points of a set A of Lebesgue area measure zero. The set A is the set of points for which $\frac{y}{x+y} = \frac{k}{n}$, $k = 0,1,\ldots,n$, and hence, A turns out to be the union of $n+1$ lines through the origin.

2.56. The integral equals $1/(2(m+1)^2)$. See the solution of Problem **2.55**.

2.57. Use symmetry and identity (2.12).

Chapter 3
A Bouquet of Series

Even if we have thousands of acts of great virtue to our credit,
our confidence in being heard must be based on God's mercy
and His love for men. Even if we stand at the very summit of
virtue, it is by mercy that we shall be saved.

St. John Chrysostom (347–407)

3.1 Single Series

An expert is a man who has made all the mistakes that can be
made in a very narrow field.

Niels Bohr (1885–1962)

3.1. Let $n, m \geq 1$ be integers. Prove that

$$\left\{ \frac{n}{m} \right\} = \frac{n}{m} - \frac{1}{m} \sum_{k=1}^{n} \sum_{l=0}^{m-1} e^{2\pi i k l / m},$$

where $\{a\}$ denotes the fractional part of a.

3.2. Let n be a positive integer. Prove that

$$\sum_{k=1}^{n-1} (-1)^k \sin^n \frac{k\pi}{n} = \frac{1 + (-1)^n}{2^n} \cdot n \cdot \cos \frac{n\pi}{2}.$$

3.3. Let $\alpha > 0$ be a real number. Discuss the convergence of the series

$$\sum_{n=2}^{\infty} \left((2 - \sqrt{e})(2 - \sqrt[3]{e}) \cdots (2 - \sqrt[n]{e}) \right)^{\alpha}.$$

O. Furdui, *Limits, Series, and Fractional Part Integrals: Problems in Mathematical
Analysis*, Problem Books in Mathematics, DOI 10.1007/978-1-4614-6762-5_3,
© Springer Science+Business Media New York 2013

3.4. Let $\beta \geq 0$ and let $\alpha > 0$. Prove that the series

$$\sum_{n=1}^{\infty} n^{\beta} (\cos 1 \cos 2 \cdots \cos n)^{\alpha} \quad \text{and} \quad \sum_{n=1}^{\infty} n^{\beta} (\sin 1 \sin 2 \cdots \sin n)^{\alpha}$$

are absolutely convergent.

3.5. Discuss the convergence of the series

$$\sum_{n=1}^{\infty} x^{\sin \frac{1}{1} + \sin \frac{1}{2} + \cdots + \sin \frac{1}{n}}, \quad x \in (0,1).$$

3.6. Prove that the series

$$\sum_{n=1}^{\infty} \cos 1 \cos^2 2 \cdots \cos^n n \quad \text{and} \quad \sum_{n=1}^{\infty} \sin 1 \sin^2 2 \cdots \sin^n n$$

are absolutely convergent.

3.7. Let f be the function defined by $f(n) = \ln 2/2^{k-1}$ if $2^{k-1} \leq n < 2^k$. Prove that

$$\sum_{n=1}^{\infty} \left(\frac{1}{n} - f(n) \right) = \gamma.$$

3.8. Let $n \geq 1$ be an integer. Calculate the sum

$$\sum_{k=1}^{\infty} \frac{1}{k(k+1)^n}.$$

3.9. Let $n \geq 1$ be an integer. Calculate the sum

$$\sum_{k=1}^{\infty} \frac{1}{k^n (k+1)}.$$

3.10. Let $n \geq 1$ be an integer. Prove that

$$\sum_{k=1}^{\infty} \frac{1}{k(k+1)(k+2) \cdots (k+n)} = \frac{1}{n \cdot n!}.$$

3.11. Prove that

$$\sum_{n=1}^{\infty} \left(n \ln \frac{2n+1}{2n-1} - 1 \right) = \frac{1}{2}(1 - \ln 2).$$

3.12. Calculate

$$\sum_{k=2}^{\infty} \left((k-1)^2 \ln \frac{k^2}{k^2-1} - \frac{k-1}{k+1} \right).$$

3.13. Integrals to series. Let $\alpha > 0$ and let $p \geq 1$ be an integer. Prove that

$$\sum_{n=1}^{\infty} \frac{\alpha^{n-1}}{(\alpha+p)(2\alpha+p)\cdots(n\alpha+p)} = e \int_0^1 x^{p-1+\alpha} e^{-x^\alpha} dx$$

and

$$\sum_{n=1}^{\infty} \frac{(-1)^{n-1}\alpha^{n-1}}{(\alpha+p)(2\alpha+p)\cdots(n\alpha+p)} = e^{-1} \int_0^1 x^{p-1+\alpha} e^{x^\alpha} dx.$$

3.14. Calculate the sum

$$\sum_{n=1}^{\infty} \frac{1}{n} \left(1 - \frac{1}{2} + \frac{1}{3} + \cdots + \frac{(-1)^{n-1}}{n} - \ln 2 \right).$$

3.15. Prove that

$$\sum_{n=1}^{\infty} \frac{1}{n} \left(\frac{1}{n+1} - \frac{1}{n+2} + \frac{1}{n+3} - \cdots \right) = \frac{1}{2}\zeta(2) - \frac{1}{2}\ln^2 2.$$

3.16. Find the value of

$$\sum_{n=1}^{\infty} \frac{1}{n^2} \left(1 - \frac{1}{2} + \frac{1}{3} + \cdots + \frac{(-1)^{n-1}}{n} \right).$$

3.17. Calculate

$$\sum_{n=1}^{\infty} \left(1 - \frac{1}{2} + \frac{1}{3} + \cdots + \frac{(-1)^{n-1}}{n} - \ln 2 \right) \cdot \ln \frac{n+1}{n}.$$

3.18. Let $p > 2$ be a real number. Find the value of

$$\sum_{n=1}^{\infty} \left(\zeta(p) - \frac{1}{1^p} - \frac{1}{2^p} - \cdots - \frac{1}{n^p} \right).$$

3.19. Let $p > 3$ be a real number. Calculate the sum

$$\sum_{n=1}^{\infty} n \left(\zeta(p) - \frac{1}{1^p} - \frac{1}{2^p} - \cdots - \frac{1}{n^p} \right).$$

3.20. An Apéry's constant series. Prove that

$$\sum_{n=1}^{\infty} \frac{1}{n} \left(\zeta(2) - \frac{1}{1^2} - \frac{1}{2^2} - \cdots - \frac{1}{n^2} \right) = \zeta(3).$$

3.21. Prove that

$$\sum_{n=1}^{\infty} \frac{1}{n^2} \left(\zeta(2) - \frac{1}{1^2} - \frac{1}{2^2} - \cdots - \frac{1}{n^2} \right) = \frac{3}{4}\zeta(4).$$

3.22. A quadratic series, $\zeta(3)$ and $\zeta(4)$. Prove that

$$\sum_{n=1}^{\infty} \left(\zeta(2) - \frac{1}{1^2} - \frac{1}{2^2} - \cdots - \frac{1}{n^2} \right)^2 = 3\zeta(3) - \frac{5}{2}\zeta(4).$$

3.23. Find the value of

$$\sum_{k=2}^{\infty} \frac{\zeta(k) - 1}{k + 1}.$$

3.24. Show that

$$\sum_{k=2}^{\infty} \frac{\zeta(k) - 1}{k + 2} = \frac{11}{6} - 2\ln A - \frac{\gamma}{3} - \frac{\ln(2\pi)}{2}.$$

3.25. Find the value of

$$\sum_{k=1}^{\infty} \frac{\zeta(2k) - 1}{k + 1}.$$

3.26. (a) Prove that $\lim_{n \to \infty} (n - \zeta(2) - \zeta(3) - \cdots - \zeta(n)) = 0$.
(b) Calculate

$$\sum_{n=2}^{\infty} n \left(n - \zeta(2) - \zeta(3) - \cdots - \zeta(n) \right).$$

(continued)

The Digamma function, ψ, is defined, for $x > 0$, as the logarithmic derivative of the Gamma function, $\psi(x) = \Gamma'(x)/\Gamma(x)$ [122, p. 13].

3.27. Let a be a real number such that $|a| < 2$. Prove that

$$\sum_{n=2}^{\infty} a^n \left(n - \zeta(2) - \zeta(3) - \cdots - \zeta(n) \right) = a \left(\frac{\psi(2-a) + \gamma}{1-a} - 1 \right).$$

3.28. A quadratic logarithmic series. Prove that

$$\sum_{n=1}^{\infty} \left(\ln 2 - \frac{1}{n+1} - \frac{1}{n+2} - \cdots - \frac{1}{2n} \right)^2 = \frac{\pi^2}{48} + \frac{\ln 2}{2} - \ln^2 2.$$

3.29. Prove that

$$\sum_{n=0}^{\infty} \left(\frac{1}{n+1} - \frac{1}{n+2} + \frac{1}{n+3} - \cdots \right)^2 = \ln 2.$$

3.2 Alternating Series

If I have been able to see further, it was only because I stood on the shoulders of giants.

Sir Isaac Newton (1642–1727)

3.30. Let $p \geq 2$ and $k \geq 1$ be integers. Find the sum

$$\sum_{n=1}^{\infty} \frac{(-1)^n}{\lfloor \sqrt[p]{n} \rfloor^k},$$

where $\lfloor a \rfloor$ denotes the greatest integer not exceeding a.

3.31. Let $k \geq 2$ be an integer. Find the sum

$$\sum_{n=1}^{\infty} \frac{(-1)^n}{n!(n+k)}.$$

3.32. Let x be a positive real number. Prove that

$$\sum_{n=1}^{\infty}(-1)^{n-1}\frac{1}{n+x} = \ln 2 - \psi(x) + \psi(x/2) + 1/x,$$

where ψ denotes the Digamma function.

3.33. Find the sum

$$\sum_{n=1}^{\infty}\frac{(-1)^{n-1}}{n}\left(1 - \frac{1}{2} + \frac{1}{3} - \cdots + \frac{(-1)^{n-1}}{n}\right).$$

3.34. Find the sum

$$\sum_{n=1}^{\infty}(-1)^n\left(\ln 2 - \frac{1}{n+1} - \frac{1}{n+2} - \cdots - \frac{1}{2n}\right).$$

3.35. Four Hardy series.

(a) Prove that

$$\sum_{n=1}^{\infty}(-1)^n\left(1 + \frac{1}{2} + \frac{1}{3} + \cdots + \frac{1}{n} - \ln n - \gamma\right) = \frac{\gamma - \ln \pi}{2},$$

$$\sum_{n=1}^{\infty}(-1)^n\left(1 + \frac{1}{3} + \cdots + \frac{1}{2n-1} - \frac{\ln n}{2} - \frac{\gamma}{2} - \ln 2\right) = \frac{\ln 8}{4} - \frac{\ln \pi}{4} - \frac{\pi}{8} + \frac{\gamma}{4}.$$

(b) Let $k \geq 1$ be an integer. Prove that

$$\sum_{n=1}^{\infty}(-1)^n\left(1 + \frac{1}{2} + \frac{1}{3} + \cdots + \frac{1}{n} - \ln(n+2k) - \gamma\right) = \frac{\gamma}{2} - \ln\frac{\sqrt{\pi}(2k-1)!!}{2^k k!},$$

$$\sum_{n=1}^{\infty}(-1)^n\left(1 + \frac{1}{2} + \frac{1}{3} + \cdots + \frac{1}{n} - \ln(n+2k+1) - \gamma\right) = \frac{\gamma}{2} - \ln\frac{2^{k+1}k!}{(2k+1)!!\sqrt{\pi}}.$$

3.36. An alternating Euler constant series. Prove that

$$\sum_{n=1}^{\infty}(-1)^{n-1}\left(\frac{1}{n} - \ln\frac{n+1}{n}\right) = \ln\frac{4}{\pi}.$$

3.37. Prove that

$$\sum_{n=1}^{\infty}(-1)^n\left(n\ln\frac{n+1}{n} - 1\right) = 1 + (\ln \pi)/2 - 6\ln A + (\ln 2)/6,$$

where A denotes the Glaisher–Kinkelin constant.

3.38. Prove that

$$\sum_{n=1}^{\infty}(-1)^n\frac{\ln n}{n^2} = \frac{\pi^2}{12}(\ln(4\pi) - 12\ln A + \gamma),$$

where A denotes the Glaisher–Kinkelin constant.

3.39. An alternating series in terms of Stieltjes constants.

(a) Prove that

$$\sum_{k=1}^{\infty}(-1)^{k-1}\frac{\ln k}{k} = \frac{\ln^2 2}{2} - \gamma\ln 2.$$

(b) More generally, if n is a nonnegative integer, show that

$$\sum_{k=1}^{\infty}(-1)^{k-1}\frac{\ln^n k}{k} = \frac{\ln^{n+1} 2}{n+1} - \sum_{j=0}^{n-1}\binom{n}{j}\gamma_j\ln^{n-j} 2.$$

3.40. Prove that

$$\sum_{n=1}^{\infty}(-1)^n\left(\zeta\left(1+\frac{1}{n}\right) - n - \gamma\right) = \gamma_1\ln 2 + \sum_{k=2}^{\infty}\frac{(-1)^k}{k!}\gamma_k(2^{1-k} - 1)\zeta(k),$$

where γ_k denotes the Stieltjes constant.

3.41. Find the value of

$$\sum_{n=1}^{\infty}\frac{(-1)^{n-1}}{n^2}\left(1 - \frac{1}{2} + \frac{1}{3} - \cdots + \frac{(-1)^{n-1}}{n}\right).$$

3.42. Prove that

$$\sum_{n=1}^{\infty}\left(1 + \frac{1}{2} + \cdots + \frac{1}{n} - \ln n - \gamma - \frac{1}{2n}\right) = \frac{1}{2}(\gamma + 1 - \ln(2\pi)).$$

3.43. Prove that

$$\sum_{k=1}^{\infty}(-1)^{k-1}k\left(1 + \frac{1}{2} + \cdots + \frac{1}{k} - \ln k - \gamma - \frac{1}{2k}\right) = -\frac{\gamma}{4} - \frac{1}{4} + 3\ln A - \frac{7}{12}\ln 2.$$

Catalan's constant, G, is defined by the series formula [43, p. 53]

$$G = \sum_{n=0}^{\infty}\frac{(-1)^n}{(2n+1)^2} = \frac{1}{1^2} - \frac{1}{3^2} + \frac{1}{5^2} - \frac{1}{7^2} + \cdots.$$

(continued)

(continued)

It is also related to *Dirichlet's beta function*

$$\beta(x) = \sum_{n=0}^{\infty} \frac{(-1)^n}{(2n+1)^x}, \quad x > 0,$$

so $G = \beta(2)$.

3.44. A quadratic alternating series. Prove that

$$\sum_{n=1}^{\infty} (-1)^n \left(\ln 2 - \frac{1}{n+1} - \frac{1}{n+2} - \cdots - \frac{1}{2n} \right)^2 = \frac{\pi^2}{48} - \frac{\pi}{8} \ln 2 - \frac{7}{8} \ln^2 2 + \frac{G}{2}.$$

3.45. Prove that

$$\sum_{n=0}^{\infty} (-1)^n \left(\frac{1}{n+1} - \frac{1}{n+2} + \frac{1}{n+3} - \cdots \right)^2 = \frac{\zeta(2)}{4}.$$

3.3 Alternating Products

> What we know is not much. What we do not know is immense.
>
> Pierre-Simon de Laplace (1749–1827)

3.46. A $\pi/2$ product. Let $p \geq 2$ be an integer. Prove that

$$\prod_{n=1}^{\infty} \left(1 + \frac{1}{\lfloor \sqrt[p]{n} \rfloor} \right)^{(-1)^{n-1}} = \frac{\pi}{2},$$

where $\lfloor a \rfloor$ denotes the greatest integer not exceeding a.

3.47. Let $p \geq 2$ be an integer. Find the value of

$$\prod_{n=2^p}^{\infty} \left(1 - \frac{1}{\lfloor \sqrt[p]{n} \rfloor} \right)^{(-1)^{n-1}},$$

where $\lfloor a \rfloor$ denotes the greatest integer not exceeding a.

3.48. Calculate

$$\prod_{n=1}^{\infty}\left(1+\frac{1}{n^2}\right)^{(-1)^{n-1}} \quad \text{and} \quad \prod_{n=2}^{\infty}\left(1-\frac{1}{n^2}\right)^{(-1)^{n-1}}.$$

3.49. Calculate

$$\prod_{n=2}^{\infty}\left(\frac{n^2+1}{n^2-1}\right)^{(-1)^{n-1}}.$$

3.50. (a) Calculate

$$\prod_{n=1}^{\infty}\left(\sqrt{\frac{\pi}{2}}\cdot\frac{(2n-1)!!\sqrt{2n+1}}{2^n n!}\right)^{(-1)^n}.$$

(b) More generally, if $x \neq n\pi$ is a real number, find the value of

$$\prod_{n=1}^{\infty}\left(\frac{x}{\sin x}\left(1-\frac{x^2}{\pi^2}\right)\cdots\left(1-\frac{x^2}{(n\pi)^2}\right)\right)^{(-1)^n}.$$

3.51. The Gamma function and an infinite product.

(a) Find the value of

$$\prod_{n=1}^{\infty}\left(\frac{2m+2n-1}{2n-1}\right)\left(\frac{2n}{2m+2n}\right),$$

where m denotes a positive integer.

(b) More generally, if $\Re(z) > 0$, find the value of

$$\prod_{n=1}^{\infty}\left(1+\frac{z}{n}\right)^{(-1)^{n-1}}.$$

3.52. A Glaisher–Kinkelin product. Prove that

$$\prod_{n=1}^{\infty}\frac{1}{e}\left(\frac{2n}{2n-1}\right)^{(4n-1)/2}=\frac{A^3}{2^{7/12}\pi^{1/4}}.$$

3.53. A Stirling product. Prove that

$$\prod_{n=1}^{\infty} \left(\frac{n!}{\sqrt{2\pi n}\,(n/e)^n} \right)^{(-1)^{n-1}} = \frac{A^3}{2^{7/12}\pi^{1/4}}.$$

3.4   Harmonic Series

> *...and I already see a way for finding the sum $\frac{1}{1} + \frac{1}{4} + \frac{1}{9} + \frac{1}{16}$ etc.*
>
> Johann Bernoulli (1667–1748)

The nth harmonic number, H_n, is the rational number defined by $H_n = 1 + 1/2 + 1/3 + \cdots + 1/n$. The first four harmonic numbers are $H_1 = 1$, $H_2 = 3/2$, $H_3 = 11/6$, $H_4 = 25/12$.

3.54. Prove that

$$\sum_{k=0}^{m} (-1)^k \binom{m+1}{k+1} \frac{1}{k+1} = H_{m+1}.$$

3.55. Prove that

$$\sum_{n=1}^{\infty} \frac{H_n}{n^2} = 2\zeta(3).$$

3.56. Prove that

$$\sum_{n=1}^{\infty} (-1)^{n-1} \frac{H_n}{n^2} = \frac{5}{8}\zeta(3).$$

3.57. Prove that

$$\sum_{n=1}^{\infty} (-1)^{n-1} \frac{H_n^2}{n} = -\frac{1}{12}(\pi^2 \ln 2 - 4\ln^3 2 - 9\zeta(3)).$$

3.58. A series of Klamkin. Prove that

$$\sum_{n=1}^{\infty} \frac{H_n}{n^3} = \frac{\pi^4}{72}.$$

3.59. (a) Prove that

$$\sum_{n=1}^{\infty} \left(1 + \frac{1}{2} + \frac{1}{3} + \cdots + \frac{1}{n+1}\right) \frac{1}{n(n+1)} = 2.$$

(b) More generally, for $k = 1, 2, \ldots$, one has that

$$\sum_{n=1}^{\infty} \left(1 + \frac{1}{2} + \cdots + \frac{1}{n+k}\right) \frac{1}{n(n+1)\cdots(n+k)} = \frac{1}{k \cdot k!} \left(\frac{1}{k} + 1 + \frac{1}{2} + \cdots + \frac{1}{k}\right).$$

3.60. Let $n \geq 2$ be an integer. Prove that

$$\sum_{k=1}^{\infty} \frac{H_k}{k(k+1)(k+2)\cdots(k+n)} = \frac{1}{n!} \left(\frac{\pi^2}{6} - 1 - \frac{1}{2^2} - \cdots - \frac{1}{(n-1)^2}\right).$$

3.61. Harmonic numbers, $\zeta(2)$ and $\zeta(3)$. Let $n \geq 2$ be an integer. Prove that

$$\sum_{k=1}^{\infty} \frac{H_k}{k^2(k+1)(k+2)\cdots(k+n)} = \frac{1}{n!} \left(2\zeta(3) - \frac{\pi^2}{6}\right)$$

$$- \frac{1}{n!} \sum_{k=2}^{n} \frac{1}{k} \left(\frac{\pi^2}{6} - 1 - \frac{1}{2^2} - \cdots - \frac{1}{(k-1)^2}\right).$$

3.62. The nth harmonic number and the tail of $\zeta(2)$. Show that

$$\sum_{n=1}^{\infty} \frac{1}{n} \left(1 + \frac{1}{2} + \cdots + \frac{1}{n}\right) \left(\zeta(2) - \sum_{k=1}^{n} \frac{1}{k^2}\right) = \frac{7}{4}\zeta(4).$$

3.63. Prove that

$$\sum_{n=1}^{\infty} \frac{H_n - \ln n - \gamma}{n} = -\frac{\gamma^2}{2} + \frac{\pi^2}{12} - \gamma_1.$$

3.64. Prove that

$$\sum_{n=1}^{\infty} \frac{1}{n} \left(\ln 2 - \frac{1}{n+1} - \frac{1}{n+2} - \cdots - \frac{1}{2n}\right) = \sum_{n=1}^{\infty} \frac{1 + \frac{1}{2} + \cdots + \frac{1}{n}}{(2n+1)(2n+2)} = \frac{\pi^2}{12} - \ln^2 2.$$

3.65. Prove that

$$\sum_{n=1}^{\infty} \frac{1}{n} \left(\ln 3 - \frac{1}{n+1} - \frac{1}{n+2} - \cdots - \frac{1}{3n}\right) = \frac{5\pi^2}{36} - \frac{3\ln^2 3}{4}.$$

3.66. Let $m \geq 2$ be an integer. Prove that

$$\sum_{n=1}^{\infty} \frac{\ln m - (H_{nm} - H_n)}{n} = \frac{(m-1)(m+2)}{24m}\pi^2 - \frac{1}{2}\ln^2 m - \frac{1}{2}\sum_{j=1}^{m-1}\ln^2\left(2\sin\frac{\pi j}{m}\right).$$

3.67. Let $k \geq 2$ be an integer. Prove that

$$\sum_{n=1}^{\infty}\left(\ln k - H_{kn} + H_n - \frac{k-1}{2kn}\right) = \frac{k-1}{2k} - \frac{\pi}{2k^2}\sum_{j=1}^{k-1}j\cot\frac{j\pi}{k} - \frac{\ln k}{2}.$$

3.68. Let $k \geq 2$ be an integer. Show that

$$\sum_{n=1}^{\infty}(-1)^{n-1}(\ln k - (H_{kn} - H_n)) = \frac{k-1}{2k}\ln 2 + \frac{\ln k}{2}$$

$$- \frac{\pi}{2k^2}\sum_{l=1}^{\lfloor k/2 \rfloor}(k+1-2l)\cot\frac{(2l-1)\pi}{2k}.$$

3.69. Let $k \geq 1$ be an integer. Prove that

$$\sum_{n=1}^{\infty}(-1)^{n-1}\frac{H_{kn}}{n} = \frac{k^2+1}{24k}\pi^2 - \frac{1}{2}\sum_{j=0}^{k-1}\ln^2\left(2\sin\frac{(2j+1)\pi}{2k}\right).$$

3.70. A quadratic series of Au–Yeung. Prove that

$$\sum_{n=1}^{\infty}\left(\frac{H_n}{n}\right)^2 = \frac{17\pi^4}{360}.$$

3.71. A special harmonic sum. Prove that

$$\sum_{n=1}^{\infty}\frac{H_n}{n}\cdot\frac{H_{n+1}}{n+1} = \zeta(2) + 2\zeta(3).$$

3.72. Nonlinear harmonic series. Let k and m be nonnegative integers and let

$$T_{k,m} = \sum_{n=1}^{\infty} \frac{H_{n+k}}{n+k} \cdot \frac{H_{n+m}}{n+m} \quad \text{and} \quad S_k = \sum_{n=1}^{\infty} \frac{H_n}{n} \cdot \frac{H_{n+k}}{n+k}.$$

(a) Let $k \geq 1$ be an integer. Prove that

$$S_k = \sum_{j=0}^{k-1} (-1)^j \binom{k-1}{j} \frac{1}{(j+1)^3} \left(\frac{3}{j+1} - \frac{2}{k} \right) + \frac{\pi^2}{6} \cdot \frac{H_k}{k} + \frac{2\zeta(3)}{k} - \frac{1}{k} \sum_{i=1}^{k} \frac{H_i}{i^2}.$$

(b) Let $1 \leq k \leq m$ be integers. Prove that

$$T_{k,m} = S_{m-k} - \sum_{i=1}^{k} \frac{H_i}{i} \cdot \frac{H_{i+m-k}}{i+m-k}.$$

3.5   Series of Functions

> *The cleverest of all, in my opinion, is the man who calls himself*
> *a fool at least once a month.*
>
> Fyodor Dostoevsky (1821–1881)

3.73. Let $x > 0$ be a real number and let $n \geq 1$ be an integer. Calculate

$$\sum_{k=1}^{\infty} \frac{1}{k^n(k+x)}.$$

3.74. Let $k \geq 1$ be an integer. Find the sum

$$\sum_{n=1}^{\infty} \left(\frac{1}{1-x} - 1 - x - x^2 - \cdots - x^n \right)^k, \quad |x| < 1.$$

3.75. Let $k \geq 1$ be an integer. Find the sum

$$\sum_{n=1}^{\infty} \frac{1}{n} \left(\frac{1}{1-x} - 1 - x - x^2 - \cdots - x^n \right)^k, \quad |x| < 1.$$

3.76. Prove that

$$\sum_{n=1}^{\infty} \frac{1}{n} \left(\sum_{k=1}^{n} \frac{x^k}{k} - \ln \frac{1}{1-x} \right) = -\frac{\ln^2(1-x)}{2}, \quad -1 \leq x < 1.$$

3.77. Let $p \geq 1$ be an integer and let x be a real number such that $|x| < 1$. Prove that

$$\sum_{n=1}^{\infty} \frac{1}{n} \left(\sum_{k=0}^{n-1} x^{pk} - \frac{1}{1-x^p} \right) = \frac{\ln(1-x^p)}{1-x^p}$$

and

$$\sum_{n=1}^{\infty} \frac{1}{n} \left(\sum_{k=0}^{n-1} (-1)^k x^{pk} - \frac{1}{1+x^p} \right) = \frac{\ln(1+x^p)}{1+x^p}.$$

3.78. Let x be a real number such that $|x| < 1$. Find the sum

$$\sum_{n=1}^{\infty} (-1)^{n-1} n \left(\ln(1-x) + x + \frac{x^2}{2} + \cdots + \frac{x^n}{n} \right).$$

3.79. Find the sum

$$\sum_{n=1}^{\infty} \left(\ln \frac{1}{1-x} - x - \frac{x^2}{2} - \cdots - \frac{x^n}{n} \right)^2, \quad |x| < 1.$$

3.80. Find the sum

$$\sum_{n=1}^{\infty} \left(\ln \frac{1}{1-x} - x - \frac{x^2}{2} - \cdots - \frac{x^n}{n} \right) \left(\frac{1}{1-x} - 1 - x - x^2 - \cdots - x^n \right), \quad |x| < 1.$$

3.81. A cosine series. Let $x \in (0, 2\pi)$, let $k \geq 1$ be an integer and let

$$S_k(x) = \sum_{n=1}^{\infty} \frac{\cos nx}{n(n+1)(n+2)\cdots(n+k)}.$$

(a) Prove that $S_k(x)$ equals

$$\frac{(2\sin \frac{x}{2})^k}{k!}$$

$$\times \left(-\cos \frac{(\pi-x)k}{2} \ln \left(2\sin \frac{x}{2} \right) - \frac{\pi-x}{2} \sin \frac{(\pi-x)k}{2} + \sum_{j=1}^{k} \frac{\cos \frac{(\pi-x)(j-k)}{2}}{j \left(2\sin \frac{x}{2} \right)^j} \right).$$

(b) Show that

$$\sum_{n=1}^{\infty} \frac{(-1)^n}{n(n+1)(n+2)\cdots(n+k)} = \frac{2^k}{k!} \left(-\ln 2 + \sum_{j=1}^{k} \frac{1}{j2^j} \right).$$

The *Cosine integral*, Ci, is the special function defined by [62, Entry 8.230(2^{10}), p. 878]

$$\text{Ci}(x) = -\int_x^\infty \frac{\cos t}{t} dt, \quad x > 0.$$

3.82. A Cosine integral series. Prove that if $0 \le a < 2\pi$, then

$$\sum_{n=1}^\infty \frac{\text{Ci}(an)}{n^2} = -\frac{\pi^2}{6} \ln \frac{2\pi}{a} + 2\pi^2 \ln A - \frac{\pi a}{2} + \frac{a^2}{8}.$$

3.83. Find the sums

$$\sum_{n=1}^\infty \sum_{m=1}^\infty \left(\frac{1}{1-x} - 1 - x - x^2 - \cdots - x^{n+m} \right), \quad |x| < 1$$

and

$$\sum_{n=1}^\infty \sum_{m=1}^\infty \left(\ln \frac{1}{1-x} - x - \frac{x^2}{2} - \cdots - \frac{x^{n+m}}{n+m} \right), \quad |x| < 1.$$

3.84. Let p be a nonnegative integer and let x be a real number. Find the sum

$$\sum_{n=1}^\infty (-1)^n \left(e^x - 1 - \frac{x}{1!} - \frac{x^2}{2!} - \cdots - \frac{x^{n+p}}{(n+p)!} \right).$$

3.85. Let x be a real number. Find the sum

$$\sum_{n=1}^\infty (-1)^{n-1} n \left(e^x - 1 - \frac{x}{1!} - \frac{x^2}{2!} - \cdots - \frac{x^n}{n!} \right).$$

3.86. Let x be a real number. Find the sum

$$\sum_{n=1}^\infty n^2 \left(e^x - 1 - \frac{x}{1!} - \frac{x^2}{2!} - \cdots - \frac{x^n}{n!} \right).$$

3.87. Let $p \ge 1$ be an integer and let x be a real number. Find the sum

$$\sum_{n=1}^\infty (-1)^n \left(\frac{x^n}{n!} + \frac{x^{n+1}}{(n+1)!} + \cdots + \frac{x^{n+p}}{(n+p)!} \right).$$

3.88. Let f be a function that has a Taylor series representation at 0 with radius of convergence R, and let $x \in (-R, R)$. Prove that

$$\sum_{n=1}^{\infty} (-1)^n f^{(n)}(0) \left(e^x - 1 - \frac{x}{1!} - \frac{x^2}{2!} - \cdots - \frac{x^n}{n!} \right) = \int_{-x}^{0} e^{x+t} f(t)\,dt.$$

3.89. An exponential series. Let f be a function which has a Taylor series representation at 0 with radius of convergence R.

(a) Prove that

$$\sum_{n=0}^{\infty} f^{(n)}(0) \left(e^x - 1 - \frac{x}{1!} - \frac{x^2}{2!} - \cdots - \frac{x^n}{n!} \right) = \int_{0}^{x} e^{x-t} f(t)\,dt, \quad |x| < R.$$

(b) Let $\alpha \in \mathbb{R}^*$. Calculate

$$\sum_{n=0}^{\infty} \alpha^n \left(e^x - 1 - \frac{x}{1!} - \frac{x^2}{2!} - \cdots - \frac{x^n}{n!} \right).$$

3.90. Let f be a function which has a Taylor series representation at 0 with radius of convergence R.

(a) Prove that

$$\sum_{n=1}^{\infty} n f^{(n)}(0) \left(e^x - 1 - \frac{x}{1!} - \frac{x^2}{2!} - \cdots - \frac{x^n}{n!} \right) = \int_{0}^{x} e^{x-t} t f'(t)\,dt, \quad |x| < R.$$

(b) Let $\alpha \in \mathbb{R}$. Calculate

$$\sum_{n=1}^{\infty} n \alpha^n \left(e^x - 1 - \frac{x}{1!} - \frac{x^2}{2!} - \cdots - \frac{x^n}{n!} \right).$$

3.91. Let f be any function that has a Taylor series representation at 0 with radius of convergence R, and let

$$T_n(x) = f(0) + \frac{f'(0)}{1!} x + \cdots + \frac{f^{(n)}(0)}{n!} x^n,$$

denote the nth-degree Taylor polynomial of f at 0. Calculate $\sum_{n=1}^{\infty} (f(x) - T_n(x))$ for $|x| < R$.

3.92. Let f be any function that has a Taylor series representation at 0 with radius of convergence R, and let T_n denote the nth-degree Taylor polynomial of f at 0. Find the sum $\sum_{n=1}^{\infty} (-1)^{n-1} (f(x) - T_n(x))$ for $|x| < R$.

3.93. An infinite sum of a function with its Taylor polynomial. Let f be any function that has a Taylor series representation at 0 with radius of convergence R, and let T_n denote the nth-degree Taylor polynomial of f at 0. Find the sum $\sum_{n=1}^{\infty} x^n (f(x) - T_n(x))$ for $|x| < R$.

3.6 Multiple Series

How can intuition deceive us at this point?

Jules Henri Poincaré (1854–1912)

3.94. A double alternating series. Find the sum

$$\sum_{n=1}^{\infty} \sum_{m=1}^{\infty} \frac{(-1)^{n+m}}{(\lfloor \sqrt{n+m} \rfloor)^3},$$

where $\lfloor a \rfloor$ denotes the greatest integer not exceeding a.

3.95. Let $\alpha > 4$ be a real number. Find, in closed form, the series

$$\sum_{n=1}^{\infty} \sum_{m=1}^{\infty} \frac{1}{(\lfloor \sqrt{n+m} \rfloor)^{\alpha}},$$

where $\lfloor a \rfloor$ denotes the greatest integer not exceeding a.

3.96. When does a double series equal a difference of two single series? Suppose that both series $\sum_{k=1}^{\infty} a_k$ and $\sum_{k=1}^{\infty} k a_k$ converge and let σ and $\tilde{\sigma}$ denote their sums, respectively. Then, the iterated series $\sum_{n=1}^{\infty} \sum_{m=1}^{\infty} a_{n+m}$ converges and its sum s equals $\tilde{\sigma} - \sigma$.

3.97. With the same notation as in the previous problem, let us suppose that the series $\sum_{n=1}^{\infty} a_n$ converges, that $\tilde{\sigma}_{2n} = o(n)$, and that the limit $\ell = \lim_{n \to \infty}(n a_{2n} + \tilde{\sigma}_{2n-1})$ exists in $\mathbb{R}$. Then, the iterated series $\sum_{n=1}^{\infty} \sum_{m=1}^{\infty} a_{n+m}$ converges and its sum s equals $\ell - \sigma$.

The logarithmic constants, δ_n, are the special constants defined by

$$\delta_n = \lim_{m \to \infty} \left(\sum_{k=1}^{m} \ln^n k - \int_1^m \ln^n x \, dx - \frac{1}{2} \ln^n m \right) = (-1)^n (\zeta^{(n)}(0) + n!)^{\bullet}.$$

(continued)

3.98. (a) Prove that

$$\sum_{n=1}^{\infty}\sum_{m=1}^{\infty}(-1)^{n+m}\frac{\ln(n+m)}{n+m} = \frac{1}{2}\left(\ln\frac{\pi}{2} + \ln^2 2 - 2\gamma\ln 2\right).$$

(b) More generally, let p be a nonnegative integer and let

$$S(p) = \sum_{n=1}^{\infty}\sum_{m=1}^{\infty}(-1)^{n+m}\frac{\ln^p(n+m)}{n+m}.$$

Prove that

$$S(p) = 2\sum_{i=0}^{p}\binom{p}{i}\delta_i\ln^{p-i}2 - \delta_p + \gamma_p - \sum_{i=0}^{p}\binom{p}{i}\gamma_i\ln^{p-i}2$$
$$+ \frac{\ln^{p+1}2}{p+1} - p!\left(2\sum_{i=1}^{p}(-1)^{p-i}\frac{\ln^i 2}{i!} + (-1)^p\right),$$

where the last sum is missing when $p = 0$.

3.99. Find the sum

$$\sum_{n=1}^{\infty}\sum_{m=1}^{\infty}(-1)^{n+m}\frac{H_{n+m}}{n\cdot m}.$$

3.100. Find the sum

$$\sum_{n=1}^{\infty}\sum_{m=1}^{\infty}(-1)^{n+m}\frac{H_n H_m}{n+m}.$$

3.101. Prove that

$$\sum_{n=1}^{\infty}\sum_{m=1}^{\infty}(-1)^{n+m}\frac{H_n H_m}{n+m+1} = -\frac{\ln^2 2}{2} - \ln 2 + 1.$$

3.102. Let $k \geq 1$ be an integer. Evaluate, in closed form, the series

$$\sum_{n_1,\dots,n_k=1}^{\infty}(-1)^{n_1+\cdots+n_k}\frac{H_{n_1}H_{n_2}\cdots H_{n_k}}{n_1+n_2+\cdots+n_k+1}.$$

3.103. Prove that

$$\sum_{n=1}^{\infty}\sum_{m=1}^{\infty}(-1)^{n+m}\frac{H_{n+m}}{n+m} = \frac{\pi^2}{12} - \frac{\ln 2}{2} - \frac{\ln^2 2}{2}.$$

3.104. Let $k \geq 1$ be a nonnegative integer and let A_k be the series defined by

$$A_k = \sum_{n_1=1}^{\infty}\sum_{n_2=1}^{\infty}\cdots\sum_{n_k=1}^{\infty}(-1)^{n_1+n_2+\cdots+n_k}\frac{H_{n_1+n_2+\cdots+n_k}}{n_1+n_2+\cdots+n_k}.$$

(a) Prove that

$$A_1 = \sum_{n=1}^{\infty}(-1)^n\frac{H_n}{n} = \frac{\ln^2 2}{2} - \frac{\pi^2}{12}.$$

(b) Show that, when $k \geq 2$, one has

$$(-1)^{k+1}A_k = \sum_{p=2}^{k}\sum_{j=0}^{p-2}\binom{p-2}{j}\frac{(-1)^j}{(j+1)^2}$$

$$\times\left(\frac{(j+1)}{2^{j+1}}\ln 2 - 1 + \left(\frac{1}{2}\right)^{j+1} + \left(\frac{1}{2}\right)^{p-1}\right) + \frac{\ln^2 2}{2} - \frac{\pi^2}{12}.$$

3.105. A logarithmic series. Let $k \geq 1$ be an integer. Calculate

$$\sum_{n_1=1}^{\infty}\cdots\sum_{n_k=1}^{\infty}(-1)^{n_1+n_2+\cdots+n_k}\frac{\ln(n_1+n_2+\cdots+n_k)}{n_1+n_2+\cdots+n_k}.$$

3.106. An alternating Hardy series. Prove that

$$\sum_{n=1}^{\infty}\sum_{m=1}^{\infty}(-1)^{n+m}\left(1+\frac{1}{2}+\cdots+\frac{1}{n+m}-\ln(n+m)-\gamma\right)$$

$$= -3\ln A + \frac{7}{12}\ln 2 + \frac{1}{2}\ln\pi - \frac{\gamma}{4}.$$

3.107. Calculate

$$\sum_{n=1}^{\infty}\sum_{m=1}^{\infty}(-1)^{n+m}\left(\ln 2-\frac{1}{n+m+1}-\frac{1}{n+m+2}-\cdots-\frac{1}{2(n+m)}\right).$$

3.108. Let p and p_i, $i=1,\ldots,n$, be positive real numbers. Prove that

$$\sum_{k_1,\ldots,k_n=1}^{\infty}\frac{1}{\left(k_1^{p_1}+k_2^{p_2}+\cdots+k_n^{p_n}\right)^p}<\infty \quad\Leftrightarrow\quad \frac{1}{p_1}+\frac{1}{p_2}+\cdots+\frac{1}{p_n}<p.$$

3.109. Let $k\geq 2$ be an integer and let $-1\leq a<1$ be a real number. Calculate the sum

$$\sum_{n_1,\ldots,n_k=1}^{\infty}\frac{a^{n_1+n_2+\cdots+n_k}}{n_1+n_2+\cdots+n_k}.$$

3.110. Let a_i, $i=1,\ldots,k$, be distinct real numbers such that $-1\leq a_i<1$. Calculate the sum

$$\sum_{n_1,\ldots,n_k=1}^{\infty}\frac{a_1^{n_1}a_2^{n_2}\cdots a_k^{n_k}}{n_1+n_2+\cdots+n_k}.$$

3.111. Let $k\geq 1$ and let $p\geq 0$ be two integers. Find the sum

$$S(p)=\sum_{m_1,\ldots,m_k=1}^{\infty}\frac{1}{m_1 m_2\cdots m_k(m_1+m_2+\cdots+m_k+p)}.$$

3.112. A factorial sum that equals 1. Prove that

$$\sum_{n=1}^{\infty}\sum_{m=1}^{\infty}\frac{1}{(n+m)!}=1.$$

3.113. Find the sum

$$\sum_{n=1}^{\infty}\sum_{m=1}^{\infty}\frac{n^2}{(n+m)!}.$$

3.114. Prove that

$$\sum_{n=1}^{\infty}\sum_{m=1}^{\infty}\frac{n\cdot m}{(n+m)!}=\frac{2}{3}e.$$

3.115. Let a and b be real numbers. Calculate

$$\sum_{n=1}^{\infty}\sum_{m=1}^{\infty}\frac{a^n b^m}{(n+m)!}\quad\text{and}\quad\sum_{n=1}^{\infty}\sum_{m=1}^{\infty}\frac{n\cdot m}{(n+m)!}a^n b^m.$$

3.116. Let a and b be real numbers. Calculate

$$\sum_{n=1}^{\infty}\sum_{m=1}^{\infty}\frac{n}{(n+m)!}a^n b^m \quad \text{and} \quad \sum_{n=1}^{\infty}\sum_{m=1}^{\infty}\frac{m}{(n+m)!}a^n b^m.$$

3.117. Find the sum

$$\sum_{n,m,p=1}^{\infty}\frac{n\cdot m}{(n+m+p)!}.$$

3.118. Prove that

$$\sum_{n,m,p=1}^{\infty}\frac{n\cdot m\cdot p}{(n+m+p)!}=\frac{31}{120}e.$$

3.119. Let $k\geq 1$ be an integer. Prove that

$$\sum_{n_1,\dots,n_k=1}^{\infty}\frac{1}{(n_1+n_2+\cdots+n_k)!}=(-1)^{k-1}\left(e\sum_{j=0}^{k-1}\frac{(-1)^j}{j!}-1\right).$$

3.120. Let k and i be integers such that $1\leq i\leq k$. Prove that

$$\sum_{n_1,\dots,n_k=1}^{\infty}\frac{n_i}{(n_1+n_2+\cdots+n_k)!}=\frac{e}{k!}.$$

3.121. Let $p>3$ be a real number. Find the value of

$$\sum_{n=1}^{\infty}\sum_{m=1}^{\infty}\frac{n}{(n+m)^p}.$$

The Stirling numbers of the first kind, $s(n,k)$, are the special numbers defined by the generating function

$$z(z-1)(z-2)\cdots(z-n+1)=\sum_{k=0}^{n}s(n,k)z^k.$$

For recurrence relations as well as interesting properties satisfied by these numbers, the reader is referred to [98, p. 624] and [122, p. 56].

3.122. (a) Let k be a fixed positive integer and let $m > k$. Then,

$$\sum_{n_1,\ldots,n_k=1}^{\infty} \frac{1}{(n_1+n_2+\cdots+n_k)^m}$$

$$= \frac{1}{(k-1)!} \sum_{i=1}^{k} s(k,i) \left(\zeta(m+1-i)-1-\frac{1}{2^{m+1-i}}-\cdots-\frac{1}{(k-1)^{m+1-i}} \right),$$

where the parentheses contain only the term $\zeta(m+1-i)$ when $k=1$.

(b) Let k and i be fixed integers such that $1 \le i \le k$, and let m be such that $m-k>1$. Then,

$$\sum_{n_1,\ldots,n_k=1}^{\infty} \frac{n_i}{(n_1+n_2+\cdots+n_k)^m}$$

$$= \frac{1}{k!} \sum_{i=1}^{k} s(k,i) \left(\zeta(m-i)-1-\frac{1}{2^{m-i}}-\cdots-\frac{1}{(k-1)^{m-i}} \right),$$

where the parentheses contain only the term $\zeta(m-i)$ when $k=1$.

3.123. A binomial series.

(a) Let $k \ge 1$ be an integer and let $m > 2k$ be a real number. Prove that

$$\sum_{n_1,\ldots,n_k=1}^{\infty} \frac{n_1 n_2 \cdots n_k}{(n_1+n_2+\cdots+n_k)^m} = \sum_{n=0}^{\infty} \frac{\binom{n+2k-1}{2k-1}}{(n+k)^m}.$$

(b) Prove that if $m > 4$, then

$$\sum_{n_1=1}^{\infty} \sum_{n_2=1}^{\infty} \frac{n_1 n_2}{(n_1+n_2)^m} = \frac{1}{6}(\zeta(m-3)-\zeta(m-1)).$$

It is a problem of *Wolstenholme*[1] to prove that, if $k \in \mathbb{N}$, the series $S_k = \sum_{n=1}^{\infty} n^k/n!$ equals an integral multiple of e, i.e., $S_k = B_k e$ [20, p. 197]. The integer B_k is known in the mathematical literature as the kth Bell number [29], [98, p. 623] and the equality $S_k = B_k e$ is known as *Dobinski's formula* [35]. For example, $B_1 = 1, B_2 = 2, B_3 = 5, B_4 = 15, B_5 = 52$. One can prove that the sequence $(B_k)_{k \in \mathbb{N}}$ verifies the recurrence relation

$$B_k = \sum_{j=1}^{k-1} \binom{k-1}{j} B_j + 1.$$

---

[1] Joseph Wolstenhome (1829–1891) was an English mathematician and the author of *Mathematical problems*.

3.124. Let m and k be positive integers. Prove that

$$\sum_{n_1,\ldots,n_k=1}^{\infty} \frac{(n_1+n_2+\cdots+n_k)^m}{(n_1+n_2+\cdots+n_k)!} = \frac{1}{(k-1)!} \sum_{i=1}^{k} s(k,i) \left(B_{m+i-1}e - \sum_{p=1}^{k-1} \frac{p^{m+i-1}}{p!} \right),$$

where the parentheses contain only the term $B_{m+i-1}e$ when $k=1$.

3.125. A double series giving Gamma. Prove that

$$\sum_{n=1}^{\infty} \sum_{m=1}^{\infty} \frac{\zeta(n+m)-1}{n+m} = \gamma.$$

3.126. Calculate

$$\sum_{n=1}^{\infty} \sum_{m=1}^{\infty} \frac{n(\zeta(n+m)-1)}{(n+m)^2}.$$

3.127. Calculate

$$\sum_{n=1}^{\infty} \sum_{m=1}^{\infty} \frac{m(\zeta(2n+2m)-1)}{(n+m)^2}.$$

3.128. Let $k \geq 2$ be an integer and let S_k be the series defined by

$$S_k = \sum_{n_1,\ldots,n_k=1}^{\infty} \frac{\zeta(n_1+\cdots+n_k)-1}{n_1+\cdots+n_k}.$$

Prove that

(a) If $l \geq 2$, then

$$S_{2l} = \frac{\zeta(2l-1)}{2l-1} - \frac{\zeta(2l-2)}{2l-2} + \cdots - \frac{\zeta(2)}{2} + \gamma.$$

(b) If $l \geq 1$, then

$$S_{2l+1} = \frac{\zeta(2l)}{2l} - \frac{\zeta(2l-1)}{2l-1} + \cdots - \frac{\zeta(3)}{3} + \frac{\zeta(2)}{2} - \gamma.$$

3.129. Prove that

$$\sum_{n_1=1}^{\infty} \sum_{n_2=1}^{\infty} (-1)^{n_1+n_2} \frac{\zeta(n_1+n_2)-1}{n_1+n_2} = \frac{3}{2} - \ln 2 - \gamma.$$

3.130. Let $k \geq 3$ be an integer and let A_k be the series defined by

$$A_k = \sum_{n_1,\ldots,n_k=1}^{\infty} (-1)^{n_1+\cdots+n_k} \frac{\zeta(n_1+\cdots+n_k)-1}{n_1+\cdots+n_k}.$$

Prove that $(-1)^{k+1}A_k$ equals

$$\frac{\zeta(k-1)-1-\frac{1}{2^{k-1}}}{k-1} + \frac{\zeta(k-2)-1-\frac{1}{2^{k-2}}}{k-2} + \cdots + \frac{\zeta(2)-1-\frac{1}{2^2}}{2} + \gamma + \ln 2 - \frac{3}{2}.$$

3.131. Two multiple zeta series. Let $k \geq 2$ be an integer. Calculate the sums

$$\sum_{n_1,\ldots,n_k=1}^{\infty} \left(\zeta(n_1+\cdots+n_k)-1\right)$$

and

$$\sum_{n_1,\ldots,n_k=1}^{\infty} (-1)^{n_1+\cdots+n_k} \left(\zeta(n_1+\cdots+n_k)-1\right).$$

3.132. (a) Prove that

$$\sum_{n=2}^{\infty} (n - \zeta(2) - \zeta(3) - \cdots - \zeta(n)) = \zeta(2) - 1.$$

(b) If $k \geq 2$ is an integer, prove that

$$\sum_{n_1,\ldots,n_k=1}^{\infty} (n_1+n_2+\cdots+n_k-\zeta(2)-\zeta(3)-\cdots-\zeta(n_1+n_2+\cdots+n_k))$$

$$= \zeta(k+1).$$

3.133. Let $k \geq 2$ be an integer, let j be such that $0 \leq j \leq k$, and let T_j be the series $\sum_{n_1,\ldots,n_k=1}^{\infty} a_{n_1,\ldots,n_k}$, with

$$a_{n_1,\ldots,n_k} = n_1 n_2 \cdots n_j \left(n_1+n_2+\cdots+n_k-\zeta(2)-\zeta(3)-\cdots-\zeta(n_1+n_2+\cdots+n_k)\right),$$

where the product $n_1 \cdots n_j$ is missing when $j = 0$. Prove that

$$T_j = \sum_{m=0}^{j} \binom{j}{m} \zeta(k+1+j-m).$$

3.7 Open Problems

> *In mathematics the art of proposing a question must be held of higher value than solving it.*
>
> Georg Ferdinand Cantor (1845–1918)

3.134. Let $k \geq 3$ be an integer. Calculate

$$\sum_{n_1,\ldots,n_k=1}^{\infty} (-1)^{n_1+n_2+\cdots+n_k} \frac{H_{n_1+n_2+\cdots+n_k}}{n_1 \cdot n_2 \cdots n_k}.$$

3.135. Let $k \geq 2$ be an integer. Evaluate, in closed form, the multiple series

$$\sum_{n_1,\ldots,n_k=1}^{\infty} (-1)^{n_1+\cdots+n_k} \frac{H_{n_1} \cdots H_{n_k}}{n_1 n_2 \cdots n_k (n_1 + n_2 + \cdots + n_k)},$$

3.136. Let $k \geq 3$ be an integer. Calculate, in closed form, the series

$$\sum_{n_1,\ldots,n_k=1}^{\infty} (-1)^{n_1+n_2+\cdots+n_k} \left(H_{n_1+n_2+\cdots+n_k} - \ln(n_1 + n_2 + \cdots + n_k) - \gamma \right).$$

3.137. Does the sum equal a rational multiple of e? Let $k \geq 4$ be an integer. Calculate the sum

$$\sum_{n_1,\ldots,n_k=1}^{\infty} \frac{n_1 n_2 \cdots n_k}{(n_1 + n_2 + \cdots + n_k)!}.$$

A conjecture. It is reasonable to conjecture that

$$\sum_{n_1,\ldots,n_k=1}^{\infty} \frac{n_1 n_2 \cdots n_k}{(n_1 + n_2 + \cdots + n_k)!} = a_k e,$$

where a_k is a rational number. We observe that $a_1 = 1$, $a_2 = 2/3$, and $a_3 = 31/120$ (see Problems **3.114** and **3.118**). If the conjecture holds true, an open problem would be to study the properties of the sequence $(a_k)_{k \in \mathbb{N}}$.

3.8   Hints

> It is not once nor twice but times without number that the same
> ideas make their appearance in the world.
>
> Aristotle (384–322 BC)

3.8.1   Single Series

> Mathematics is an art, and as an art chooses beauty and
> freedom.
>
> Marston Morse (1892–1977)

3.2. Use the formula $\sin x = (e^{ix} - e^{-ix})/2i$ and the Binomial Theorem.

3.3. Use the inequality $1 - e^{1/x} + 1/(x-1) > 0$ which holds for $x > 1$.

3.4. Apply the AM–GM inequality to the product $\prod_{k=1}^{n} \cos^2 k$.

3.5. Apply Raabe's test for series.

3.6. $|\cos 1 \cos^2 2 \cdots \cos^n n| \le |\cos 1 \cos 2 \cdots \cos n|$ and use Problem **3.4**.

3.7. Write $\sum_{n=1}^{\infty} (1/n - f(n)) = \sum_{k=1}^{\infty} \left(\sum_{2^{k-1} \le n < 2^k} (1/n - f(n)) \right)$.

3.8. Let $S_n = \sum_{k=1}^{\infty} 1/k(k+1)^n$ and prove that $S_n = S_{n-1} - (\zeta(n) - 1)$ by using the formula $\frac{1}{k(k+1)^n} = \frac{1}{k(k+1)^{n-1}} - \frac{1}{(k+1)^n}$.

3.9. Use the formula $\frac{1}{k^n(k+1)} = \frac{1}{k^n} - \frac{1}{k^{n-1}(k+1)}$.

3.10. Use that $\frac{1}{k(k+1)\cdots(k+n)} = \frac{1}{n} \left(\frac{1}{k(k+1)\cdots(k+n-1)} - \frac{1}{(k+1)\cdots(k+n)} \right)$.

3.12. Calculate the nth partial sum of the series.

3.13. Use partial fractions and prove that

$$\frac{\alpha^{n-1}}{(\alpha+p)(2\alpha+p)\cdots(n\alpha+p)} = \int_0^1 x^{p-1+\alpha} \left(\sum_{m=0}^{n-1} \frac{(-1)^m x^{\alpha m}}{m!(n-1-m)!} \right) dx.$$

3.14. Use Abel's summation formula (A.2).

3.15. Use that $1/(n+1) - 1/(n+2) + 1/(n+3) - \cdots = \int_0^1 x^n/(1+x)dx$.

3.16. Write $1 - 1/2 + 1/3 + \cdots + (-1)^{n-1}/n = \int_0^1 (1 - (-x)^n)/(1+x)dx$.

3.17. Calculate the $2n$th partial sum of the series.

3.18. Calculate the nth partial sum of the series by Abel's summation formula (A.1).

3.19. Calculate directly the nth partial sum of the series.

3.20. Use Abel's summation formula (A.2).

3.23. and 3.24. Use that $\zeta(k) - 1 = \sum_{m=2}^{\infty} 1/m^k$.

3.25. Use that $\zeta(2k) - 1 = \sum_{m=2}^{\infty} 1/m^{2k}$ and change the order of summation.

3.26. and 3.27. Use that $\sum_{k=1}^{\infty} 1/k(k+1)^n = n - \zeta(2) - \zeta(3) - \cdots - \zeta(n)$ and change the order of summation.

3.28. Use Abel's summation formula (A.2) and the power series

$$\ln(1+x)\ln(1-x) = \sum_{n=1}^{\infty} \frac{1}{n}\left(H_n - H_{2n} - \frac{1}{2n}\right)x^{2n}, \quad x^2 < 1.$$

3.29. Use that $1/(n+1) - 1/(n+2) + 1/(n+3) - \cdots = \int_0^1 x^n/(1+x)\mathrm{d}x$.

3.8.2 Alternating Series

> *The profound study of nature is the most fertile source of mathematical discoveries.*
>
> Joseph Fourier (1768–1830)

3.30. Sum the terms of the series according to the cases when $n = k^p$ and $k^p < n < (k+1)^p$.

3.31. Use that $\frac{1}{n!(n+k)} = \frac{1}{k}\left(\frac{1}{n!} - \frac{1}{(n-1)!(n+k)}\right)$.

3.33. Use Abel's summation formula (A.2).

3.34. and 3.35. Calculate the nth partial sum of the series by using Abel's summation formula (A.1).

3.37. Calculate the $2n$th partial sum of the series by using both Stirling's formula and the limit definition of the Glaisher–Kinkelin constant.

3.38. Calculate the $2n$th partial sum of the series and use the Glaisher product $\prod_{k=1}^{\infty} k^{1/k^2} = \left(A^{12}/(2\pi e^\gamma)\right)^{\pi^2/6}$.

3.41. Use that $1 - 1/2 + 1/3 - \cdots + (-1)^{n-1}/n = \int_0^1 (1 - (-x)^n)/(1+x)\mathrm{d}x$.

3.42. Use Abel's summation formula (A.2).

3.43. Calculate the $2n$th partial sum of the series and use the definition of the Glaisher–Kinkelin constant.

3.45. Use that $\sum_{k=1}^{\infty} (-1)^{k-1}/(n+k) = \int_0^1 x^n/(1+x)\mathrm{d}x$.

3.8.3 Alternating Products

> *We are servants rather than masters in mathematics.*
>
> Charles Hermite (1822–1901)

3.46. and **3.47.** Apply the logarithm to the product and sum the terms according to the cases when $n = k^p$ and $k^p < n < (k+1)^p$.

3.48. and **3.49.** Use Euler's infinite product formula for the sine function $\sin \pi z = \pi z \prod_{n=1}^{\infty} \left(1 - z^2/n^2\right)$ for $z \in \mathbb{C}$.

3.50. (b) Apply the logarithm to the product, calculate the $2n$th partial sum of the series, and use the infinite product formula for the sine function.

3.51. (b) Calculate the $2n$th partial sum of the logarithmic series and use Weierstrass product for the Gamma function $1/\Gamma(z) = z e^{\gamma z} \prod_{k=1}^{\infty} \left(1 + z/k\right) e^{-z/k}$.

3.52. Take the logarithm of the product and calculate directly the nth partial sum of the logarithmic series.

3.53. Transform the product into a series and use Abel's summation formula (A.2).

3.8.4 Harmonic Series

> *In mathematics you don't understand things. You just get used to them.*
>
> John von Neumann (1903–1957)

3.54. Prove the identity by induction on m.

3.55. —3.58. Use that $\int_0^1 x^n \ln(1-x)\,dx = -H_{n+1}/(n+1)$.

3.59. Use that $\frac{1}{n(n+1)\cdots(n+k)} = \frac{1}{k}\left(\frac{1}{n(n+1)\cdots(n+k-1)} - \frac{1}{(n+1)\cdots(n+k)}\right)$.

3.60. Use that $\frac{1}{k(k+1)\cdots(k+n)} = \frac{1}{n}\left(\frac{1}{k(k+1)\cdots(k+n-1)} - \frac{1}{(k+1)\cdots(k+n)}\right)$.

3.61. Denote the series by V_n and prove that $V_n = \frac{1}{n}V_{n-1} - \frac{1}{n}a_n$, where $a_n = \sum_{k=1}^{\infty} H_k/k(k+1)(k+2)\cdots(k+n)$.

3.62. Calculate the sum by using the logarithmic integrals $\int_0^1 x^n \ln(1-x)\,dx = -H_{n+1}/(n+1)$ and $\int_0^1 y^{k-1} \ln y\,dy = -1/k^2$.

3.64. Use Abel's summation formula (A.1).

3.70. Use the integral $\int_0^1 x^n \ln(1-x)\,dx = -H_{n+1}/(n+1)$.

3.8.5 Series of Functions

3.74. and **3.75.** Observe that $1 + x + \cdots + x^n = (1 - x^{n+1})/(1 - x)$.

3.76. Use that $\sum_{k=1}^{n} x^k/k - \ln 1/(1-x) = -\int_0^x t^n/(1-t)dt$.

3.77. Calculate the inner sum.

3.78. Calculate directly the $2n$th partial sum of the series.

3.79. $\ln 1/(1-x) - x - x^2/2 - \cdots - x^n/n = \sum_{k=1}^{\infty} x^{k+n}/(k+n) = \int_0^x t^n/(1-t)dt$.

3.80. Use the formulae $\ln 1/(1-x) - x - x^2/2 - \cdots - x^n/n = \sum_{k=1}^{\infty} x^{k+n}/(k+n)$ and $1/(1-x) - 1 - x - \cdots - x^n = \sum_{p=1}^{\infty} x^{p+n}$.

3.81. Consider the series $T_k(x) = \sum_{n=1}^{\infty} e^{inx}/n(n+1)(n+2)\cdots(n+k)$ and prove it satisfies the recurrence $T_k(x) = (1 - e^{-ix})T_{k-1}(x)/k + 1/k \cdot k!$.

3.83. Use the hint of Problem **3.80**.

3.84. Calculate the nth partial sum of the series by Abel's summation formula.

3.85. and **3.86.** Prove that the sum of the series, viewed as a function of x, verifies a linear differential equation.

3.87. Calculate the nth partial sum of the series by Abel's summation formula.

3.89. and **3.90.** Show that the sum of the series, viewed as a function of x, verifies a linear differential equation.

3.92. Calculate the nth partial sum of the series by Abel's summation formula.

3.8.6 Multiple Series

3.95. Evaluate the series by adding over the intervals $k^2 \le n + m < (k+1)^2$.

3.99. Use that $\int_0^1 x^{k-1} \ln(1-x)dx = -H_k/k$.

3.100.—3.102. Use the power series $-\ln(1+x)/(1+x) = \sum_{n=1}^{\infty}(-1)^n H_n x^n$ for $-1 < x < 1$ combined with $\int_0^1 x^{k-1} dx = 1/k$.

3.106. First, apply Abel's summation formula (A.2) to prove that the infinite series $\sum_{m=1}^{\infty}(-1)^m \left(1 + \frac{1}{2} + \cdots + \frac{1}{n+m} - \ln(n+m) - \gamma\right)$ equals the logarithmic series $-\sum_{m=1}^{\infty}\left(\ln(n+2m) - \ln(n+2m-1) - 1/(n+2m)\right)$. Second, calculate the double sum $\sum_{n=1}^{\infty}(-1)^n \sum_{m=1}^{\infty}\left(\ln(n+2m) - \ln(n+2m-1) - 1/(n+2m)\right)$ by applying one more time Abel's summation formula (A.2).

3.107. Show, by using Abel's summation formula, that

$$\sum_{m=1}^{\infty}(-1)^m \left(\ln 2 - \frac{1}{n+m+1} - \frac{1}{n+m+2} - \cdots - \frac{1}{2(n+m)}\right) = \int_0^1 \frac{-x^{2n+2} dx}{1+x+x^2+x^3}.$$

3.108. The series behaves the same like the integral $\int_0^{\infty} \cdots \int_0^{\infty} \frac{dx_1 \cdots dx_n}{(1+x_1^{p_1}+\cdots+x_n^{p_n})^p}$.

3.109. Write the general term of the sum as an integral.

3.110. Use that $1/(n_1 + \cdots + n_k) = \int_0^1 x^{n_1+\cdots+n_k-1} dx$.

3.111. Use that $\int_0^1 x^{m_1+\cdots+m_k+p-1} dx = 1/(m_1 + \cdots + m_k + p)$.

3.112. Sum the series by diagonals.

3.113.—3.115. Use that $\int_0^1 x^n (1-x)^{m-1} dx = n!(m-1)!/(n+m)!$.

3.116. Differentiate the first series from Problem **3.115**.

3.117. Observe that $\frac{nm}{(n+m+p)!} = \frac{nm}{(n+m)!(p-1)!} \int_0^1 x^{n+m}(1-x)^{p-1} dx$.

3.118. Observe that $\frac{nmp}{(n+m+p)!} = \frac{nmp}{(n+m)!(p-1)!} \int_0^1 x^{n+m}(1-x)^{p-1} dx$.

3.119., 3.122. and **3.124.** The series can be calculated by using that

$$\sum_{n_1,n_2,\ldots,n_k=1}^{\infty} a_{n_1+n_2+\cdots+n_k} = \sum_{j=k}^{\infty}\left(\sum_{n_1+n_2+\cdots+n_k=j} a_{n_1+n_2+\cdots+n_k}\right).$$

3.120. and **3.121.** Use symmetry and the hint from Problem **3.119**.

3.123. Use that $1/a^m = 1/\Gamma(m) \int_0^{\infty} e^{-at} t^{m-1} dt$ combined with the power series

$$\frac{1}{(1-x)^{\lambda}} = \sum_{n=0}^{\infty} \frac{\Gamma(n+\lambda)}{n!\Gamma(\lambda)} x^n, \quad -1 < x < 1, \quad \lambda > 0.$$

3.125. Use that $\sum_{n,m=1}^{\infty} a_{n+m} = \sum_{k=2}^{\infty}\left(\sum_{n+m=k} a_{n+m}\right)$.

3.126. Use symmetry and Problem **3.125**.

3.127. Use symmetry and the hint of Problem **3.125**.

3.128., 3.130. and **3.131.** Observe that $\zeta(n_1 + \cdots + n_k) - 1 = \sum_{p=2}^{\infty} 1/p^{n_1+\cdots+n_k}$.

3.132. and **3.133.** Use that $n - \zeta(2) - \zeta(3) - \cdots - \zeta(n) = \sum_{k=1}^{\infty} 1/k(k+1)^n$.

3.9 Sololutions

> If I am given a formula, and I am ignorant of its meaning, it
> cannot teach me anything, but if I already know it what does the
> formula teach me?
>
> Blessed Augustine (354–430)

This section contains the solutions of the problems from Chap. 3.

3.9.1 Single Series

> I recognize the lion by his paw. (After reading an anonymous
> solution to a problem that he realized was Newton's solution)
>
> Jakob Bernoulli (1654–1705)

3.1. The solution of the problem and other properties concerning the fractional part of special real numbers are discussed by M. Th. Rassias in [107, Sect. 5, p. 12].

3.2. Using the Binomial Theorem we have, since $\sin(k\pi/n) = \frac{1}{2i}\left(e^{k\pi i/n} - e^{-k\pi i/n}\right)$, that

$$\sin^n(k\pi/n) = \frac{1}{(2i)^n}\left(e^{k\pi i/n} - e^{-k\pi i/n}\right)^n$$

$$= \frac{1}{(2i)^n}\left(e^{-k\pi i/n}\left(e^{2k\pi i/n} - 1\right)\right)^n$$

$$= \frac{(-1)^k}{(2i)^n}\left(e^{2k\pi i/n} - 1\right)^n$$

$$= \frac{(-1)^k}{(2i)^n}\sum_{m=0}^{n}\binom{n}{m}e^{2km\pi i/n}(-1)^{n-m}.$$

It follows, since $(-1)^{-m} = (-1)^m$, that

$$\sum_{k=1}^{n-1}(-1)^k\sin^n(k\pi/n) = \sum_{k=0}^{n-1}(-1)^k\sin^n(k\pi/n)$$

$$= \frac{(-1)^n}{(2i)^n}\sum_{k=0}^{n-1}\left(\sum_{m=0}^{n}\binom{n}{m}(-1)^m e^{2km\pi i/n}\right)$$

$$= \frac{(-1)^n}{(2i)^n}\sum_{m=0}^{n}\binom{n}{m}(-1)^m\left(\sum_{k=0}^{n-1}e^{2mk\pi i/n}\right)$$

$$= \frac{(-1)^n}{(2i)^n}\sum_{m=0}^{n}\binom{n}{m}(-1)^m\sum_{k=0}^{n-1}\left(e^{2m\pi i/n}\right)^k. \qquad (3.1)$$

On the other hand, we have that

$$\sum_{k=0}^{n-1}\left(e^{2m\pi i/n}\right)^{k}=\begin{cases}n & \text{if } m=0 \quad \text{or} \quad m=n,\\ 0 & \text{if } m\neq 0,n.\end{cases} \tag{3.2}$$

Combining (3.1) and (3.2), we get that

$$\sum_{k=0}^{n-1}(-1)^{k}\sin^{n}(k\pi/n)=\frac{(-1)^{n}}{(2i)^{n}}\left((-1)^{0}n+(-1)^{n}n\right).$$

It follows that

$$\sum_{k=0}^{n-1}(-1)^{k}\sin^{n}(k\pi/n)=\Re\left(\frac{(-1)^{n}}{(2i)^{n}}\left((-1)^{0}n+(-1)^{n}n\right)\right)$$

$$=\Re\left(\frac{i^{n}}{2^{n}}\cdot n\left(1+(-1)^{n}\right)\right)$$

$$=\frac{1+(-1)^{n}}{2^{n}}\cdot n\cdot\cos\frac{n\pi}{2}.$$

3.3. The series converges if $\alpha>1$ and diverges otherwise. See the comment on Problem **1.80**.

Remark. We mention that the case when $\alpha=1$ is recorded as Problem 3.80 in [14]. Another interesting problem, which appears in Pólya and Szegö [104, Problem 159, p. 29], is about studying the convergence of the series

$$\sum_{n=1}^{\infty}\left(2-e^{\alpha}\right)\left(2-e^{\frac{\alpha}{2}}\right)\cdots\left(2-e^{\frac{\alpha}{n}}\right),$$

where α is a positive real number.

3.4. We will be using the formula

$$\cos 2+\cos 4+\cdots+\cos(2n)=\frac{\sin n\cos(n+1)}{\sin 1},\quad n\in\mathbb{N}.$$

Let $x_{n}=\prod_{k=1}^{n}\cos k$. We have, based on the Arithmetic Mean–Geometric Mean Inequality, that

$$x_{n}^{2}\leq\left(\frac{\sum_{k=1}^{n}\cos^{2}k}{n}\right)^{n}=\left(\frac{n+\sum_{k=1}^{n}\cos(2k)}{2n}\right)^{n}$$

$$=\frac{1}{2^{n}}\left(1+\frac{\sin n\cos(n+1)}{n\sin 1}\right)^{n}$$

$$=\frac{1}{(2\sin 1)^{n}}\left(\sin 1+\frac{\sin n\cos(n+1)}{n}\right)^{n}.$$

Let $0 < \varepsilon < 1 - \sin 1 < \sin 1$. Since $\lim_{n\to\infty}(\sin n \cos(n+1))/n = 0$, there is N_ε such that for all $n \geq N_\varepsilon$, we have

$$-\varepsilon < \frac{\sin n \cos(n+1)}{n} < \varepsilon.$$

It follows that

$$0 < (\sin 1 - \varepsilon)^n \leq \left(\sin 1 + \frac{\sin n \cos(n+1)}{n}\right)^n \leq (\varepsilon + \sin 1)^n < \varepsilon + \sin 1 < 1.$$

Thus, for all $n \geq N_\varepsilon$, we have that

$$0 < |\cos 1 \cos 2 \cdots \cos n| \leq \frac{1}{(2\sin 1)^{n/2}}, \tag{3.3}$$

and, hence,

$$n^\beta |\cos 1 \cos 2 \cdots \cos n|^\alpha \leq \frac{n^\beta}{((2\sin 1)^{\alpha/2})^n}, \quad n \geq N_\varepsilon.$$

We have that $2\sin 1 > 1$ and $\sum_{n=1}^{\infty} n^\beta \left(1/(2\sin 1)^{\alpha/2}\right)^n$ converges, and the result follows from the comparison theorem.

For an alternative solution of this problem, as well as comments, generalizations, and remarks, see [66].

3.5. The series converges for $x < 1/e$ and diverges for $x \in [1/e, 1)$. Let

$$a_n = x^{\sin\frac{1}{1} + \sin\frac{1}{2} + \cdots + \sin\frac{1}{n}}.$$

A calculation shows that $n\left(\frac{a_{n+1}}{a_n} - 1\right) = n\left(x^{\sin\frac{1}{n+1}} - 1\right)$, and hence

$$\lim_{n\to\infty} n\left(\frac{a_{n+1}}{a_n} - 1\right) = \lim_{n\to\infty} n\left(x^{\sin\frac{1}{n+1}} - 1\right) = \lim_{n\to\infty} \frac{x^{\sin\frac{1}{n+1}} - 1}{\sin\frac{1}{n+1}} \cdot \lim_{n\to\infty} n\sin\frac{1}{n+1} = \ln x.$$

Thus, the series converges for $x < 1/e$ and diverges when $x > 1/e$.

Now we study the case when $x = 1/e$. One can prove that the sequence

$$y_n = \sin\frac{1}{1} + \sin\frac{1}{2} + \cdots + \sin\frac{1}{n} - \ln n$$

converges. Thus,

$$\left(\frac{1}{e}\right)^{\sin\frac{1}{1} + \sin\frac{1}{2} + \cdots + \sin\frac{1}{n}} = \left(\frac{1}{e}\right)^{\sin\frac{1}{1} + \sin\frac{1}{2} + \cdots + \sin\frac{1}{n} - \ln n} \cdot \frac{1}{n} \sim \frac{1}{n},$$

and, hence, the series diverges.

3.6. Since $|\cos 1 \cos^2 2 \cdots \cos^n n| \leq |\cos 1 \cos 2 \cdots \cos n|$, the result follows from Problem **3.4** with $\beta = 0$ and $\alpha = 1$.

3.7. We have

$$\sum_{n=1}^{\infty}\left(\frac{1}{n}-f(n)\right)=\sum_{k=1}^{\infty}\left(\sum_{2^{k-1}\le n<2^k}\left(\frac{1}{n}-f(n)\right)\right)$$

$$=\sum_{k=1}^{\infty}\left(\sum_{2^{k-1}\le n<2^k}\frac{1}{n}-\sum_{2^{k-1}\le n<2^k}f(n)\right)$$

$$=\sum_{k=1}^{\infty}\left(\sum_{2^{k-1}\le n<2^k}\frac{1}{n}-\ln 2\right)$$

$$=\sum_{k=1}^{\infty}\left(\frac{1}{2^{k-1}}+\frac{1}{2^{k-1}+1}+\cdots+\frac{1}{2^k-1}-\ln 2\right)$$

$$=\gamma,$$

where the last equality follows by calculating the nth partial sum of the series and by using the limit definition of the Euler–Mascheroni constant.

3.8. We have that

$$S_n=\sum_{k=1}^{\infty}\frac{1}{k(k+1)^n}=n-\zeta(2)-\zeta(3)-\cdots-\zeta(n).$$

See the solution of Problem **1.31**.

3.9. The series equals

$$(-1)^{n+1}+\sum_{j=2}^{n}(-1)^{n+j}\zeta(j).$$

Let $S_n=\sum_{k=1}^{\infty}1/k^n(k+1)$. We have, since

$$\frac{1}{k^n(k+1)}=\frac{1}{k^n}-\frac{1}{k^{n-1}(k+1)},$$

that $S_n=\zeta(n)-S_{n-1}$, and it follows that $(-1)^nS_n=(-1)^n\zeta(n)+(-1)^{n-1}S_{n-1}$. Thus, $S_n=(-1)^{n+1}+\sum_{j=2}^{n}(-1)^{n+j}\zeta(j)$, and the problem is solved.

Remark. We have, as a consequence of this problem, that

$$\int_0^1 \mathrm{Li}_n(x)\mathrm{d}x=(-1)^{n+1}+\sum_{j=2}^{n}(-1)^{n+j}\zeta(j),$$

where $\mathrm{Li}_n(z)$ denotes the Polylogarithm function.

3.10. Let $x_n=\sum_{k=1}^{\infty}\frac{1}{k(k+1)\cdots(k+n)}$. We have

$$\frac{1}{k(k+1)\cdots(k+n)}=\frac{1}{n}\left(\frac{1}{k(k+1)\cdots(k+n-1)}-\frac{1}{(k+1)\cdots(k+n)}\right),$$

and it follows that

$$x_n = \frac{1}{n}\left(x_{n-1} - \sum_{k=1}^{\infty} \frac{1}{(k+1)\cdots(k+n)}\right)$$

$$\overset{k+1=m}{=} \frac{1}{n}\left(x_{n-1} - \left(\sum_{m=1}^{\infty} \frac{1}{m(m+1)\cdots(m+n-1)} - \frac{1}{n!}\right)\right)$$

$$= \frac{1}{n}\left(x_{n-1} - x_{n-1} + \frac{1}{n!}\right)$$

$$= \frac{1}{n \cdot n!}.$$

Remark. This problem is due to Andreoli (see [5]).

3.11. A solution, which is based on summing the first n terms of the series, is given in [108, Problem 2.2.4, p. 69]. Another approach would be to express the general term of the series in the form $\int_0^1 x^2/(4n^2 - x^2)dx$ (see [20, Problem 1, p. 526]).

3.12. The series equals $2\gamma + 1/2 - \ln(2\pi)$. Let S_n be the nth partial sum of the series. A calculation shows that

$$S_n = \sum_{k=2}^{n}\left((k-1)^2 \ln\frac{k^2}{k^2-1} - \frac{k-1}{k+1}\right)$$

$$= 2\sum_{k=2}^{n}(k-1)^2 \ln k - \sum_{k=2}^{n}(k-1)^2 \ln(k-1)$$

$$- \sum_{k=2}^{n}(k-1)^2 \ln(k+1) - \sum_{k=2}^{n}\left(1 - \frac{2}{k+1}\right).$$

We change the limits of summation in the second and the third sum, and we get that

$$S_n = -2\ln n! + (n^2+2)\ln n - (n-1)^2\ln(n+1) - n + 1 + 2\left(\sum_{k=2}^{n}\frac{1}{k+1} - \ln n\right).$$

Using Stirling's formula, $\ln n! = (1/2)\ln(2\pi) + (n+1/2)\ln n - n + O(1/n)$, we get that

$$S_n = -\ln(2\pi) + \left((n-1)^2\ln\frac{n}{n+1} + n\right) + 1 + 2\left(\sum_{k=2}^{n}\frac{1}{k+1} - \ln n\right) + O\left(\frac{1}{n}\right).$$

Since $\lim_{n\to\infty}\left((n-1)^2\ln\frac{n}{n+1} + n\right) = 5/2$, one has that $\lim_{n\to\infty} S_n = 2\gamma + 1/2 - \ln(2\pi)$.

3.13. Using partial fractions we get that

$$\frac{\alpha^{n-1}}{(\alpha+p)(2\alpha+p)\cdots(n\alpha+p)} = \frac{A_1}{\alpha+p} + \frac{A_2}{2\alpha+p} + \cdots + \frac{A_n}{n\alpha+p},$$

where $A_k = \frac{(-1)^{k-1}}{(k-1)!(n-k)!}$. Thus,

$$
\frac{\alpha^{n-1}}{(\alpha+p)(2\alpha+p)\cdots(n\alpha+p)} = \sum_{k=1}^{n} \frac{(-1)^{k-1}}{(k-1)!(n-k)!(k\alpha+p)}
$$

$$
= \sum_{k=1}^{n} \frac{(-1)^{k-1}}{(k-1)!(n-k)!} \int_0^1 x^{\alpha k + p - 1}\,dx
$$

$$
= \int_0^1 x^{p-1}\left(\sum_{k=1}^{n} \frac{(-1)^{k-1}x^{\alpha k}}{(k-1)!(n-k)!} \right) dx
$$

$$
\overset{k-1=m}{=} \int_0^1 x^{p-1+\alpha}\left(\sum_{m=0}^{n-1} \frac{(-1)^m x^{\alpha m}}{m!(n-1-m)!} \right) dx.
$$

It follows that

$$
\sum_{n=1}^{\infty} \frac{\alpha^{n-1}}{(\alpha+p)(2\alpha+p)\cdots(n\alpha+p)} = \int_0^1 x^{p-1+\alpha} \sum_{n=1}^{\infty} \left(\sum_{m=0}^{n-1} \frac{(-1)^m x^{\alpha m}}{m!(n-1-m)!} \right) dx.
$$

We prove that

$$
\sum_{n=1}^{\infty} \left(\sum_{m=0}^{n-1} \frac{(-1)^m x^{\alpha m}}{m!(n-1-m)!} \right) = e \cdot e^{-x^\alpha}. \tag{3.4}
$$

To see this, we consider the following power series expansions:

$$
e^t = \sum_{k=0}^{\infty} \frac{t^k}{k!} \quad \text{and} \quad e^{-x^\alpha t} = \sum_{k=0}^{\infty} (-1)^k \cdot \frac{x^{\alpha k}}{k!} \cdot t^k, \quad t \in \mathbb{R}.
$$

Multiplying these two power series we obtain that

$$
e^t \cdot e^{-x^\alpha t} = \sum_{n=1}^{\infty} \left(\sum_{m=0}^{n-1} \frac{(-1)^m x^{\alpha m}}{m!(n-1-m)!} \right) t^{n-1},
$$

and we get, for $t = 1$, that (3.4) holds. This completes the proof. The second part of the problem is solved similarly.

An alternative solution is given in [10].

3.14. The sum equals $\ln^2 2/2$. We need the following power series expansion [16, Entry (5.2.9), p. 76]:

$$
-\frac{\ln(1-z)}{1-z} = \sum_{n=1}^{\infty} H_n z^n, \quad |z| < 1, \tag{3.5}
$$

Let S be the value of the series. Using Abel's summation formula (A.2), with

$$
a_k = \frac{1}{k} \quad \text{and} \quad b_k = 1 - \frac{1}{2} + \frac{1}{3} + \cdots + \frac{(-1)^{k-1}}{k} - \ln 2,
$$

we get that

$$S = \lim_{n \to \infty} H_n \left(1 - \frac{1}{2} + \frac{1}{3} + \cdots + \frac{(-1)^n}{n+1} - \ln 2 \right) + \sum_{k=1}^{\infty} \frac{(-1)^{k+1}}{k+1} H_k$$

$$= \sum_{k=1}^{\infty} \frac{(-1)^{k+1}}{k+1} H_k,$$

since

$$\lim_{n \to \infty} H_n \left(1 - \frac{1}{2} + \frac{1}{3} + \cdots + \frac{(-1)^n}{n+1} - \ln 2 \right) = 0.$$

Integrating (3.5) we get that

$$\frac{\ln^2(1-z)}{2} = \sum_{n=1}^{\infty} \frac{H_n}{n+1} \cdot z^{n+1}, \quad |z| < 1,$$

and, when $z = -1$, one has that $\ln^2 2/2 = \sum_{n=1}^{\infty} (-1)^{n+1} H_n/(n+1)$.

3.15. We have

$$\frac{1}{n+1} - \frac{1}{n+2} + \frac{1}{n+3} - \cdots = \int_0^1 \left(x^n - x^{n+1} + x^{n+2} - \cdots \right) dx = \int_0^1 \frac{x^n}{1+x} dx.$$

It follows that

$$\sum_{n=1}^{\infty} \frac{1}{n} \left(\frac{1}{n+1} - \frac{1}{n+2} + \frac{1}{n+3} - \cdots \right) = \sum_{n=1}^{\infty} \frac{1}{n} \int_0^1 \frac{x^n}{1+x} dx = \int_0^1 \frac{1}{1+x} \left(\sum_{n=1}^{\infty} \frac{x^n}{n} \right) dx$$

$$= -\int_0^1 \frac{\ln(1-x)}{1+x} dx \overset{1-x=y}{=} -\int_0^1 \frac{\ln y}{2-y} dy$$

$$= -\int_0^1 \frac{\ln(y/2) + \ln 2}{2-y} dy$$

$$= -\int_0^1 \frac{\ln(y/2)}{2-y} dy - \ln^2 2$$

$$\overset{y/2=t}{=} -\int_0^{1/2} \frac{\ln t}{1-t} dt - \ln^2 2$$

$$= -\int_{1/2}^1 \frac{\ln(1-t)}{t} dt - \ln^2 2$$

$$= -\left(\int_0^1 - \int_0^{1/2} \right) \frac{\ln(1-t)}{t} dt - \ln^2 2$$

$$= \text{Li}_2(1) - \text{Li}_2(1/2) - \ln^2 2$$

$$= 1/2 \left(\zeta(2) - \ln^2 2 \right).$$

3.16. See the solution of Problem **3.41**.

3.17. The sum equals $\gamma \ln 2 - \frac{1}{2}\ln^2 2$. Let

$$S_{2n} = \sum_{k=1}^{2n} \left(1 - \frac{1}{2} + \frac{1}{3} - \cdots + \frac{(-1)^{k-1}}{k} - \ln 2\right) \cdot \ln \frac{k+1}{k},$$

be the $2n$th partial sum of the series. A calculation shows that

$$S_{2n} = \sum_{k=2}^{2n} (-1)^k \frac{\ln k}{k} + \left(1 - \frac{1}{2} + \frac{1}{3} - \cdots - \frac{1}{2n}\right) \ln(2n+1) - \ln 2 \cdot \ln(2n+1)$$

$$= 2\sum_{k=1}^{n} \frac{\ln(2k)}{2k} - \sum_{k=2}^{2n} \frac{\ln k}{k} + (H_{2n} - H_n)\ln(2n+1) - \ln 2 \cdot \ln(2n+1)$$

$$= \ln 2 \cdot H_n + \sum_{k=1}^{n} \frac{\ln k}{k} - \sum_{k=2}^{2n} \frac{\ln k}{k} + (H_{2n} - \ln(2n) - \gamma)\ln(2n+1)$$

$$\quad + (\ln n + \gamma - H_n)\ln(2n+1)$$

$$= \ln 2\,(H_n - \ln n) + \ln 2 \ln n + \left(\sum_{k=2}^{n} \frac{\ln k}{k} - \frac{\ln^2 n}{2}\right) + \frac{\ln^2 n}{2}$$

$$\quad - \left(\sum_{k=2}^{2n} \frac{\ln k}{k} - \frac{\ln^2(2n)}{2}\right) - \frac{\ln^2(2n)}{2}$$

$$\quad + (H_{2n} - \ln(2n) - \gamma)\ln(2n+1) + (\ln n + \gamma - H_n)\ln(2n+1)$$

$$= \ln 2\,(H_n - \ln n) + \left(\sum_{k=2}^{n} \frac{\ln k}{k} - \frac{\ln^2 n}{2}\right) - \left(\sum_{k=2}^{2n} \frac{\ln k}{k} - \frac{\ln^2(2n)}{2}\right)$$

$$\quad + (H_{2n} - \ln(2n) - \gamma)\ln(2n+1) + (\ln n + \gamma - H_n)\ln(2n+1) - \frac{1}{2}\ln^2 2.$$

Recall that the Stieltjes constant γ_1 is defined by the limit

$$\gamma_1 = \lim_{n \to \infty} \left(\sum_{k=1}^{n} \frac{\ln k}{k} - \frac{\ln^2 n}{2}\right).$$

Also, in [134] the following inequalities are proved:

$$\frac{1}{2(n+1)} < \gamma - \sum_{k=1}^{n} \frac{1}{k} + \ln(n+1) < \frac{1}{2n}, \quad n \in \mathbb{N},$$

from which it follows that

$$\lim_{n\to\infty} (H_{2n} - \ln(2n) - \gamma)\ln(2n+1) = 0 \quad \text{and} \quad \lim_{n\to\infty} (\ln n + \gamma - H_n)\ln(2n+1) = 0.$$

This implies that

$$\lim_{n\to\infty} S_{2n} = \gamma\ln 2 + \gamma_1 - \gamma_1 - \frac{1}{2}\ln^2 2 = \gamma\ln 2 - \frac{1}{2}\ln^2 2.$$

3.18. The series equals $\zeta(p-1) - \zeta(p)$. Let S be the value of the series and let S_n denote the nth partial sum of the series. An application of Abel's summation formula (A.1), with $a_k = 1$ and $b_k = \zeta(p) - 1 - \frac{1}{2^p} - \cdots - \frac{1}{k^p}$, shows that

$$S_n = \sum_{k=1}^{n}\left(\zeta(p) - 1 - \frac{1}{2^p} - \cdots - \frac{1}{k^p}\right)$$

$$= n\left(\zeta(p) - 1 - \frac{1}{2^p} - \cdots - \frac{1}{(n+1)^p}\right) + \sum_{k=1}^{n}\frac{k}{(k+1)^p}. \qquad (3.6)$$

Using Stolz–Cesàro lemma (the $0/0$ case), we obtain that

$$\lim_{n\to\infty} n\left(\zeta(p) - 1 - \frac{1}{2^p} - \cdots - \frac{1}{(n+1)^p}\right) = \lim_{n\to\infty}\frac{\zeta(p) - 1 - \frac{1}{2^p} - \cdots - \frac{1}{(n+1)^p}}{\frac{1}{n}}$$

$$= \lim_{n\to\infty}\frac{n(n+1)}{(n+2)^p}$$

$$= 0.$$

Letting $n \to \infty$ in (3.6), and using the preceding limit, we get that

$$S = \sum_{k=1}^{\infty}\frac{k}{(k+1)^p} = \sum_{k=1}^{\infty}\frac{1}{(k+1)^{p-1}} - \sum_{k=1}^{\infty}\frac{1}{(k+1)^p} = \zeta(p-1) - \zeta(p).$$

For an alternative solution one can use the definition of the Riemann zeta function and part (a) of Problem **3.122**.

3.19. The series equals $(\zeta(p-2) - \zeta(p-1))/2$. Let S be the value of the series and let S_n denote the nth partial sum of the series. Direct calculations show that

$$S_n = \sum_{k=1}^{n} k\left(\zeta(p) - \frac{1}{1^p} - \frac{1}{2^p} - \cdots - \frac{1}{k^p}\right)$$

$$= \frac{n(n+1)}{2}\zeta(p) - \sum_{k=1}^{n}\left(\frac{n(n+1)}{2} - \frac{(k-1)k}{2}\right)\frac{1}{k^p}$$

$$= \frac{n(n+1)}{2}\left(\zeta(p) - \sum_{k=1}^{n}\frac{1}{k^p}\right) + \frac{1}{2}\sum_{k=1}^{n}\frac{k-1}{k^{p-1}}. \qquad (3.7)$$

On the other hand,

$$0 \le \zeta(p) - \sum_{k=1}^{n} \frac{1}{k^p} = \sum_{k=n+1}^{\infty} \frac{1}{k^p} \le \int_{n}^{\infty} \frac{dx}{x^p} = \frac{1}{(p-1)n^{p-1}},$$

and hence, for $p > 3$, one has that

$$\lim_{n \to \infty} \frac{n(n+1)}{2} \left(\zeta(p) - \sum_{k=1}^{n} \frac{1}{k^p} \right) = 0.$$

Letting $n \to \infty$ in (3.7), and using the preceding limit, we obtain that

$$S = \frac{1}{2} \sum_{k=1}^{\infty} \frac{k-1}{k^{p-1}} = \frac{1}{2} \sum_{k=1}^{\infty} \frac{1}{k^{p-2}} - \frac{1}{2} \sum_{k=1}^{\infty} \frac{1}{k^{p-1}} = \frac{1}{2} \zeta(p-2) - \frac{1}{2} \zeta(p-1).$$

For an alternative solution one can use the definition of the Riemann zeta function and part (b) of Problem **3.122**.

3.20. We apply Abel's summation formula (A.2) and we get that

$$\sum_{n=1}^{\infty} \frac{1}{n} \left(\zeta(2) - \frac{1}{1^2} - \frac{1}{2^2} - \cdots - \frac{1}{n^2} \right) = \lim_{n \to \infty} H_n \left(\zeta(2) - \frac{1}{1^2} - \frac{1}{2^2} - \cdots - \frac{1}{(n+1)^2} \right)$$

$$+ \sum_{n=1}^{\infty} \frac{H_n}{(n+1)^2}$$

$$= \sum_{n=1}^{\infty} \left(\frac{H_{n+1}}{(n+1)^2} - \frac{1}{(n+1)^3} \right)$$

$$= \zeta(3),$$

since $\sum_{n=1}^{\infty} H_n/n^2 = 2\zeta(3)$ (see Problem **3.55**).

Remark. More generally, one can prove that if $k \ge 2$ is an integer, then

$$\sum_{n=1}^{\infty} \frac{1}{n} \left(\zeta(k) - \frac{1}{1^k} - \frac{1}{2^k} - \cdots - \frac{1}{n^k} \right) = \frac{1}{2} k \zeta(k+1) - \frac{1}{2} \sum_{i=1}^{k-2} \zeta(k-i)\zeta(i+1),$$

where the second sum is missing when $k = 2$.

In particular, when $k = 3$, we have that

$$\sum_{n=1}^{\infty} \frac{1}{n} \left(\zeta(3) - \frac{1}{1^3} - \frac{1}{2^3} - \cdots - \frac{1}{n^3} \right) = \frac{\zeta(4)}{4}.$$

3.21. We have, since $\int_0^1 x^{k-1} \ln x \, dx = -1/k^2$, that

$$\sum_{n=1}^{\infty} \frac{1}{n^2}\left(\zeta(2) - \frac{1}{1^2} - \cdots - \frac{1}{n^2}\right) = \sum_{n=1}^{\infty} \frac{1}{n^2}\left(\sum_{k=n+1}^{\infty} \frac{1}{k^2}\right)$$

$$= -\sum_{n=1}^{\infty} \frac{1}{n^2} \int_0^1 \left(\sum_{k=n+1}^{\infty} x^{k-1}\right) \ln x\, dx$$

$$= -\int_0^1 \frac{\ln x}{1-x}\left(\sum_{n=1}^{\infty} \frac{x^n}{n^2}\right) dx$$

$$= -\int_0^1 \frac{\ln x \operatorname{Li}_2(x)}{1-x} dx$$

$$= \frac{3}{4}\zeta(4),$$

the last integral being recorded in [44, Table 5, p. 1436].

An alternative solution is based on an application of Abel's summation formula in which case the series reduces to a special Euler sum that can be calculated by integrals involving polylogarithmic functions [44].

3.22. We have, by Abel's summation formula (A.2), that

$$\sum_{n=1}^{\infty} \left(\zeta(2) - \frac{1}{1^2} - \cdots - \frac{1}{n^2}\right)^2 = \sum_{n=1}^{\infty} \frac{n}{(n+1)^2}\left(2\zeta(2) - \frac{2}{1^2} - \cdots - \frac{2}{n^2} - \frac{1}{(n+1)^2}\right)$$

$$= 2\sum_{n=1}^{\infty} \frac{n}{(n+1)^2}\left(\zeta(2) - \frac{1}{1^2} - \cdots - \frac{1}{n^2}\right)$$

$$- \sum_{n=1}^{\infty} \frac{n}{(n+1)^4}$$

$$= 2\sum_{n=1}^{\infty} \frac{1}{n+1}\left(\zeta(2) - \frac{1}{1^2} - \cdots - \frac{1}{n^2}\right)$$

$$- 2\sum_{n=1}^{\infty} \frac{1}{(n+1)^2}\left(\zeta(2) - \frac{1}{1^2} - \cdots - \frac{1}{n^2}\right)$$

$$- \zeta(3) + \zeta(4).$$

We have, based on Problem **3.20**, that

$$\sum_{n=1}^{\infty} \frac{1}{n+1}\left(\zeta(2) - \frac{1}{1^2} - \cdots - \frac{1}{n^2}\right) = \sum_{n=1}^{\infty} \frac{1}{n+1}\left(\zeta(2) - \frac{1}{1^2} - \cdots - \frac{1}{(n+1)^2}\right)$$

$$+ \sum_{n=1}^{\infty} \frac{1}{(n+1)^3}$$

$$= 2\zeta(3) - \zeta(2).$$

On the other hand, a calculation similar to the one in the solution of Problem **3.21** shows that

$$\sum_{n=1}^{\infty} \frac{1}{(n+1)^2} \left(\zeta(2) - \frac{1}{1^2} - \cdots - \frac{1}{n^2} \right) = - \int_0^1 \frac{\ln x}{x(1-x)} \left(\text{Li}_2(x) - x \right) dx$$

$$= \frac{7}{4} \zeta(4) - \zeta(2),$$

since $\int_0^1 \ln x/(1-x) dx = -\zeta(2)$ and

$$\int_0^1 \frac{\ln x \, \text{Li}_2(x)}{x} dx = -\zeta(4), \quad \int_0^1 \frac{\ln x \, \text{Li}_2(x)}{1-x} dx = -\frac{3}{4} \zeta(4).$$

Putting all these together we get that the sum is calculated and the problem is solved.

3.23. The series equals $3/2 - \gamma/2 - \ln(2\pi)/2$. We have

$$\sum_{k=2}^{\infty} \frac{\zeta(k)-1}{k+1} = \sum_{k=2}^{\infty} \frac{1}{k+1} \left(\sum_{m=2}^{\infty} \left(\frac{1}{m} \right)^k \right)$$

$$= \sum_{m=2}^{\infty} \left(\sum_{k=2}^{\infty} \frac{1}{k+1} \left(\frac{1}{m} \right)^k \right)$$

$$= \sum_{m=2}^{\infty} m \left(\sum_{k=2}^{\infty} \frac{1}{k+1} \left(\frac{1}{m} \right)^{k+1} \right)$$

$$= \sum_{m=2}^{\infty} m \left(\ln \frac{m}{m-1} - \frac{1}{m} - \frac{1}{2m^2} \right).$$

Let S_n denote the nth partial sum of the series. A calculation shows that

$$S_n = \sum_{m=2}^{n} m \left(\ln \frac{m}{m-1} - \frac{1}{m} - \frac{1}{2m^2} \right)$$

$$= \left(n + \frac{1}{2} \right) \ln n - \ln n! - n + 1 - \frac{1}{2} \left(\sum_{m=2}^{n} \frac{1}{m} - \ln n \right).$$

Using Stirling's formula, $\ln n! = (1/2)\ln(2\pi) + (n+1/2)\ln n - n + O(1/n)$, we get that

$$S_n = -\frac{\ln(2\pi)}{2} + 1 - \frac{1}{2}\left(\sum_{m=2}^n \frac{1}{m} - \ln n\right) - O\left(\frac{1}{n}\right),$$

and it follows that $\lim_{n\to\infty} S_n = 3/2 - \gamma/2 - \ln(2\pi)/2$.

3.24. We have

$$\sum_{k=2}^\infty \frac{\zeta(k)-1}{k+2} = \sum_{k=2}^\infty \frac{1}{k+2}\left(\sum_{m=2}^\infty \left(\frac{1}{m}\right)^k\right)$$

$$= \sum_{m=2}^\infty \left(\sum_{k=2}^\infty \frac{1}{k+2}\left(\frac{1}{m}\right)^k\right)$$

$$= \sum_{m=2}^\infty m^2\left(\sum_{k=2}^\infty \frac{1}{k+2}\left(\frac{1}{m}\right)^{k+2}\right)$$

$$= \sum_{m=2}^\infty m^2\left(\ln\frac{m}{m-1} - \frac{1}{m} - \frac{1}{2m^2} - \frac{1}{3m^3}\right).$$

Let S_n denote the nth partial sum of the series. A calculation shows that

$$S_n = \sum_{k=2}^n k^2\left(\ln\frac{k}{k-1} - \frac{1}{k} - \frac{1}{2k^2} - \frac{1}{3k^3}\right)$$

$$= (n+1)^2\ln n - \ln n! - 2\sum_{k=2}^n k\ln k - \frac{n^2+2n-3}{2} - \frac{1}{3}\sum_{k=2}^n \frac{1}{k}.$$

Let $A_n = n^{-n^2/2 - n/2 - 1/12} \cdot e^{n^2/4} \cdot \prod_{k=1}^n k^k$. We have

$$2\sum_{k=2}^n k\ln k = 2\ln A_n + \left(n^2 + n + \frac{1}{6}\right)\ln n - \frac{n^2}{2}.$$

Using Stirling's formula, $\ln n! = (1/2)\ln(2\pi) + (n+1/2)\ln n - n + O(1/n)$, we get that

$$S_n = -\frac{\ln(2\pi)}{2} - 2\ln A_n + \frac{3}{2} - \frac{1}{3}\left(\sum_{k=2}^n \frac{1}{k} - \ln n\right) - O\left(\frac{1}{n}\right),$$

and this implies that $\lim_{n\to\infty} S_n = 11/6 - 2\ln A - \gamma/3 - \ln(2\pi)/2$.

3.25. The series equals $3/2 - \ln \pi$. We have

$$\sum_{k=1}^{\infty} \frac{\zeta(2k)-1}{k+1} = \sum_{k=1}^{\infty} \left(\sum_{m=2}^{\infty} \left(\frac{1}{m} \right)^{2k} \right)$$

$$= \sum_{m=2}^{\infty} \left(\sum_{k=1}^{\infty} \frac{1}{k+1} \left(\frac{1}{m^2} \right)^{k} \right)$$

$$= \sum_{m=2}^{\infty} m^2 \left(\sum_{k=1}^{\infty} \frac{1}{k+1} \left(\frac{1}{m^2} \right)^{k+1} \right)$$

$$= \sum_{m=2}^{\infty} m^2 \left(\ln \frac{m^2}{m^2-1} - \frac{1}{m^2} \right).$$

Let S_n denote the nth partial sum of the series. We have

$$S_n = \sum_{m=2}^{n} m^2 \left(\ln \frac{m^2}{m^2-1} - \frac{1}{m^2} \right)$$

$$= -2\ln n! + (n+1)^2 \ln n - n^2 \ln(n+1) - n + \ln 2 + 1.$$

Using Stirling's formula, $\ln n! = (1/2)\ln(2\pi) + (n+1/2)\ln n - n + O(1/n)$, one has that

$$S_n = -\ln \pi + 1 + n^2 \ln \frac{n}{n+1} + n - O\left(\frac{1}{n} \right).$$

A calculation shows that $\lim_{n\to\infty}(n^2 \ln(n/(n+1)) + n) = 1/2$, and this implies that $\lim_{n\to\infty} S_n = 3/2 - \ln \pi$.

Remark. It is worth mentioning that an excellent book which contains a systematic collection of various families of series associated with the Riemann and Hurwitz zeta functions is [122].

3.26. (a) We have, based on Problem **3.8**, that

$$S_n = \sum_{k=1}^{\infty} \frac{1}{k(k+1)^n} = n - \zeta(2) - \zeta(3) - \cdots - \zeta(n).$$

Clearly $S_n > 0$ and hence $n - \zeta(2) - \zeta(3) - \cdots - \zeta(n) > 0$. On the other hand, we have, based on Bernoulli's inequality, that $(1+k)^n \geq 1 + kn$, and it follows that

$$0 < S_n \leq \sum_{k=1}^{\infty} \frac{1}{k(1+kn)} = \gamma + \psi\left(1 + \frac{1}{n} \right),$$

where ψ denotes the Digamma function. Letting $n \to \infty$, we obtain, since $\psi(1) = -\gamma$, that $\lim_{n\to\infty} S_n = 0$, and the first part of the problem is solved.

(b) The series equals $\zeta(2) + \zeta(3) - 1$. We need the following series formula:

$$\sum_{n=2}^{\infty} n a^n = \frac{2a^2 - a^3}{(1-a)^2}, \quad |a| < 1.$$

We have

$$\sum_{n=2}^{\infty} n\left(n - \zeta(2) - \zeta(3) - \cdots - \zeta(n)\right) = \sum_{n=2}^{\infty} n \left(\sum_{k=1}^{\infty} \frac{1}{k(k+1)^n}\right)$$

$$= \sum_{k=1}^{\infty} \frac{1}{k}\left(\sum_{n=2}^{\infty} \frac{n}{(k+1)^n}\right)$$

$$= \sum_{k=1}^{\infty} \frac{1}{k}\left(\frac{2k+1}{(k+1)k^2}\right)$$

$$= 2\sum_{k=1}^{\infty} \frac{1}{(k+1)k^2} + \sum_{k=1}^{\infty} \frac{1}{(k+1)k^3}$$

$$= \zeta(2) + \zeta(3) - 1.$$

We used the following two series formulae:

$$\sum_{k=1}^{\infty} \frac{1}{(k+1)k^2} = \zeta(2) - 1 \quad \text{and} \quad \sum_{k=1}^{\infty} \frac{1}{(k+1)k^3} = \zeta(3) - \zeta(2) + 1,$$

which can be proved by direct calculations.

Similarly, one can prove that

$$\sum_{n=2}^{\infty} (-1)^n n(n - \zeta(2) - \zeta(3) - \cdots - \zeta(n)) = \frac{5}{4} - \frac{\pi^2}{12}.$$

3.27. Clearly when $a = 0$, there is nothing to prove, so we consider the case when $a \neq 0$. We have

$$\sum_{n=2}^{\infty} a^n \left(n - \zeta(2) - \zeta(3) - \cdots - \zeta(n)\right) = \sum_{n=2}^{\infty} a^n \left(\sum_{k=1}^{\infty} \frac{1}{k(k+1)^n}\right)$$

$$= \sum_{k=1}^{\infty} \frac{1}{k}\left(\sum_{n=2}^{\infty} \left(\frac{a}{k+1}\right)^n\right)$$

$$= a^2 \sum_{k=1}^{\infty} \frac{1}{k(k+1)(k+1-a)}.$$

We distinguish here two cases.

Case $a = 1$. We have, based on the preceding calculations, that

$$\sum_{n=2}^{\infty} (n - \zeta(2) - \zeta(3) - \cdots - \zeta(n)) = \sum_{k=1}^{\infty} \frac{1}{k^2(k+1)} = \zeta(2) - 1.$$

Case $a \neq 1$. We have

$$\sum_{n=2}^{\infty} a^n (n - \zeta(2) - \zeta(3) - \cdots - \zeta(n)) = a^2 \sum_{k=1}^{\infty} \frac{1}{k(k+1)(k+1-a)}$$

$$= a^2 \sum_{k=1}^{\infty} \left(\frac{1}{k(k+1-a)} - \frac{1}{(k+1)(k+1-a)} \right)$$

$$= a^2 \left(\sum_{k=1}^{\infty} \frac{1}{1-a} \left(\frac{1}{k} - \frac{1}{k+1-a} \right) - \sum_{k=1}^{\infty} \frac{1}{a} \left(\frac{1}{k+1-a} - \frac{1}{k+1} \right) \right)$$

$$= a^2 \left(\sum_{k=1}^{\infty} \frac{1}{1-a} \left(\frac{1}{k} - \frac{1}{k+1-a} \right) - \sum_{k=1}^{\infty} \frac{1}{a} \left(\frac{1}{k+1-a} - \frac{1}{k} + \frac{1}{k} - \frac{1}{k+1} \right) \right)$$

$$= a^2 \left(\frac{1}{a(1-a)} \sum_{k=1}^{\infty} \left(\frac{1}{k} - \frac{1}{k+1-a} \right) - \frac{1}{a} \sum_{k=1}^{\infty} \left(\frac{1}{k} - \frac{1}{k+1} \right) \right)$$

$$= a \left(\frac{\psi(2-a) + \gamma}{1-a} - 1 \right).$$

This formula agrees even with the case when $a = 1$, since

$$\lim_{a \to 1} a \left(\frac{\psi(2-a) + \gamma}{1-a} - 1 \right) = \psi'(1) - 1 = \zeta(2) - 1.$$

3.28. Using Abel's summation formula (A.2), with $a_n = 1$ and

$$b_n = \left(\ln 2 - \frac{1}{n+1} - \frac{1}{n+2} - \cdots - \frac{1}{2n} \right)^2 = (\ln 2 - H_{2n} + H_n)^2,$$

we get that

$$S = \sum_{n=1}^{\infty} \left(\ln 2 - \frac{1}{n+1} - \frac{1}{n+2} - \cdots - \frac{1}{2n} \right)^2 = \lim_{n \to \infty} n (\ln 2 - H_{2n+2} + H_{n+1})^2$$

$$+ \sum_{n=1}^{\infty} n \left((\ln 2 - H_{2n} + H_n)^2 - (\ln 2 - H_{2n+2} + H_{n+1})^2 \right)$$

$$= \sum_{n=1}^{\infty} n \left(\frac{1}{2n+1} - \frac{1}{2n+2} \right) \left(2\ln 2 - 2H_{2n} + 2H_n + \frac{1}{2n+2} - \frac{1}{2n+1} \right)$$

$$= 2 \sum_{n=1}^{\infty} n \left(\frac{1}{2n+1} - \frac{1}{2n+2} \right) (\ln 2 - H_{2n} + H_n) - \sum_{n=1}^{\infty} n \left(\frac{1}{2n+1} - \frac{1}{2n+2} \right)^2,$$

since $\lim_{n \to \infty} n \, (\ln 2 - H_{2n+1} + H_{n+1})^2 = 0$. Now,

$$2n \left(\frac{1}{2n+1} - \frac{1}{2n+2} \right) = \frac{1}{n+1} - \frac{1}{2n+1}$$

and

$$\sum_{n=1}^{\infty} n \left(\frac{1}{2n+1} - \frac{1}{2n+2} \right)^2 = \frac{3\ln 2}{2} - \frac{5\pi^2}{48}.$$

Thus,

$$S = \sum_{n=1}^{\infty} \left(\frac{1}{n+1} - \frac{1}{2n+1} \right) (\ln 2 - H_{2n} + H_n) - \frac{3\ln 2}{2} + \frac{5\pi^2}{48}. \tag{3.8}$$

Next, we need the following series [62, Formula 1.516 (3), p. 52]:

$$\ln(1+x)\ln(1-x) = \sum_{n=1}^{\infty} \frac{1}{n} \left(H_n - H_{2n} - \frac{1}{2n} \right) x^{2n}, \quad x^2 < 1.$$

It follows, by differentiation, that

$$\frac{\ln(1-x)}{1+x} - \frac{\ln(1+x)}{1-x} - \frac{\ln(1-x^2)}{x} + \frac{(2\ln 2)x}{1-x^2} = 2 \sum_{n=1}^{\infty} (H_n - H_{2n} + \ln 2) x^{2n-1}.$$

This implies that

$$\int_0^1 x^2 \left(\frac{\ln(1-x)}{1+x} - \frac{\ln(1+x)}{1-x} - \frac{\ln(1-x^2)}{x} + \frac{(2\ln 2)x}{1-x^2} \right) dx = \sum_{n=1}^{\infty} \frac{H_n - H_{2n} + \ln 2}{n+1}$$

and

$$\int_0^1 \frac{x}{2} \left(\frac{\ln(1-x)}{1+x} - \frac{\ln(1+x)}{1-x} - \frac{\ln(1-x^2)}{x} + \frac{(2\ln 2)x}{1-x^2} \right) dx = \sum_{n=1}^{\infty} \frac{H_n - H_{2n} + \ln 2}{2n+1}.$$

Hence,

$$\int_0^1 \left(x^2 - \frac{x}{2} \right) \left(\frac{\ln(1-x)}{1+x} - \frac{\ln(1+x)}{1-x} - \frac{\ln(1-x^2)}{x} + \frac{(2\ln 2)x}{1-x^2} \right) dx$$

$$= \sum_{n=1}^{\infty} \left(\frac{1}{n+1} - \frac{1}{2n+1} \right) (H_n - H_{2n} + \ln 2). \tag{3.9}$$

A calculation shows that

$$\left(x^2-\frac{x}{2}\right)\left(\frac{\ln(1-x)}{1+x}-\frac{\ln(1+x)}{1-x}-\frac{\ln(1-x^2)}{x}+\frac{(2\ln2)x}{1-x^2}\right)$$

$$=\left(x-\frac{1}{2}\right)\left(\frac{(2\ln2)x^2}{1-x^2}-\frac{\ln(1-x)}{1+x}-\frac{\ln(1+x)}{1-x}\right).$$

We have, by Maple calculations, that

$$\int_0^1\left(x-\frac{1}{2}\right)\left(\frac{(2\ln2)x^2}{1-x^2}-\frac{\ln(1-x)}{1+x}-\frac{\ln(1+x)}{1-x}\right)dx=-\ln^2 2+2\ln2-\frac{\pi^2}{12}.$$

$$(3.10)$$

Combining (3.8)–(3.10) we get that $S=\pi^2/48+(\ln2)/2-\ln^2 2$.

Remark. Since $\ln2-\frac{1}{n+1}-\frac{1}{n+2}-\cdots-\frac{1}{2n}=O(1/n)$, it is reasonable to expect that the value of the series can be expressed in terms of $\zeta(2)$.

3.29. We have

$$\frac{1}{n+1}-\frac{1}{n+2}+\frac{1}{n+3}-\cdots=\int_0^1\left(x^n-x^{n+1}+x^{n+2}-\cdots\right)dx=\int_0^1\frac{x^n}{1+x}dx.$$

It follows that

$$\sum_{n=0}^{\infty}\left(\frac{1}{n+1}-\frac{1}{n+2}+\frac{1}{n+3}-\cdots\right)^2=\sum_{n=0}^{\infty}\left(\int_0^1\frac{x^n}{1+x}dx\right)\left(\int_0^1\frac{y^n}{1+y}dy\right)$$

$$=\int_0^1\int_0^1\frac{1}{(1+x)(1+y)}\left(\sum_{n=0}^{\infty}(xy)^n\right)dxdy$$

$$=\int_0^1\int_0^1\frac{1}{(1+x)(1+y)(1-xy)}dxdy$$

$$=\int_0^1\frac{1}{1+x}\left(\int_0^1\frac{1}{(1+y)(1-xy)}dy\right)dx$$

$$=\int_0^1\frac{1}{1+x}\left(\frac{\ln2-\ln(1-x)}{1+x}\right)dx$$

$$=\left(\frac{(1-x)\ln(1-x)}{2(1+x)}+\frac{\ln(1+x)}{2}-\frac{\ln2}{1+x}\right)\Big|_0^1$$

$$=\ln2.$$

3.9.2 Alternating Series

> *I have tried to avoid long numerical computations, thereby following Riemann's postulate that proofs should be given through ideas and not voluminous computations.*
>
> David Hilbert (1862–1943)

3.30. The series equals $-\ln 2$. First we note that for all $k \in \mathbb{N}$, one has that

$$(k+1)^p - k^p - 1 \equiv 0 \ (\mathrm{mod}\, 2).$$

It follows, since there are $(k+1)^p - k^p - 1$ integers in the interval $(k^p, (k+1)^p)$, that

$$\sum_{k^p < n < (k+1)^p} (-1)^n = 0.$$

Also, we note that $(-1)^{k^p} = (-1)^k$. We have

$$\sum_{n=1}^{\infty} \frac{(-1)^n}{\lfloor \sqrt[p]{n} \rfloor} = \sum_{k=1}^{\infty} \left(\sum_{n=k^p} \frac{(-1)^n}{\lfloor \sqrt[p]{n} \rfloor} \right) + \sum_{k=1}^{\infty} \left(\sum_{k^p < n < (k+1)^p} \frac{(-1)^n}{\lfloor \sqrt[p]{n} \rfloor} \right)$$

$$= \sum_{k=1}^{\infty} \frac{(-1)^{k^p}}{k} + \sum_{k=1}^{\infty} \left(\sum_{k^p < n < (k+1)^p} \frac{(-1)^n}{\lfloor \sqrt[p]{n} \rfloor} \right)$$

$$= \sum_{k=1}^{\infty} \frac{(-1)^k}{k} + \sum_{k=1}^{\infty} \left(\sum_{k^p < n < (k+1)^p} \frac{(-1)^n}{k} \right)$$

$$= -\ln 2 + \sum_{k=1}^{\infty} \frac{1}{k} \left(\sum_{k^p < n < (k+1)^p} (-1)^n \right)$$

$$= -\ln 2.$$

For an alternative solution see [64].

3.31. The sum equals

$$(k-1)! \left(-\frac{1}{e} - \frac{1}{e} \left(\frac{1}{1!} + \frac{1}{2!} + \cdots + \frac{1}{(k-1)!} \right) + 1 - \frac{1}{k!} \right).$$

Let $S_k = \sum_{n=1}^{\infty} (-1)^n / (n!(n+k))$. We have, since

$$\frac{1}{n!(n+k)} = \frac{1}{k} \left(\frac{1}{n!} - \frac{1}{(n-1)!(n+k)} \right),$$

that

$$S_k = \frac{1}{k} \sum_{n=1}^{\infty} \frac{(-1)^n}{n!} - \frac{1}{k} \sum_{n=1}^{\infty} \frac{(-1)^n}{(n-1)!(n+k)}$$

$$\stackrel{n-1=m}{=} \frac{1}{k}\left(\frac{1}{e}-1\right) + \frac{1}{k} \sum_{m=0}^{\infty} \frac{(-1)^m}{m!(m+k+1)}$$

$$= \frac{1}{k}\left(\frac{1}{e}-1\right) + \frac{1}{k}\left(S_{k+1}+\frac{1}{k+1}\right).$$

This implies that

$$\frac{S_{k+1}}{k!} = \frac{S_k}{(k-1)!} - \frac{1}{e}\cdot\frac{1}{k!} + \frac{1}{k!} - \frac{1}{(k+1)!},$$

and the result follows.

3.32. We need the following lemma.

Lemma 3.1. *Let x be a positive real number. The following equality holds*

$$\sum_{n=1}^{\infty} (-1)^{n-1} \frac{1}{n(n+x)} = \frac{1}{x}\left(\psi(x) - \psi\left(\frac{x}{2}\right) - \frac{1}{x}\right).$$

Proof. We need the following product formula due to Weierstrass [7, Formula (2.8), p. 16]:

$$\Gamma(x) = \frac{e^{-\gamma x}}{x} \prod_{n=1}^{\infty} \frac{e^{x/n}}{1+x/n}, \quad x > 0,$$

from which it follows, by logarithmic differentiation, that

$$\psi(x) = \frac{\Gamma'(x)}{\Gamma(x)} = -\gamma - \frac{1}{x} + \sum_{n=1}^{\infty}\left(\frac{1}{n} - \frac{1}{n+x}\right), \quad x > 0.$$

Let $S_{2n} = \sum_{k=1}^{2n} (-1)^{k-1} \frac{1}{k(k+x)}$. We have that

$$S_{2n} = \frac{1}{x} \sum_{k=1}^{2n} (-1)^{k-1}\left(\frac{1}{k} - \frac{1}{k+x}\right)$$

$$= \frac{1}{x}\left[\sum_{k=1}^{2n}\left(\frac{1}{k} - \frac{1}{k+x}\right) - 2\sum_{k=1}^{n}\left(\frac{1}{2k} - \frac{1}{2k+x}\right)\right]$$

$$= \frac{1}{x}\left[\sum_{k=1}^{2n}\left(\frac{1}{k} - \frac{1}{k+x}\right) - \sum_{k=1}^{n}\left(\frac{1}{k} - \frac{1}{k+x/2}\right)\right].$$

Thus,

$$\lim_{n\to\infty} S_{2n} = \frac{1}{x}\left(\psi(x)+\gamma+\frac{1}{x}-\psi\left(\frac{x}{2}\right)-\gamma-\frac{2}{x}\right)$$

$$= \frac{1}{x}\left(\psi(x)-\frac{1}{x}-\psi\left(\frac{x}{2}\right)\right),$$

and the lemma is proved.

We have, based on Lemma 3.1, that

$$\sum_{n=1}^{\infty}(-1)^{n-1}\frac{1}{n+x} = \sum_{n=1}^{\infty}(-1)^{n-1}\left(\frac{1}{n+x}-\frac{1}{n}+\frac{1}{n}\right)$$

$$= \sum_{n=1}^{\infty}\frac{(-1)^{n-1}}{n}-\sum_{n=1}^{\infty}(-1)^{n-1}\left(\frac{1}{n}-\frac{1}{n+x}\right)$$

$$= \ln 2 - x\sum_{n=1}^{\infty}(-1)^{n-1}\frac{1}{n(n+x)}$$

$$= \ln 2 - \psi(x) + \psi\left(\frac{x}{2}\right)+\frac{1}{x}.$$

3.33. The series equals $\pi^2/12+\ln^2 2/2$. Let S be the value of the series. Using Abel's summation formula (A.2), with $a_k=(-1)^{k-1}/k$ and $b_k=\sum_{j=1}^{k}(-1)^{j-1}/j$, we have that

$$S = \lim_{n\to\infty}\left(\sum_{k=1}^{n}\frac{(-1)^{k-1}}{k}\right)\cdot\left(\sum_{j=1}^{n+1}\frac{(-1)^{j-1}}{j}\right)$$

$$+\sum_{k=1}^{\infty}\left(1-\frac{1}{2}+\frac{1}{3}-\cdots+\frac{(-1)^{k-1}}{k}\right)\frac{(-1)^{k+1}}{k+1}$$

$$= \ln^2 2+\sum_{k=1}^{\infty}\left(1-\frac{1}{2}+\frac{1}{3}-\cdots+\frac{(-1)^{k-1}}{k}\right)\frac{(-1)^{k+1}}{k+1}$$

$$= \ln^2 2+\sum_{k=1}^{\infty}\left(1-\frac{1}{2}+\frac{1}{3}-\cdots+\frac{(-1)^{k-1}}{k}+\frac{(-1)^k}{k+1}-\frac{(-1)^k}{k+1}\right)\frac{(-1)^{k+1}}{k+1}$$

$$= \ln^2 2-\sum_{k=1}^{\infty}\left(1-\frac{1}{2}+\frac{1}{3}-\cdots+\frac{(-1)^k}{k+1}\right)\frac{(-1)^k}{k+1}+\sum_{k=1}^{\infty}\frac{1}{(k+1)^2}$$

$$\overset{k+1=m}{=}\ln^2 2-\sum_{m=2}^{\infty}\left(1-\frac{1}{2}+\cdots+\frac{(-1)^{m-1}}{m}\right)\frac{(-1)^{m-1}}{m}+\frac{\pi^2}{6}-1$$

$$= \ln^2 2 - (S-1) + \frac{\pi^2}{6} - 1$$

$$= \ln^2 2 - S + \frac{\pi^2}{6}.$$

3.34. The sum equals $\pi/8 - 3\ln 2/4$. Let

$$S_n = \sum_{k=1}^{n} (-1)^k \left(\ln 2 - \frac{1}{k+1} - \frac{1}{k+2} - \cdots - \frac{1}{2k} \right).$$

An application of Abel's summation formula (A.1), with

$$a_k = (-1)^k \quad \text{and} \quad b_k = \ln 2 - \frac{1}{k+1} - \frac{1}{k+2} - \cdots - \frac{1}{2k},$$

shows that

$$S_n = \left((-1) + (-1)^2 + \cdots + (-1)^n \right) \left(\ln 2 - \frac{1}{n+2} - \frac{1}{n+3} - \cdots - \frac{1}{2n+2} \right)$$

$$+ \sum_{k=1}^{n} \left((-1) + (-1)^2 + \cdots + (-1)^k \right) \frac{1}{(2k+1)(2k+2)}. \tag{3.11}$$

Passing to the limit, as n tends to infinity, in (3.11), we get that

$$\lim_{n \to \infty} S_n = - \sum_{\substack{k=1 \\ k=\text{odd}}}^{\infty} \frac{1}{(2k+1)(2k+2)} = - \sum_{l=0}^{\infty} \frac{1}{(4l+3)(4l+4)}$$

$$= - \sum_{l=0}^{\infty} \left(\frac{1}{4l+3} - \frac{1}{4l+4} \right) = - \int_0^1 \sum_{l=0}^{\infty} \left(x^{4l+2} - x^{4l+3} \right) dx$$

$$= - \int_0^1 \frac{x^2 dx}{(1+x)(1+x^2)} = - \frac{3\ln 2}{4} + \frac{\pi}{8}.$$

For an alternative solution see [80].

3.35. All series can be calculated by an application of Abel's summation formula. We include below, for the sake of completeness, the calculations of the second series.

Let S be the sum of the series and let S_n be its nth partial sum. We have

$$S_n = \sum_{k=1}^{n} (-1)^k \left(1 + \frac{1}{3} + \frac{1}{5} + \cdots + \frac{1}{2k-1} - \frac{\ln k}{2} - \frac{\gamma}{2} - \ln 2 \right).$$

An application of Abel's summation formula (A.1), with

$$a_k = (-1)^k \quad \text{and} \quad b_k = 1 + \frac{1}{3} + \frac{1}{5} + \cdots + \frac{1}{2k-1} - \frac{\ln k}{2} - \frac{\gamma}{2} - \ln 2,$$

shows that

$$S_n = \left((-1) + (-1)^2 + \cdots + (-1)^n \right) \left(1 + \frac{1}{3} + \cdots + \frac{1}{2n+1} - \frac{\ln(n+1)}{2} - \frac{\gamma}{2} - \ln 2 \right)$$

$$+ \sum_{k=1}^{n} \left((-1) + (-1)^2 + \cdots + (-1)^k \right) \left(\frac{\ln(k+1)}{2} - \frac{\ln k}{2} - \frac{1}{2k+1} \right).$$

It follows that

$$S = \sum_{n=1}^{\infty} \left((-1) + (-1)^2 + \cdots + (-1)^n \right) \left(\frac{\ln(n+1)}{2} - \frac{\ln n}{2} - \frac{1}{2n+1} \right)$$

$$= \sum_{p=0}^{\infty} \left(\frac{1}{4p+3} + \frac{\ln(2p+1)}{2} - \frac{\ln(2p+2)}{2} \right).$$

Let T_n be the nth partial sum of the preceding series. A calculation shows that

$$T_n = \sum_{p=0}^{n} \left(\frac{1}{4p+3} + \frac{\ln(2p+1)}{2} - \frac{\ln(2p+2)}{2} \right)$$

$$= \sum_{p=0}^{n} \frac{1}{4p+3} + \frac{1}{2} \ln(2n+2)! - (n+1)\ln 2 - \ln(n+1)!.$$

Using Stirling's formula, $\ln n! = (1/2)\ln(2\pi) + (n+1/2)\ln n - n + O(1/n)$, one has that

$$T_n = \sum_{p=0}^{n} \frac{1}{4p+3} - \frac{\ln \pi}{4} - \frac{\ln(n+1)}{4} + O\left(\frac{1}{n}\right)$$

$$= \sum_{p=0}^{n} \left(\frac{1}{4p+3} - \frac{1}{4p+4} \right) - \frac{\ln \pi}{4} + \frac{1}{4} \left(\sum_{p=0}^{n} \frac{1}{p+1} - \ln(n+1) \right) + O\left(\frac{1}{n}\right).$$

It follows, since

$$\sum_{p=0}^{\infty} \left(\frac{1}{4p+3} - \frac{1}{4p+4} \right) = \frac{3\ln 2}{4} - \frac{\pi}{8} \quad \text{(see the solution of Problem 3.34),}$$

that $\lim_{n\to\infty} T_n = (\ln 8)/4 - \pi/8 - (\ln \pi)/4 + \gamma/4$.

Remark. We mention that the first series is an old problem due to Hardy [20, Problem 45, p. 277].

3.36. Euler's constant γ is defined by the series formula

$$\gamma = \sum_{n=1}^{\infty} \left(\frac{1}{n} - \ln \frac{n+1}{n} \right),$$

and its *alternating counterpart* (this problem) is a series of Sondow [120].

3.37. Let S_{2n} denote the $2n$th partial sum of the series. We have

$$S_{2n} = \sum_{k=1}^{2n} (-1)^k \left(k \ln \frac{k+1}{k} - 1 \right)$$

$$= \sum_{k=1}^{2n} (-1)^k k \ln(k+1) - \sum_{k=1}^{2n} (-1)^k k \ln k$$

$$= \sum_{k=2}^{2n+1} (-1)^{k-1} (k-1) \ln k + \sum_{k=1}^{2n} (-1)^{k-1} k \ln k$$

$$= 2n \ln(2n+1) + \sum_{k=2}^{2n} (-1)^{k-1} (2k-1) \ln k$$

$$= 2n \ln(2n+1) - 2 \sum_{k=2}^{2n} (-1)^k k \ln k + \sum_{k=2}^{2n} (-1)^k \ln k.$$

Using Stirling's formula, $\ln n! = (1/2) \ln(2\pi) + (n+1/2) \ln n - n + O(1/n)$, we obtain that

$$\sum_{k=2}^{2n} (-1)^k \ln k = \ln \frac{2^{2n}(n!)^2}{(2n)!} = \frac{1}{2} \ln \pi + \frac{1}{2} \ln n + O\left(\frac{1}{n} \right)$$

and

$$\sum_{k=2}^{2n} (-1)^{k-1} k \ln k = 2 \ln 2 + 3 \ln 3 + \cdots + 2n \ln(2n) - 2 \sum_{k=1}^{n} 2k \ln(2k)$$

$$= \ln \left(2^2 3^3 \cdots (2n)^{2n} \right) - 2n(n+1) \ln 2 - 4 \ln \left(2^2 3^3 \cdots n^n \right)$$

$$= \ln \frac{2^2 3^3 \cdots (2n)^{2n}}{(2n)^{2n^2+n+1/12} e^{-n^2}} - 4 \ln \frac{2^2 3^3 \cdots n^n}{n^{n^2/2+n/2+1/12} e^{-n^2/4}}$$

$$+ \left(\frac{1}{12} - n \right) \ln 2 - \left(n + \frac{1}{4} \right) \ln n.$$

It follows that

$$S_{2n} = 2\ln \frac{2^2 3^3 \cdots (2n)^{2n}}{(2n)^{2n^2+n+1/12} e^{-n^2}} - 8\ln \frac{2^2 3^3 \cdots n^n}{n^{n^2/2+n/2+1/12} e^{-n^2/4}}$$

$$+ \frac{\ln 2}{6} + \frac{\ln \pi}{2} + \ln \left(\frac{2n+1}{2n}\right)^{2n} + O\left(\frac{1}{n}\right).$$

Thus, $\lim_{n\to\infty} S_{2n} = 1 + (\ln \pi)/2 - 6\ln A + (\ln 2)/6$.

Remark. The problem has an equivalent formulation as follows:

$$\prod_{n=1}^{\infty} \left(\frac{1}{e}\left(\frac{n+1}{n}\right)^n\right)^{(-1)^n} = \frac{e \cdot \sqrt{\pi} \cdot \sqrt[6]{2}}{A^6}.$$

For an alternative solution see [127].

3.38. Calculate the $2n$th partial sum of the series and use the *Glaisher product* [43, p. 135]

$$\prod_{k=1}^{\infty} k^{1/k^2} = \left(\frac{A^{12}}{2\pi e^\gamma}\right)^{\frac{\pi^2}{6}}.$$

3.39. See [128].

3.40. This problem, which is due to Coffey, is solved in [27].

3.16. and **3.41.** We prove that

$$\sum_{n=1}^{\infty} \frac{1}{n^2}\left(1 - \frac{1}{2} + \frac{1}{3} - \cdots + \frac{(-1)^{n-1}}{n}\right) = \frac{\pi^2}{4}\ln 2 - \frac{\zeta(3)}{4}$$

and

$$\sum_{n=1}^{\infty} \frac{(-1)^{n-1}}{n^2}\left(1 - \frac{1}{2} + \frac{1}{3} - \cdots + \frac{(-1)^{n-1}}{n}\right) = \frac{\pi^2}{4}\ln 2 - \frac{5\zeta(3)}{8}.$$

We have

$$\sum_{n=1}^{\infty} \frac{1}{n^2}\left(1 - \frac{1}{2} + \frac{1}{3} - \cdots + \frac{(-1)^{n-1}}{n}\right) = \sum_{n=1}^{\infty} \frac{1}{n^2}\int_0^1 \left(1 + (-x) + \cdots + (-x)^{n-1}\right) dx$$

$$= \sum_{n=1}^{\infty} \frac{1}{n^2}\int_0^1 \frac{1 - (-x)^n}{1+x} dx$$

$$= \int_0^1 \frac{1}{1+x} \left(\sum_{n=1}^{\infty} \frac{1}{n^2} - \sum_{n=1}^{\infty} \frac{(-x)^n}{n^2} \right) dx$$

$$= \int_0^1 \frac{1}{1+x} \left(\zeta(2) - \text{Li}_2(-x) \right) dx,$$

where $\text{Li}_2(z)$ denotes the Dilogarithm function defined, for $|z| < 1$, by

$$\text{Li}_2(z) = \sum_{n=1}^{\infty} \frac{z^n}{n^2} = - \int_0^z \frac{\ln(1-t)}{t} dt.$$

We calculate the integral by parts, with

$$f(x) = \zeta(2) - \text{Li}_2(-x), \quad f'(x) = \frac{\ln(1+x)}{x}, \quad g'(x) = \frac{1}{1+x}, \quad g(x) = \ln(1+x),$$

and we have that

$$\int_0^1 \frac{1}{1+x} \left(\zeta(2) - \text{Li}_2(-x) \right) dx = \left. \left(\zeta(2) - \text{Li}_2(-x) \right) \ln(1+x) \right|_0^1 - \int_0^1 \frac{\ln^2(1+x)}{x} dx$$

$$= \left(\zeta(2) - \text{Li}_2(-1) \right) \ln 2 - \int_0^1 \frac{\ln^2(1+x)}{x} dx$$

$$= \frac{\pi^2}{4} \ln 2 - \frac{\zeta(3)}{4},$$

since

$$\int_0^1 \frac{\ln^2(1+x)}{x} dx = \frac{\zeta(3)}{4} \quad \text{and} \quad \text{Li}_2(-1) = -\frac{\pi^2}{12}.$$

Similarly,

$$\sum_{n=1}^{\infty} \frac{(-1)^{n-1}}{n^2} \left(1 - \frac{1}{2} + \frac{1}{3} - \cdots + \frac{(-1)^{n-1}}{n} \right)$$

$$= \left. \left(\sum_{n=1}^{\infty} \frac{(-1)^{n-1}}{n^2} + \text{Li}_2(x) \right) \ln(1+x) \right|_0^1$$

$$+ \int_0^1 \frac{\ln(1+x)\ln(1-x)}{x} dx$$

$$= \frac{\pi^2}{4} \ln 2 - \frac{5\zeta(3)}{8},$$

since

$$\int_0^1 \frac{\ln(1+x)\ln(1-x)}{x} dx = -\frac{5\zeta(3)}{8}.$$

3.42. We have, based on Abel's summation formula (A.2), that

$$\sum_{n=1}^{\infty}\left(1+\frac{1}{2}+\cdots+\frac{1}{n}-\ln n-\gamma-\frac{1}{2n}\right)=\lim_{n\to\infty}n\left(H_{n+1}-\ln(n+1)-\gamma-\frac{1}{2(n+1)}\right)$$

$$+\sum_{n=1}^{\infty}n\left(\ln\frac{n+1}{n}-\frac{1}{2(n+1)}-\frac{1}{2n}\right)$$

$$=\sum_{n=1}^{\infty}n\left(\ln\frac{n+1}{n}-\frac{1}{2(n+1)}-\frac{1}{2n}\right),$$

since $\lim_{n\to\infty}n\left(H_{n+1}-\ln(n+1)-\gamma-1/(2n+2)\right)=0$. Let S_n be the nth partial sum of the series. A calculation shows that

$$S_n=n\ln(n+1)-\ln n!-n+\frac{1}{2}(H_{n+1}-1),$$

and by Stirling's formula, $\lim_{n\to\infty}S_n=(\gamma+1-\ln(2\pi))/2$.

3.43. Let S_{2n} be the $2n$th partial sum of the series. We have

$$S_{2n}=\sum_{k=1}^{2n}(-1)^{k-1}k\left(1+\frac{1}{2}+\cdots+\frac{1}{k}-\ln k-\gamma-\frac{1}{2k}\right)$$

$$=\sum_{k=1}^{2n}(-1)^{k-1}kH_k-\sum_{k=1}^{2n}(-1)^{k-1}k\ln k-\gamma\sum_{k=1}^{2n}(-1)^{k-1}k-\frac{1}{2}\sum_{k=1}^{2n}(-1)^{k-1}$$

$$=\sum_{k=1}^{2n}(-1)^{k-1}kH_k-\sum_{k=1}^{2n}(-1)^{k-1}k\ln k+n\gamma.$$

Let $A_n=n^{-n^2/2-n/2-1/12}e^{n^2/4}\prod_{k=1}^{n}k^k$. A calculation shows (see the solution of Problem **3.37**) that

$$\sum_{k=1}^{2n}(-1)^{k-1}k\ln k=-4\ln A_n+\ln A_{2n}-\left(n+\frac{1}{4}\right)\ln n-\left(n-\frac{1}{12}\right)\ln 2.$$

On the other hand,

$$\sum_{k=1}^{2n}(-1)^{k-1}kH_k=\frac{1}{2}+\frac{1}{4}H_n-\left(n+\frac{1}{2}\right)H_{2n+1}.$$

The preceding equality can be proved either by induction on n or by Abel's summation formula. It follows that

$$S_{2n} = \frac{1}{2} + \frac{1}{4}H_n - \left(n + \frac{1}{2}\right)H_{2n+1}$$

$$+ 4\ln A_n - \ln A_{2n} + \left(n + \frac{1}{4}\right)\ln n + \left(n - \frac{1}{12}\right)\ln 2 + n\gamma$$

$$= \frac{1}{2} + \frac{1}{4}(H_n - \ln n) - \left(n + \frac{1}{2}\right)(H_{2n+1} - \ln(2n+1) - \gamma) - \frac{\gamma}{2}$$

$$- \frac{7\ln 2}{12} + 4\ln A_n - \ln A_{2n} + \left(n + \frac{1}{2}\right)\ln \frac{2n}{2n+1}.$$

Since

$$\lim_{n \to \infty} \left(n + \frac{1}{2}\right)(H_{2n+1} - \ln(2n+1) - \gamma) = \frac{1}{4} \quad \text{(prove it!)},$$

we get that $\lim_{n \to \infty} S_{2n} = -\gamma/4 - 1/4 + 3\ln A - (7\ln 2)/12$.

3.44. Let S be the value of the series and let $y_m = (-1) + (-1)^2 + \cdots + (-1)^m$. We have, based on Abel's summation formula (A.2) with

$$a_m = (-1)^m \quad \text{and} \quad b_m = \left(\ln 2 - \sum_{i=1}^{2m} \frac{1}{i} + \sum_{i=1}^{m} \frac{1}{i}\right)^2,$$

that

$$S = \lim_{m \to \infty} y_m \left(\ln 2 - \sum_{i=1}^{2m+2} \frac{1}{i} + \sum_{i=1}^{m+1} \frac{1}{i}\right)^2$$

$$+ \sum_{m=1}^{\infty} y_m \left[\left(\ln 2 - \sum_{i=1}^{2m} \frac{1}{i} + \sum_{i=1}^{m} \frac{1}{i}\right)^2 - \left(\ln 2 - \sum_{i=1}^{2m+2} \frac{1}{i} + \sum_{i=1}^{m+1} \frac{1}{i}\right)^2\right]$$

$$\overset{m:=2m-1}{=} \sum_{m=1}^{\infty} \left[\left(\ln 2 - \sum_{i=1}^{4m} \frac{1}{i} + \sum_{i=1}^{2m} \frac{1}{i}\right)^2 - \left(\ln 2 - \sum_{i=1}^{4m-2} \frac{1}{i} + \sum_{i=1}^{2m-1} \frac{1}{i}\right)^2\right]$$

$$= \sum_{m=1}^{\infty} \left(\frac{1}{4m} - \frac{1}{4m-1}\right)\left(2\ln 2 - 2\sum_{i=2m}^{4m} \frac{1}{i} + \frac{1}{4m-1} + \frac{3}{4m}\right)$$

$$= -2\ln 2 \sum_{m=1}^{\infty} \frac{1}{4m(4m-1)} + 2\sum_{m=1}^{\infty} \left[\left(\frac{1}{4m-1} - \frac{1}{4m}\right) \cdot \sum_{i=2m}^{4m} \frac{1}{i}\right]$$

$$+ \sum_{m=1}^{\infty} \left(\frac{3}{16m^2} - \frac{1}{2m(4m-1)} - \frac{1}{(4m-1)^2} \right)$$

$$= -2(\ln 2 + 1) \sum_{m=1}^{\infty} \frac{1}{4m(4m-1)} + 2C + \frac{\pi^2}{32} - \sum_{m=1}^{\infty} \frac{1}{(4m-1)^2},$$

where

$$C = \sum_{m=1}^{\infty} \left[\left(\frac{1}{4m-1} - \frac{1}{4m} \right) \cdot \sum_{i=2m}^{4m} \frac{1}{i} \right].$$

One can check that

$$\sum_{m=1}^{\infty} \frac{1}{4m(4m-1)} = \frac{3\ln 2}{4} - \frac{\pi}{8} \quad \text{and} \quad \sum_{m=1}^{\infty} \frac{1}{(4m-1)^2} = \frac{\pi^2}{16} - \frac{1}{2}G.$$

It follows that

$$S = \frac{\pi}{4}\ln 2 + \frac{\pi}{4} - \frac{\pi^2}{32} - \frac{3}{2}\ln^2 2 - \frac{3}{2}\ln 2 + 2C + \frac{1}{2}G. \tag{3.12}$$

Now we calculate C. We have

$$C = \sum_{m=1}^{\infty} \left[\left(\frac{1}{4m-1} - \frac{1}{4m} \right) \cdot \left(\sum_{i=1}^{4m} \frac{1}{i} - \sum_{i=1}^{2m-1} \frac{1}{i} \right) \right]$$

$$= \sum_{m=1}^{\infty} \left(\frac{1}{4m-1} - \frac{1}{4m} \right) \left(\sum_{i=1}^{m} \left(\frac{1}{4i-3} - \frac{1}{4i-2} + \frac{1}{4i-1} - \frac{1}{4i} \right) + \frac{1}{2m} \right)$$

$$= \sum_{m=1}^{\infty} \frac{1}{2m(4m-1)} - \frac{1}{8} \sum_{m=1}^{\infty} \frac{1}{m^2}$$

$$+ \sum_{m=1}^{\infty} \left[\left(\frac{1}{4m-1} - \frac{1}{4m} \right) \cdot \sum_{i=1}^{m} \left(\frac{1}{4i-3} - \frac{1}{4i-2} + \frac{1}{4i-1} - \frac{1}{4i} \right) \right]$$

$$= \frac{3\ln 2}{2} - \frac{\pi}{4} - \frac{\pi^2}{48} + D,$$

where

$$D = \sum_{m=1}^{\infty} \left[\left(\frac{1}{4m-1} - \frac{1}{4m} \right) \cdot \sum_{i=1}^{m} \left(\frac{1}{4i-3} - \frac{1}{4i-2} + \frac{1}{4i-1} - \frac{1}{4i} \right) \right]$$

$$= \sum_{i=1}^{\infty} \left[\left(\frac{1}{4i-3} - \frac{1}{4i-2} + \frac{1}{4i-1} - \frac{1}{4i} \right) \cdot \sum_{m=i}^{\infty} \left(\frac{1}{4m-1} - \frac{1}{4m} \right) \right].$$

Let

$$D(x) = \sum_{i=1}^{\infty} \left(\frac{1}{4i-3} - \frac{1}{4i-2} + \frac{1}{4i-1} - \frac{1}{4i} \right) \sum_{m=i}^{\infty} \left(\frac{x^{4m-1}}{4m-1} - \frac{x^{4m}}{4m} \right).$$

It follows that

$$D'(x) = \sum_{i=1}^{\infty} \left(\frac{1}{4i-3} - \frac{1}{4i-2} + \frac{1}{4i-1} - \frac{1}{4i} \right) \frac{x^{4i-2}}{1+x+x^2+x^3},$$

and this implies that $\left(1+x+x^2+x^3\right) D'(x)$ is equal to

$$\sum_{i=1}^{\infty} \left(x \int_0^x t^{4i-4}\,dt - \int_0^x t^{4i-3}\,dt + \frac{1}{x} \int_0^x t^{4i-2}\,dt - \frac{1}{x^2} \int_0^x t^{4i-1}\,dt \right)$$

$$= x \int_0^x \frac{1}{1-t^4}\,dt - \int_0^x \frac{t}{1-t^4}\,dt + \frac{1}{x} \int_0^x \frac{t^2}{1-t^4}\,dt - \frac{1}{x^2} \int_0^x \frac{t^3}{1-t^4}\,dt$$

$$= \frac{x}{4} \ln \frac{1+x}{1-x} + \frac{x}{2} \arctan x - \frac{1}{4} \ln \frac{1+x^2}{1-x^2} + \frac{1}{4x} \ln \frac{1+x}{1-x} - \frac{\arctan x}{2x}$$

$$+ \frac{1}{4x^2} \ln(1-x^4).$$

Hence,

$$D'(x) = \frac{x}{4(1+x)(1+x^2)} \ln \frac{1+x}{1-x} + \frac{x \cdot \arctan x}{2(1+x)(1+x^2)}$$

$$- \frac{1}{4(1+x)(1+x^2)} \ln \frac{1+x^2}{1-x^2} + \frac{1}{4x(1+x)(1+x^2)} \ln \frac{1+x}{1-x}$$

$$- \frac{\arctan x}{2x(1+x)(1+x^2)} + \frac{\ln(1-x^4)}{4x^2(1+x)(1+x^2)}$$

$$= \frac{1}{8} \left(\frac{1+x}{1+x^2} - \frac{1}{1+x} \right) \ln \frac{1+x}{1-x} + \frac{1}{4} \left(\frac{1+x}{1+x^2} - \frac{1}{1+x} \right) \arctan x$$

$$- \frac{1}{8} \left(\frac{1}{1+x} - \frac{x-1}{1+x^2} \right) \ln \frac{1+x^2}{1-x^2} + \frac{1}{8} \left(\frac{2}{x} - \frac{1}{1+x} - \frac{x+1}{1+x^2} \right) \ln \frac{1+x}{1-x}$$

$$- \frac{1}{4} \left(\frac{2}{x} - \frac{1}{1+x} - \frac{x+1}{1+x^2} \right) \arctan x$$

$$+ \frac{1}{8} \left(\frac{2}{x^2} - \frac{2}{x} + \frac{1}{1+x} + \frac{x-1}{1+x^2} \right) \ln(1-x^4)$$

$$= \frac{\ln(1-x^2)}{4x^2} + \frac{1}{2} \left(\frac{1}{1+x} - \frac{1}{x} \right) \ln(1-x)$$

$$+\frac{1}{4}\left(\frac{1}{x^2}-\frac{1}{x}\right)\ln(1+x^2)+\frac{\arctan x}{2(1+x^2)}+\frac{x\ln(1+x^2)}{4(1+x^2)}$$

$$+\frac{1}{2}\left(\frac{x}{1+x^2}-\frac{1}{x}\right)\arctan x-\frac{\ln(1+x^2)}{4(1+x^2)}.$$

Thus,

$$D=D(1)=\int_0^1 D'(x)dx$$

and, since all integrals can be calculated exactly, one has that

$$D=\frac{\pi}{8}+\frac{3\pi^2}{64}-\frac{3}{4}\ln 2-\frac{3\pi}{16}\ln 2+\frac{5}{16}\ln^2 2.$$

This implies that

$$C=\frac{5}{192}\pi^2-\frac{\pi}{8}+\frac{3}{4}\ln 2-\frac{3\pi}{16}\ln 2+\frac{5}{16}\ln^2 2,$$

which in turn implies, based on (3.12), that

$$S=\frac{\pi^2}{48}-\frac{\pi}{8}\ln 2-\frac{7}{8}\ln^2 2+\frac{1}{2}G.$$

Remark. An alternative solution of this problem was published in [18] (although the term $13\pi^2/192$ from the expression of σ is incorrect) with a correction in [19].

3.45. We have

$$\frac{1}{n+1}-\frac{1}{n+2}+\frac{1}{n+3}-\cdots=\int_0^1(x^n-x^{n+1}+x^{n+2}-\cdots)dx=\int_0^1\frac{x^n}{1+x}dx.$$

Let $S=\sum_{n=0}^\infty(-1)^n\left(\frac{1}{n+1}-\frac{1}{n+2}+\frac{1}{n+3}-\cdots\right)^2$. It follows that

$$S=\sum_{n=0}^\infty(-1)^n\left(\int_0^1\frac{x^n}{1+x}dx\right)\left(\int_0^1\frac{y^n}{1+y}dy\right)$$

$$=\int_0^1\int_0^1\frac{1}{(1+x)(1+y)}\left(\sum_{n=0}^\infty(-xy)^n\right)dxdy$$

$$=\int_0^1\int_0^1\frac{dxdy}{(1+x)(1+y)(1+xy)}$$

$$=\int_0^1\frac{1}{1+x}\left(\int_0^1\frac{1}{(1+y)(1+xy)}dy\right)dx$$

$$= \int_0^1 \frac{1}{1+x}\left(\frac{\ln 2 - \ln(1+x)}{1-x}\right)dx$$

$$= \int_0^1 \frac{\ln 2 - \ln(1+x)}{1-x^2}dx.$$

A calculation shows that an antiderivative of the function $f(x) = \frac{\ln 2 - \ln(1+x)}{1-x^2}$ is

$$F(x) = \frac{1}{2}\ln(1+x)\cdot\ln\left(\frac{2}{1+x}\right) + \frac{1}{4}\ln^2(1+x) - \frac{1}{2}\cdot\text{Li}_2\left(\frac{1-x}{2}\right),$$

where $\text{Li}_2(z)$ denotes the Dilogarithm function. Thus,

$$\sum_{n=0}^{\infty}(-1)^n\left(\frac{1}{n+1} - \frac{1}{n+2} + \frac{1}{n+3} - \cdots\right)^2 = F(1) - F(0)$$

$$= \frac{\ln^2 2}{4} + \frac{1}{2}\cdot\text{Li}_2\left(\frac{1}{2}\right)$$

$$= \frac{\pi^2}{24}.$$

We used that $\text{Li}_2(1/2) = \pi^2/12 - \ln^2 2/2$ (see Appendix A).

3.9.3   Alternating Products

> What is it indeed that gives us the feeling of elegance in a
> solution, in a demonstration? It is the harmony of the diverse
> parts, their symmetry, their happy balance; in a word it is all
> that introduces order, all that gives unity, that permits us to see
> clearly and to comprehend at once both the ensemble and the
> details.
>
>   Jules Henri Poincaré (1854–1912)

3.46. First, we note that for $k \in \mathbb{N}$, one has that

$$(k+1)^p - k^p - 1 \equiv 0 \,(\text{mod}\, 2).$$

It follows, since there are $(k+1)^p - k^p - 1$ integers in the interval $(k^p, (k+1)^p)$, that

$$\sum_{k^p < n < (k+1)^p}(-1)^{n-1} = 0.$$

Also, we have that $(-1)^{k^p} = (-1)^k$. A calculation shows that

$$
\begin{aligned}
S &= \sum_{n=1}^{\infty} (-1)^{n-1} \ln\left(1 + \frac{1}{\lfloor \sqrt[p]{n} \rfloor}\right) \\
&= \sum_{k=1}^{\infty} \left(\sum_{k^p \le n < (k+1)^p} (-1)^{n-1} \ln\left(1 + \frac{1}{\lfloor \sqrt[p]{n} \rfloor}\right) \right) \\
&= \sum_{k=1}^{\infty} \left((-1)^{k^p-1} \ln\left(1 + \frac{1}{k}\right) + \sum_{k^p < n < (k+1)^p} (-1)^{n-1} \ln\left(1 + \frac{1}{\lfloor \sqrt[p]{n} \rfloor}\right) \right) \\
&= \sum_{k=1}^{\infty} (-1)^{k^p-1} \ln\left(1 + \frac{1}{k}\right) + \sum_{k=1}^{\infty} \ln\left(1 + \frac{1}{k}\right) \left(\sum_{k^p < n < (k+1)^p} (-1)^{n-1} \right) \\
&= \sum_{k=1}^{\infty} (-1)^{k-1} \ln\left(1 + \frac{1}{k}\right).
\end{aligned}
$$

Let $S_{2n} = \sum_{k=1}^{2n} (-1)^{k-1} \ln(1 + 1/k)$ be the $2n$th partial sum of the series. A calculation shows that

$$
S_{2n} = \ln\left(\frac{(2 \cdot 4 \cdots (2n))^2}{(3 \cdot 5 \cdots (2n-1))^2(2n+1)} \right) \xrightarrow{n \to \infty} \ln \frac{\pi}{2},
$$

where the limit follows in view of the Wallis product formula.

3.47. The product equals $\pi/2$. See the solution of the previous problem. It is worth mentioning that the following product formulae hold

$$
\prod_{n=1}^{\infty} \left(1 + \frac{1}{\lfloor \sqrt[p]{n} \rfloor}\right)^{(-1)^{n-1}} = \prod_{n=1}^{\infty} \left(1 + \frac{1}{n}\right)^{(-1)^{n-1}} = \prod_{n=2^p}^{\infty} \left(1 - \frac{1}{\lfloor \sqrt[p]{n} \rfloor}\right)^{(-1)^{n-1}} = \frac{\pi}{2}.
$$

3.48. and **3.49.** We prove that

$$
\prod_{n=1}^{\infty} \left(1 + \frac{1}{n^2}\right)^{(-1)^{n-1}} = \frac{\pi}{2} \tanh \frac{\pi}{2}, \qquad \prod_{n=2}^{\infty} \left(1 - \frac{1}{n^2}\right)^{(-1)^{n-1}} = \frac{\pi^2}{8}
$$

and

$$
\prod_{n=2}^{\infty} \left(\frac{n^2+1}{n^2-1}\right)^{(-1)^{n-1}} = \frac{2}{\pi} \tanh \frac{\pi}{2}.
$$

We need Euler's product formula for the sine function [109, Sect. 3, p. 12]

$$
\sin \pi z = \pi z \prod_{n=1}^{\infty} \left(1 - \frac{z^2}{n^2}\right), \quad z \in \mathbb{C}. \tag{3.13}
$$

Replacing z by iz in (3.13) and using Euler's formula, $\sin\alpha = (e^{i\alpha} - e^{-i\alpha})/2i$, we obtain that

$$\frac{e^{\pi z} - e^{-\pi z}}{2\pi z} = \frac{\sin\pi iz}{\pi iz} = \prod_{n=1}^{\infty}\left(1 + \frac{z^2}{n^2}\right), \quad z \in \mathbb{C}. \qquad (3.14)$$

When $z = 1$ and $z = 1/2$, we obtain, based on (3.14), that

$$\frac{e^{\pi} - e^{-\pi}}{2\pi} = \prod_{n=1}^{\infty}\left(1 + \frac{1}{n^2}\right) \quad \text{and} \quad \frac{e^{\frac{\pi}{2}} - e^{-\frac{\pi}{2}}}{\pi} = \prod_{n=1}^{\infty}\left(1 + \frac{1}{4n^2}\right), \qquad (3.15)$$

respectively. We have, based on (3.15), that

$$\prod_{n=1}^{\infty}\left(1 + \frac{1}{n^2}\right)^{(-1)^{n-1}} = \left(1 + \frac{1}{1^2}\right)\left(1 + \frac{1}{2^2}\right)^{-1}\cdots$$

$$= \frac{\prod_{n=1}^{\infty}\left(1 + \frac{1}{n^2}\right)}{\prod_{n=1}^{\infty}\left(1 + \frac{1}{4n^2}\right)^2}$$

$$= \frac{\pi}{2} \cdot \frac{e^{\frac{\pi}{2}} + e^{-\frac{\pi}{2}}}{e^{\frac{\pi}{2}} - e^{-\frac{\pi}{2}}}.$$

Also,

$$\frac{\sin\pi z}{\pi z(1 - z^2)} = \prod_{n=2}^{\infty}\left(1 - \frac{z^2}{n^2}\right).$$

Letting $z \to 1$ in the preceding formula, we obtain that $1/2 = \prod_{n=2}^{\infty}\left(1 - 1/n^2\right)$. On the other hand, when $z = 1/2$ one has that $2/\pi = \prod_{n=1}^{\infty}\left(1 - 1/4n^2\right)$. Thus,

$$\prod_{n=2}^{\infty}\left(1 - \frac{1}{n^2}\right)^{(-1)^{n-1}} = \left(1 - \frac{1}{2^2}\right)^{-1}\left(1 - \frac{1}{3^2}\right)\cdots$$

$$= \frac{\prod_{n=2}^{\infty}\left(1 - \frac{1}{n^2}\right)}{\prod_{n=1}^{\infty}\left(1 - \frac{1}{4n^2}\right)^2}$$

$$= \frac{\pi^2}{8}.$$

The third product can be calculated by dividing the two product formulae.

3.50. The first product equals $\sqrt[4]{8}/\sqrt{\pi}$ and the second product equals $2/x\sin(x/2)$. Recall the infinite product representation for the sine function

$$\sin x = x \prod_{n=1}^{\infty} \left(1 - \frac{x^2}{n^2\pi^2}\right).$$

Since the first product can be obtained from the second product, when $x = \pi/2$, we concentrate on the calculation of the second product. We have

$$S_{2n} = \sum_{k=1}^{2n} (-1)^k \left(\ln\left(1 - \frac{x^2}{\pi^2}\right) + \cdots + \ln\left(1 - \frac{x^2}{k^2\pi^2}\right) + \ln\frac{x}{\sin x}\right)$$

$$= \ln\left(\left(1 - \frac{(x/2)^2}{\pi^2}\right)\left(1 - \frac{(x/2)^2}{(2\pi)^2}\right)\cdots\left(1 - \frac{(x/2)^2}{(n\pi)^2}\right)\right).$$

Letting n tend to ∞ in the preceding equality, we get that $\lim_{n\to\infty} S_{2n} = \ln\frac{2\sin\frac{x}{2}}{x}$, and the problem is solved.

3.51. (a) Part (a) of the problem follows from part (b) by letting $z = 2m$ and, by the solution of part (b), the product equals

$$\frac{2m\Gamma^2(m)2^{2m-2}}{\Gamma(2m)} = 2^{2m}\binom{2m}{m}^{-1}.$$

(b) The product equals

$$\frac{z\Gamma^2\left(\frac{z}{2}\right)2^{z-2}}{\Gamma(z)}.$$

We prove that

$$\sum_{n=1}^{\infty} (-1)^{n-1}\ln\left(1 + \frac{z}{n}\right) = \ln\frac{z\Gamma^2\left(\frac{z}{2}\right)2^{z-2}}{\Gamma(z)}.$$

We need *Weierstrass product* for the Gamma function [122, Entry 2, p.1]

$$\frac{1}{\Gamma(z)} = ze^{\gamma z}\prod_{k=1}^{\infty}\left(1 + \frac{z}{k}\right)e^{-\frac{z}{k}}, \quad z \neq 0, -1, -2, \ldots \tag{3.16}$$

Let S_{2n} be the $2n$th partial sum of the series. We have

$$S_{2n} = \sum_{k=1}^{2n} (-1)^{k-1}\ln\left(1 + \frac{z}{k}\right)$$

$$= \ln\left(1 + \frac{z}{1}\right) - \ln\left(1 + \frac{z}{2}\right) + \cdots + \ln\left(1 + \frac{z}{2n-1}\right) - \ln\left(1 + \frac{z}{2n}\right)$$

$$= \ln\left(1 + \frac{z}{1}\right) + \ln\left(1 + \frac{z}{2}\right) + \cdots + \ln\left(1 + \frac{z}{2n}\right)$$

$$- 2\left[\ln\left(1 + \frac{z}{2}\right) + \ln\left(1 + \frac{z}{4}\right) + \cdots + \ln\left(1 + \frac{z}{2n}\right)\right]$$

$$= \ln\left(1+\frac{z}{1}\right) + \ln\left(1+\frac{z}{2}\right) + \cdots + \ln\left(1+\frac{z}{2n}\right) - z\left(1+\frac{1}{2}+\cdots+\frac{1}{2n}\right)$$

$$- 2\left[\ln\left(1+\frac{z/2}{1}\right) + \cdots + \ln\left(1+\frac{z/2}{n}\right) - \frac{z}{2}\left(1+\frac{1}{2}+\cdots+\frac{1}{n}\right)\right]$$

$$+ z\left(\frac{1}{n+1}+\frac{1}{n+2}+\cdots+\frac{1}{2n}\right).$$

Letting n tend to ∞ in the preceding equality and using (3.16), we obtain that

$$\lim_{n\to\infty} S_{2n} = \ln\frac{1}{z\Gamma(z)e^{\gamma z}} - 2\ln\frac{1}{(z/2)\Gamma(z/2)e^{\gamma(z/2)}} + z\ln 2 = \ln\frac{z\Gamma^2\left(\frac{z}{2}\right)2^{z-2}}{\Gamma(z)}.$$

For an alternative solution see [115].

3.52. The problem is equivalent to proving that

$$\sum_{n=1}^{\infty}\left(\frac{4n-1}{2}\ln\frac{2n}{2n-1}-1\right) = 3\ln A - \frac{7}{12}\ln 2 - \frac{1}{4}\ln\pi.$$

Let T_n denote the nth partial sum of the series. We have

$$T_n = \sum_{p=1}^{n}\left(\frac{4p-1}{2}\ln\frac{2p}{2p-1}-1\right)$$

$$= \sum_{p=1}^{n}2p\ln(2p) - \sum_{p=1}^{n}2p\ln(2p-1) - \frac{1}{2}\sum_{p=1}^{n}\ln\frac{2p}{2p-1} - n$$

$$= \sum_{p=1}^{n}2p\ln(2p) - \sum_{p=1}^{n}(2p-1)\ln(2p-1) - \sum_{p=1}^{n}\ln(2p-1) - \frac{1}{2}\ln\frac{2^{2n}(n!)^2}{(2n)!} - n$$

$$= \sum_{p=1}^{n}2p\ln(2p) - \left(\sum_{p=1}^{2n}p\ln p - \sum_{p=1}^{n}2p\ln(2p)\right) - \ln\frac{(2n)!}{2^n n!} - \ln\frac{2^n n!}{\sqrt{(2n)!}} - n$$

$$= 2\sum_{p=1}^{n}2p\ln(2p) - \sum_{p=1}^{2n}p\ln p - \ln\sqrt{(2n)!} - n$$

$$= 2n(n+1)\ln 2 + 4\sum_{p=1}^{n}p\ln p - \sum_{p=1}^{2n}p\ln p - \ln\sqrt{(2n)!} - n.$$

Let

$$x_n = 1\ln 1 + 2\ln 2 + \cdots + n\ln n - \left(\frac{n^2}{2}+\frac{n}{2}+\frac{1}{12}\right)\ln n + \frac{n^2}{4},$$

and we note that $x_n \to \ln A$. Straightforward calculations show that

$$T_n = 4x_n - x_{2n} + \left(n - \frac{1}{12}\right) \ln 2 + \left(n + \frac{1}{4}\right) \ln n - \frac{1}{2} \ln(2n)! - n.$$

Using Stirling's formula

$$\ln(2n)! = \frac{\ln(2\pi)}{2} + \left(2n + \frac{1}{2}\right) \ln(2n) - 2n + O\left(\frac{1}{n}\right),$$

we obtain that

$$T_n = 4x_n - x_{2n} - \frac{7}{12} \ln 2 - \frac{1}{4} \ln \pi - O\left(\frac{1}{n}\right) \overset{n \to \infty}{\to} 3 \ln A - \frac{7}{12} \ln 2 - \frac{1}{4} \ln \pi.$$

For an alternative solution see [89].

3.53. The problem is equivalent to proving that

$$\sum_{n=1}^{\infty} (-1)^{n-1} \left(\ln n! - \left(n + \frac{1}{2}\right) \ln n + n - \ln \sqrt{2\pi}\right) = 3 \ln A - \frac{7}{12} \ln 2 - \frac{1}{4} \ln 4.$$

Let S be the sum of the series. An application of Abel's summation formula (A.2), with

$$a_k = (-1)^{k-1} \quad \text{and} \quad b_k = \ln k! - \left(k + \frac{1}{2}\right) \ln k + k - \ln \sqrt{2\pi},$$

shows that

$$S = \lim_{n \to \infty} \left(\sum_{k=1}^{n} (-1)^{k-1}\right) \cdot \left(\ln(n+1)! - \left(n + \frac{3}{2}\right) \ln(n+1) + n + 1 - \ln \sqrt{2\pi}\right)$$

$$+ \sum_{k=1}^{\infty} \left((-1)^{1-1} + \cdots + (-1)^{k-1}\right) \left(\frac{2k+1}{2} \ln \frac{k+1}{k} - 1\right)$$

$$= \sum_{k=1}^{\infty} \left((-1)^{1-1} + \cdots + (-1)^{k-1}\right) \left(\frac{2k+1}{2} \ln \frac{k+1}{k} - 1\right)$$

$$\overset{k=2p-1}{=} \sum_{p=1}^{\infty} \left(\frac{4p-1}{2} \ln \frac{2p}{2p-1} - 1\right),$$

and this is Problem **3.52**.

For an alternative solution see [130].

Remark. The problem gives an unexpected connection between a product involving the classical Stirling term and the Glaisher–Kinkelin constant.

3.9.4   Harmonic Series

> *Truth is ever to be found in the simplicity, and not in the*
> *multiplicity and confusion of things.*
>
> Sir Isaac Newton (1642–1727)

3.54. The identity can be proved by induction on m.

3.55. We will be using the following formula:

$$\int_0^1 x^n \ln(1-x)dx = -\frac{H_{n+1}}{n+1}. \tag{3.17}$$

This formula, which is quite old, is recorded in various tables of definite integrals. It appears as formula 865.5 in [38], where a reference is given to *Nouvelles Tables d'Intégrales Définies*, by B. de Haan: P. Engels, Leyden, 1867, and it also appears, in thin disguise, as entry 4.293(8) of [62].

We have

$$\sum_{n=1}^{\infty} \frac{H_n}{n^2} = -\sum_{n=1}^{\infty} \frac{1}{n} \int_0^1 x^{n-1} \ln(1-x)dx$$

$$= -\int_0^1 \ln(1-x) \left(\sum_{n=1}^{\infty} \frac{x^{n-1}}{n} \right) dx$$

$$= \int_0^1 \frac{\ln^2(1-x)}{x} dx$$

$$= \int_0^1 \frac{\ln^2 x}{1-x} dx$$

$$= \sum_{m=0}^{\infty} \int_0^1 x^m \ln^2 x \, dx$$

$$= 2 \sum_{m=0}^{\infty} \frac{1}{(m+1)^3}$$

$$= 2\zeta(3).$$

Remark. We mention that the series from Problems **3.55**, **3.58**, and **3.59** (part (b)) are due to Klamkin [73–75].

3.56. A calculation, based on formula (3.17), shows that

$$\sum_{n=1}^{\infty} (-1)^{n-1} \frac{H_n}{n^2} = -\int_0^1 \frac{\ln(1-x)\ln(1+x)}{x} dx = \frac{5}{8}\zeta(3),$$

where the last equality can be checked with Maple.

3.57. We will be using the following identity[2] [16, Formula 5.2.9, p. 76], [62, Formula 1.513(6), p. 52]:

$$-\frac{\ln(1-x)}{1-x} = \sum_{n=1}^{\infty} H_n \cdot x^n, \quad -1 < x < 1. \tag{3.18}$$

We have, based on (3.17) and (3.18), that

$$\sum_{n=1}^{\infty} (-1)^{n-1} \frac{H_n^2}{n} = \sum_{n=1}^{\infty} (-1)^n H_n \int_0^1 x^{n-1} \ln(1-x)dx$$

$$= \int_0^1 \frac{\ln(1-x)}{x} \left(\sum_{n=1}^{\infty} (-x)^n H_n \right) dx$$

$$= -\int_0^1 \frac{\ln(1-x)\ln(1+x)}{x(1+x)} dx$$

$$= \int_0^1 \frac{\ln(1-x)\ln(1+x)}{1+x} dx - \int_0^1 \frac{\ln(1-x)\ln(1+x)}{x} dx$$

$$= -\frac{1}{12} \left(\pi^2 \ln 2 - 4\ln^3 2 - 9\zeta(3) \right),$$

since

$$\int_0^1 \frac{\ln(1-x)\ln(1+x)}{1+x} dx = \frac{1}{24} (8\ln^3 2 - \pi^2 \ln 4 + 3\zeta(3)),$$

the preceding integral being calculated with Mathematica.

3.58. Recall that the Dilogarithm function [84, 122], denoted by $\mathrm{Li}_2(z)$, is the special function defined by

$$\mathrm{Li}_2(z) = \sum_{n=1}^{\infty} \frac{z^n}{n^2} = -\int_0^z \frac{\ln(1-t)}{t} dt, \quad |z| \le 1. \tag{3.19}$$

A special value associated with this function is obtained when $z = 1$ for which we have that $\mathrm{Li}_2(1) = \sum_{k=1}^{\infty} 1/k^2 = \pi^2/6$.

We have, based on (3.17) and (3.19), that

$$\sum_{n=1}^{\infty} \frac{H_n}{n^3} = -\sum_{n=1}^{\infty} \frac{1}{n^2} \int_0^1 x^{n-1} \ln(1-x)dx$$

$$= -\int_0^1 \frac{\ln(1-x)}{x} \left(\sum_{n=1}^{\infty} \frac{x^n}{n^2} \right) dx$$

---

[2] The function $f(x) = -\frac{\ln(1-x)}{1-x}$ is known in the mathematical literature as the generating function for the nth harmonic number.

$$= -\int_0^1 \frac{\ln(1-x)}{x} \text{Li}_2(x) dx$$

$$= \left. \frac{\text{Li}_2^2(x)}{2} \right|_0^1$$

$$= \frac{\pi^4}{72}.$$

Remarks and Further Comments. The series from the previous problems, in which a single harmonic number is involved, are called *linear Euler sums.* Some of these series are special cases of the following classical series formula due to Euler

$$2\sum_{k=1}^{\infty} \frac{H_k}{k^n} = (n+2)\zeta(n+1) - \sum_{k=1}^{n-2} \zeta(n-k)\zeta(k+1) \quad (n \in \mathbb{N} \setminus \{1\}).$$

Many papers involving the calculation of linear Euler sums, as well as new proofs of the above formula, have been published in the mathematical literature; see [8, 24, 25, 70, 97] and the references given therein.

3.59. Let

$$S_k = \sum_{n=1}^{\infty} \frac{H_{n+k}}{n(n+1)(n+2)\cdots(n+k)}.$$

Since

$$\frac{1}{n(n+1)(n+2)\cdots(n+k)} = \frac{1}{k} \left(\frac{1}{n(n+1)\cdots(n+k-1)} - \frac{1}{(n+1)\cdots(n+k)} \right),$$

we get that

$$\frac{H_{n+k}}{n(n+1)(n+2)\cdots(n+k)} = \frac{1}{k} \cdot \frac{H_{n+k}}{n(n+1)\cdots(n+k-1)} - \frac{1}{k} \cdot \frac{H_{n+k}}{(n+1)\cdots(n+k)}$$

$$= \frac{1}{k} \cdot \frac{H_{n+k-1}}{n(n+1)\cdots(n+k-1)} + \frac{1}{k} \cdot \frac{1}{n(n+1)\cdots(n+k)}$$

$$- \frac{1}{k} \cdot \frac{H_{n+k}}{(n+1)\cdots(n+k)}.$$

It follows, in view of Problem **3.10**, that

$$S_k = \frac{1}{k} S_{k-1} + \frac{1}{k} \sum_{n=1}^{\infty} \frac{1}{n(n+1)\cdots(n+k)} - \frac{1}{k} \sum_{n=1}^{\infty} \frac{H_{n+k}}{(n+1)\cdots(n+k)}$$

$$= \frac{1}{k} S_{k-1} + \frac{1}{k^2 k!} - \frac{1}{k} \sum_{m=2}^{\infty} \frac{H_{m+k-1}}{m(m+1)\cdots(m+k-1)}$$

$$= \frac{1}{k}S_{k-1} + \frac{1}{k^2 k!} - \frac{1}{k}\left(S_{k-1} - \frac{H_k}{k!}\right)$$

$$= \frac{1}{k \cdot k!}\left(\frac{1}{k} + H_k\right).$$

3.60. Let

$$S_n = \sum_{k=1}^{\infty} \frac{H_k}{k(k+1)(k+2)\cdots(k+n)}.$$

We have

$$\frac{H_k}{k(k+1)\cdots(k+n)} = \frac{1}{n}\left(\frac{H_k}{k(k+1)\cdots(k+n-1)} - \frac{H_k}{(k+1)\cdots(k+n)}\right),$$

and it follows that

$$S_n = \frac{1}{n}\sum_{k=1}^{\infty} \frac{H_k}{k(k+1)(k+2)\cdots(k+n-1)} - \frac{1}{n}\sum_{k=1}^{\infty} \frac{H_k}{(k+1)(k+2)\cdots(k+n)}$$

$$= \frac{1}{n}S_{n-1} - \frac{1}{n}\sum_{m=2}^{\infty} \frac{H_{m-1}}{m(m+1)\cdots(m+n-1)}.$$

On the other hand,

$$\sum_{m=2}^{\infty} \frac{H_{m-1}}{m(m+1)\cdots(m+n-1)} = \sum_{m=2}^{\infty} \frac{H_m - \frac{1}{m}}{m(m+1)\cdots(m+n-1)}$$

$$= \sum_{m=2}^{\infty} \frac{H_m}{m(m+1)\cdots(m+n-1)}$$

$$- \sum_{m=2}^{\infty} \frac{1}{m^2(m+1)\cdots(m+n-1)}$$

$$= \sum_{m=1}^{\infty} \frac{H_m}{m(m+1)\cdots(m+n-1)} - \frac{1}{n!}$$

$$- \sum_{m=2}^{\infty} \frac{1}{m^2(m+1)\cdots(m+n-1)}$$

$$= S_{n-1} - \frac{1}{n!} - T_{n-1},$$

where

$$T_{n-1} = \sum_{m=2}^{\infty} \frac{1}{m^2(m+1)\cdots(m+n-1)}.$$

Putting all these together we get that

$$S_n = \frac{1}{n \cdot n!} + \frac{1}{n}T_{n-1}. \tag{3.20}$$

To calculate T_{n-1} we note that for $n \geq 3$, one has

$$\frac{1}{m^2(m+1)\cdots(m+n-1)}$$
$$= \frac{1}{n-1}\left(\frac{1}{m^2(m+1)\cdots(m+n-2)} - \frac{1}{m(m+1)\cdots(m+n-1)}\right).$$

Thus,

$$T_{n-1} = \frac{1}{n-1}T_{n-2} - \frac{1}{n-1}\sum_{m=2}^{\infty}\frac{1}{m(m+1)\cdots(m+n-1)}.$$

On the other hand,

$$\sum_{m=2}^{\infty}\frac{1}{m(m+1)\cdots(m+n-1)} = \sum_{m=1}^{\infty}\frac{1}{m(m+1)\cdots(m+n-1)} - \frac{1}{n!}$$
$$= \frac{1}{(n-1)\cdot(n-1)!} - \frac{1}{n!},$$

where the last equality follows based on Problem **3.10**. Thus, for $n \geq 3$, one has

$$T_{n-1} = \frac{T_{n-2}}{n-1} - \frac{1}{(n-1)^2 \cdot (n-1)!} + \frac{1}{(n-1) \cdot n!}.$$

This implies that

$$(n-1)!T_{n-1} = (n-2)!T_{n-2} - \frac{1}{(n-1)^2} + \frac{1}{n(n-1)},$$

and it follows, since

$$T_1 = \sum_{m=2}^{\infty}\frac{1}{m^2(m+1)} = \frac{\pi^2}{6} - \frac{3}{2},$$

that

$$T_{n-1} = \frac{1}{(n-1)!}\left(\frac{\pi^2}{6} - \frac{1}{n} - \left(\frac{1}{1^2} + \frac{1}{2^2} + \cdots + \frac{1}{(n-1)^2}\right)\right). \tag{3.21}$$

Combining (3.20) and (3.21), we get that the desired result holds and the problem is solved.

When $n = 1$ one has that

$$S_1 = \sum_{k=1}^{\infty} \frac{H_k}{k(k+1)} = \sum_{k=1}^{\infty} \left(\frac{H_k}{k} - \frac{H_{k+1}}{k+1} + \frac{1}{(k+1)^2} \right) = \frac{\pi^2}{6}.$$

Remark. The problem has been generalized in the following way [23,40]: For every natural number $n \geq 2$,

$$n! \sum_{k=1}^{\infty} \frac{H_k}{k(k+1)\cdots(k+n)} x^{k+n} = -\int_0^x \frac{(x-t)^n \ln(1-t)}{t(1-t)} dt.$$

3.61. Let

$$V_n = \sum_{k=1}^{\infty} \frac{H_k}{k^2(k+1)(k+2)\cdots(k+n)}.$$

We have, for $n \geq 2$, that

$$\frac{H_k}{k^2(k+1)(k+2)\cdots(k+n)}$$

$$= \frac{1}{n} \cdot \frac{H_k}{k^2(k+1)(k+2)\cdots(k+n-1)} - \frac{1}{n} \cdot \frac{H_k}{k(k+1)(k+2)\cdots(k+n)}$$

and it follows that $V_n = \frac{1}{n} V_{n-1} - \frac{1}{n} a_n$, where $a_n = \sum_{k=1}^{\infty} \frac{H_k}{k(k+1)(k+2)\cdots(k+n)}$. This implies that $n! \cdot V_n = (n-1)! \cdot V_{n-1} - (n-1)! \cdot a_n$. Thus,

$$n! \cdot V_n = V_1 - (1! \cdot a_2 + 2! \cdot a_3 + 3! \cdot a_4 + \cdots + (n-1)! \cdot a_n).$$

On the other hand,

$$V_1 = \sum_{k=1}^{\infty} \frac{H_k}{k^2(k+1)} = \sum_{k=1}^{\infty} \frac{H_k}{k^2} - \sum_{k=1}^{\infty} \frac{H_k}{k(k+1)} = 2\zeta(3) - \frac{\pi^2}{6}$$

and (see Problem **3.60**)

$$a_n = \frac{1}{n!} \left(\frac{\pi^2}{6} - \left(1 + \frac{1}{2^2} + \cdots + \frac{1}{(n-1)^2} \right) \right).$$

Putting all these together we get that the desired result holds and the problem is solved.

For an alternative solution, see [116].

3.62. We need the following integral [44, Table 5]:

$$\int_0^1 \frac{\ln x \cdot \mathrm{Li}_2(x)}{1-x} dx = -\frac{3}{4}\zeta(4).$$

Since

$$\int_0^1 y^{n+m-1}\ln y\,dy = -\frac{1}{(n+m)^2} \quad\text{and}\quad \zeta(2) - \sum_{k=1}^n \frac{1}{k^2} = \sum_{m=1}^\infty \frac{1}{(n+m)^2},$$

one has, based on (3.17), that

$$\frac{H_n\left(\zeta(2) - \sum_{k=1}^n 1/k^2\right)}{n} = \left(\int_0^1 x^{n-1}\ln(1-x)dx\right)\left(\sum_{m=1}^\infty \int_0^1 y^{n+m-1}\ln y\,dy\right)$$

$$= \int_0^1 x^{n-1}\ln(1-x)\,dx \int_0^1 \left(\sum_{m=1}^\infty y^{n+m-1}\right)\ln y\,dy$$

$$= \int_0^1 \int_0^1 x^{n-1}y^n \frac{\ln(1-x)\ln y}{1-y}dxdy.$$

Thus,

$$S = \sum_{n=1}^\infty \frac{H_n\left(\zeta(2) - \sum_{k=1}^n 1/k^2\right)}{n} = \sum_{n=1}^\infty \int_0^1 \int_0^1 x^{n-1}y^n \frac{\ln(1-x)\ln y}{1-y}dxdy$$

$$= \int_0^1 \int_0^1 \frac{y\ln y\ln(1-x)}{(1-y)(1-xy)}dxdy = \int_0^1 \frac{y\ln y}{1-y}\left(\int_0^1 \frac{\ln(1-x)}{1-xy}dx\right)dy.$$

To calculate the inner integral we note that an antiderivative of $\frac{\ln(1-x)}{1-xy}$ is the function

$$-\frac{\ln(1-x)\ln\left(\frac{xy-1}{y-1}\right) + \text{Li}_2\left(\frac{y(1-x)}{y-1}\right)}{y}.$$

It follows that

$$\int_0^1 \frac{\ln(1-x)}{1-xy}dx = \frac{\text{Li}_2\left(\frac{y}{y-1}\right)}{y}.$$

On the other hand, $\text{Li}_2\left(\frac{y}{y-1}\right) = -\frac{1}{2}\ln^2(1-y) - \text{Li}_2(y)$ [84], and it follows that

$$S = \int_0^1 \frac{y\ln y}{1-y}\cdot\frac{-\frac{1}{2}\ln^2(1-y) - \text{Li}_2(y)}{y}dy = -\frac{1}{2}\int_0^1 \frac{\ln y\ln^2(1-y)}{1-y}dy$$

$$- \int_0^1 \frac{\ln y\,\text{Li}_2(y)}{1-y}dy = -\frac{1}{2}\int_0^1 \frac{\ln y\ln^2(1-y)}{1-y}dy + \frac{3}{4}\zeta(4).$$

Let $I = \int_0^1 \frac{\ln y \ln^2(1-y)}{1-y} dy$. We integrate by parts and we have that

$$I = \frac{1}{3}\int_0^1 \frac{\ln^3(1-y)}{y}dy = \frac{1}{3}\int_0^1 \frac{\ln^3 x}{1-x}dx = \frac{1}{3}\int_0^1\left(\ln^3 x \sum_{m=0}^\infty x^m\right)dx$$

$$= \frac{1}{3}\sum_{m=0}^\infty \int_0^1 x^m \ln^3 x\,dx = -\frac{1}{3}\sum_{m=0}^\infty \frac{3!}{(m+1)^4} = -2\zeta(4),$$

since $\int_0^1 x^m \ln^3 x\,dx = -3!/(m+1)^4$. Thus, $S = -\frac{1}{2}I + \frac{3}{4}\zeta(4) = \frac{7}{4}\zeta(4)$.
For an alternative solution see [131].

3.63. See [11].

3.64. Let

$$S_n = \sum_{k=1}^n \frac{1}{k}\left(\ln 2 - \frac{1}{k+1} - \frac{1}{k+2} - \cdots - \frac{1}{2k}\right).$$

Using Abel's summation formula (A.1), with

$$a_k = \frac{1}{k} \quad \text{and} \quad b_k = \ln 2 - \frac{1}{k+1} - \frac{1}{k+2} - \cdots - \frac{1}{2k},$$

one has that

$$S_n = H_n \cdot \left(\ln 2 - \frac{1}{n+2} - \frac{1}{n+3} - \cdots - \frac{1}{2n+2}\right)$$

$$+ \sum_{k=1}^n\left(1 + \frac{1}{2} + \cdots + \frac{1}{k}\right)\frac{1}{(2k+1)(2k+2)}.$$

This implies, since

$$\lim_{n\to\infty} H_n\left(\ln 2 - \frac{1}{n+1} - \frac{1}{n+2} - \cdots - \frac{1}{2n}\right) = 0 \quad \text{(prove it!)},$$

that

$$\sum_{n=1}^\infty \frac{1}{n}\left(\ln 2 - \frac{1}{n+1} - \frac{1}{n+2} - \cdots - \frac{1}{2n}\right) = \sum_{n=1}^\infty \frac{1 + \frac{1}{2} + \cdots + \frac{1}{n}}{(2n+1)(2n+2)}.$$

To prove the second part of the problem we have, based on (3.18), that

$$-\frac{\ln(1-x^2)}{1-x^2} = \sum_{n=1}^\infty H_n \cdot x^{2n}, \quad -1 < x < 1.$$

It follows that

$$-\int_0^z\left(\int_0^t \frac{\ln(1-x^2)}{1-x^2}dx\right)dt = \sum_{n=1}^\infty \frac{H_n}{(2n+1)(2n+2)}z^{2n+2}.$$

When $z = 1$, this implies

$$-\int_0^1 \left(\int_0^t \frac{\ln(1-x^2)}{1-x^2} \, dx \right) dt = \sum_{n=1}^{\infty} \frac{H_n}{(2n+1)(2n+2)}.$$

We calculate the integral by parts with

$$f(t) = \int_0^t \frac{\ln(1-x^2)}{1-x^2} \, dx, \quad f'(t) = \frac{\ln(1-t^2)}{1-t^2}, \quad g'(t) = 1, \quad g(t) = t,$$

and we get that

$$\int_0^1 \left(\int_0^t \frac{\ln(1-x^2)}{1-x^2} \, dx \right) dt = \int_0^1 \frac{\ln(1-x^2)}{1-x^2} \, dx - \int_0^1 \frac{x\ln(1-x^2)}{1-x^2} \, dx$$

$$= \int_0^1 \frac{\ln(1-x^2)}{1+x} \, dx$$

$$= \int_0^1 \frac{\ln(1+x)}{1+x} \, dx + \int_0^1 \frac{\ln(1-x)}{1+x} \, dx$$

$$= \frac{\ln^2 2}{2} + \int_0^1 \frac{\ln(1-x)}{1+x} \, dx.$$

Thus,

$$\sum_{n=1}^{\infty} \frac{H_n}{(2n+1)(2n+2)} = -\frac{\ln^2 2}{2} - \int_0^1 \frac{\ln(1-x)}{1+x} \, dx = \frac{\pi^2}{12} - \ln^2 2,$$

since

$$\int_0^1 \frac{\ln(1-x)}{1+x} \, dx = -\frac{\pi^2}{12} + \frac{\ln^2 2}{2}.$$

For an alternative solution see [12].

3.65. and 3.66. See [79].

3.67. Let $S(k) = \sum_{n=1}^{\infty} \left(\ln k - H_{kn} + H_n - \frac{k-1}{2kn} \right)$. One can prove, based on [62, Formula 1.352(2), p. 37] and [62, Formula 1.392(1), p. 40], that the following formulae hold

$$\sum_{j=1}^{k-1} j \cos \frac{2m\pi j}{k} = -\frac{k}{2} \quad \text{and} \quad \prod_{m=1}^{k-1} \sin \frac{m\pi}{k} = \frac{k}{2^{k-1}}. \tag{3.22}$$

Recall that the Digamma function, ψ, verifies the identity [122, Formula 3, p. 14]

$$\psi(x) = \frac{\Gamma'(x)}{\Gamma(x)} = -\gamma - \frac{1}{x} + \sum_{n=1}^{\infty} \left(\frac{1}{n} - \frac{1}{n+x} \right), \quad x > 0.$$

For p and q positive integers, with $p < q$, one has the *Gauss formula*, [122, Formula 47, p. 19], which allows the calculation of the Digamma function at positive rationals less than 1

$$\psi\left(\frac{p}{q}\right) = -\gamma - \frac{\pi}{2}\cot\frac{p\pi}{q} - \ln q + \sum_{k=1}^{q-1}\cos\frac{2k\pi p}{q}\ln\left(2\sin\frac{k\pi}{q}\right). \qquad (3.23)$$

We have, based on Abel's summation formula (A.2), that

$$S(k) = \sum_{n=1}^{\infty} n\left(\ln k - H_{kn} + H_n - \frac{k-1}{2kn} - \ln k + H_{k(n+1)} - H_{n+1} + \frac{k-1}{2k(n+1)}\right)$$

$$= \sum_{n=1}^{\infty} n\left(\frac{1}{kn+1} + \frac{1}{kn+2} + \cdots + \frac{1}{kn+k-1} - \frac{k-1}{kn+k} - \frac{k-1}{2kn(n+1)}\right)$$

$$= \sum_{n=1}^{\infty} n\left(\sum_{j=1}^{k-1}\left(\frac{1}{kn+j} - \frac{1}{kn+k}\right) - \frac{k-1}{2kn(n+1)}\right)$$

$$= \sum_{n=1}^{\infty}\left(\sum_{j=1}^{k-1}\left(\frac{n}{kn+j} - \frac{n}{kn+k}\right) - \frac{k-1}{2k(n+1)}\right)$$

$$= \sum_{n=1}^{\infty}\left(\frac{1}{k}\sum_{j=1}^{k-1}\left(\frac{1}{n+1} - \frac{j/k}{n+j/k}\right) - \frac{k-1}{2k(n+1)}\right)$$

$$= \sum_{n=1}^{\infty}\left(\frac{1}{k}\sum_{j=1}^{k-1}\left(\frac{j/k}{n} - \frac{j/k}{n+j/k}\right) + \frac{1}{k}\sum_{j=1}^{k-1}\left(\frac{1}{n+1} - \frac{j/k}{n}\right) - \frac{k-1}{2k(n+1)}\right)$$

$$= \sum_{n=1}^{\infty}\left(\frac{1}{k}\sum_{j=1}^{k-1}\left(\frac{j/k}{n} - \frac{j/k}{n+j/k}\right) + \frac{k-1}{2k}\left(\frac{1}{n+1} - \frac{1}{n}\right)\right)$$

$$= \frac{1}{k}\sum_{j=1}^{k-1}\frac{j}{k}\sum_{n=1}^{\infty}\left(\frac{1}{n} - \frac{1}{n+j/k}\right) - \frac{k-1}{2k}$$

$$= \frac{1}{k^2}\sum_{j=1}^{k-1} j\left(\psi\left(\frac{j}{k}\right) + \gamma + \frac{k}{j}\right) - \frac{k-1}{2k}$$

$$= \frac{1}{k^2}\sum_{j=1}^{k-1} j\left(\psi\left(\frac{j}{k}\right) + \gamma\right) + \frac{k-1}{2k}.$$

Using (3.23) we get that

$$\psi\left(\frac{j}{k}\right) + \gamma = -\frac{\pi}{2}\cot\frac{j\pi}{k} - \ln k + \sum_{m=1}^{k-1}\cos\frac{2mj\pi}{k}\ln\left(2\sin\frac{m\pi}{k}\right).$$

It follows, based on (3.22), that

$$\frac{1}{k^2}\sum_{j=1}^{k-1}j\left(\psi\left(\frac{j}{k}\right)+\gamma\right) = -\frac{\pi}{2k^2}\sum_{j=1}^{k-1}j\cot\frac{j\pi}{k} - \frac{k-1}{2k}\ln k$$

$$+\frac{1}{k^2}\sum_{m=1}^{k-1}\ln\left(2\sin\frac{m\pi}{k}\right)\sum_{j=1}^{k-1}j\cos\frac{2m\pi j}{k}$$

$$= -\frac{\pi}{2k^2}\sum_{j=1}^{k-1}j\cot\frac{j\pi}{k} - \frac{k-1}{2k}\ln k - \frac{1}{2k}\sum_{m=1}^{k-1}\ln\left(2\sin\frac{m\pi}{k}\right)$$

$$= -\frac{\pi}{2k^2}\sum_{j=1}^{k-1}j\cot\frac{j\pi}{k} - \frac{k-1}{2k}\ln k$$

$$-\frac{1}{2k}\ln\left(2^{k-1}\prod_{m=1}^{k-1}\sin\frac{m\pi}{k}\right)$$

$$= -\frac{\pi}{2k^2}\sum_{j=1}^{k-1}j\cot\frac{j\pi}{k} - \frac{k-1}{2k}\ln k - \frac{\ln k}{2k}$$

$$= -\frac{\pi}{2k^2}\sum_{j=1}^{k-1}j\cot\frac{j\pi}{k} - \frac{\ln k}{2}.$$

This implies that

$$S(k) = \frac{k-1}{2k} - \frac{\pi}{2k^2}\sum_{j=1}^{k-1}j\cot\frac{j\pi}{k} - \frac{\ln k}{2}.$$

When $k = 2$ or $k = 3$, we get the following series formulae:

$$\sum_{n=1}^{\infty}\left(\ln 2 - H_{2n} + H_n - \frac{1}{4n}\right) = \frac{1}{4} - \frac{\ln 2}{2}$$

and

$$\sum_{n=1}^{\infty}\left(\ln 3 - H_{3n} + H_n - \frac{1}{3n}\right) = \frac{1}{3} + \frac{\pi\sqrt{3}}{54} - \frac{\ln 3}{2}.$$

Remark. It is not hard to check that the series $\sum_{n=1}^{\infty}(\ln k - H_{kn} + H_n)$ diverges, hence the correctional term $(k-1)/2kn$ which makes the series convergent.

3.68. and **3.69.** These problems, which are due to Kouba, are solved in [78].

3.70. Recall that the Dilogarithm function [84, 122] denoted by $\mathrm{Li}_2(z)$ is the special function defined, for $|z| \le 1$, by

$$\mathrm{Li}_2(z) = \sum_{n=1}^{\infty} \frac{z^n}{n^2} = -\int_0^z \frac{\ln(1-t)}{t} dt.$$

When $z = 1$ we have that $\mathrm{Li}_2(1) = \sum_{k=1}^{\infty} 1/k^2 = \pi^2/6$.

If $x \in (0,1)$, the following identity holds

$$\ln x \ln(1-x) - \frac{1}{2}\ln^2(1-x) + \mathrm{Li}_2(x) + \int_1^{1/(1-x)} \frac{\ln(u-1)}{u} du = 0. \qquad (3.24)$$

To prove this identity we let $f : (0,1) \to \mathbb{R}$ be the function defined by the left hand side of (3.24). Then, a straightforward calculation shows that $f'(x) = 0$. Thus, f is a constant function and hence $f(x) = \lim_{x \to 0} f(x) = 0$.

We also need the following integral (see entry 4.262(2) of [62]):

$$\int_0^1 \frac{\ln^3(1-x)}{x} dx = -\frac{\pi^4}{15}.$$

Now we are ready to solve the problem. We have, based on (3.17), that

$$\sum_{n=1}^{\infty} \left(\frac{H_n}{n}\right)^2 = \sum_{n=1}^{\infty} \int_0^1 x^{n-1}\ln(1-x)dx \int_0^1 y^{n-1}\ln(1-y)dy$$

$$= \int_0^1 \int_0^1 \left(\sum_{n=1}^{\infty}(xy)^{n-1}\right) \ln(1-x)\ln(1-y)dxdy$$

$$= \int_0^1 \int_0^1 \frac{\ln(1-x)\ln(1-y)}{1-xy} dxdy$$

$$= \int_0^1 \ln(1-x)\left(\int_0^1 \frac{\ln(1-y)}{1-xy}dy\right) dx.$$

We calculate the inner integral, by making the substitution $1 - xy = t$, and we have

$$\int_0^1 \frac{\ln(1-y)}{1-xy}dy = \frac{1}{x}\int_{1-x}^1 \frac{\ln(1-1/x+t/x)}{t}dt$$

$$= \frac{1}{x}\int_{1-x}^1 \frac{\ln(1/x)}{t}dt + \frac{1}{x}\int_{1-x}^1 \frac{\ln(x-1+t)}{t}dt$$

$$= \frac{1}{x}\ln x \cdot \ln(1-x) + \frac{1}{x}\int_{1-x}^1 \frac{\ln(x-1+t)}{t}dt.$$

The substitution $t = (1-x)u$, in the preceding integral, combined with (3.24) implies that

$$\int_0^1 \frac{\ln(1-y)}{1-xy}\,dy = \frac{1}{x}\ln x\ln(1-x) + \frac{1}{x}\int_1^{1/(1-x)} \frac{\ln(1-x)(u-1)}{u}\,du$$

$$= \frac{1}{x}\left(\ln x\ln(1-x) - \ln^2(1-x) + \int_1^{1/(1-x)} \frac{\ln(u-1)}{u}\,du\right)$$

$$= \frac{1}{x}\left(-\frac{1}{2}\ln^2(1-x) - \operatorname{Li}_2(x)\right).$$

It follows that

$$\sum_{n=1}^{\infty}\left(\frac{H_n}{n}\right)^2 = -\frac{1}{2}\int_0^1 \frac{\ln^3(1-x)}{x}\,dx - \int_0^1 \frac{\ln(1-x)\operatorname{Li}_2(x)}{x}\,dx$$

$$= \frac{\pi^4}{30} + \frac{1}{2}\left(\operatorname{Li}_2(x)\right)^2 \Big|_{x=0}^{x=1}$$

$$= \frac{\pi^4}{30} + \frac{\pi^4}{72}$$

$$= \frac{17\pi^4}{360}.$$

Remark. Historically, this series identity was discovered numerically by Enrico Au–Yeung, an undergraduate student in the Faculty of Mathematics in Waterloo, and proved rigorously by Borwein and Borwein in [17]. This intriguing series, involving the square of the nth harmonic number, has become a classic in the theory of *nonlinear harmonic series*. It appears recorded as an independent study problem in [108, Problem 2.6.1, p. 110] and, as a formula, in [98, Entry 25.16.13, p. 614]. We mention that a nonlinear harmonic series is a series involving products of at least two harmonic numbers. For reference materials on this type of series, as well as surprising evaluations of sums involving harmonic numbers, the reader should refer to the work of Srivastava and Choi [25], [123, pp. 363, 364].

3.71. and **3.72.** These problems are solved in [47].

3.9.5   Series of Functions

> *There are things which seem incredible to most men who have not studied Mathematics.*
>
> Archimedes of Syracuse (287–212 BC)

3.73. The series equals

$$\frac{1}{x^n}\left(\sum_{j=2}^{n}(-1)^{j+n}x^{j-1}\zeta(j) + (-1)^{n+1}\left(\psi(x) + \frac{1}{x} + \gamma\right)\right),$$

where the parenthesis contains only the term $(-1)^{n+1}(\psi(x)+1/x+\gamma)$ when $n=1$. See the solution of Problem **3.9**.

3.74. The series equals $\frac{x^{2k}}{(1-x)^k(1-x^k)}$. We have

$$\sum_{n=1}^{\infty}\left(\frac{1}{1-x}-1-x-x^2-\cdots-x^n\right)^k = \sum_{n=1}^{\infty}\left(\frac{x^{n+1}}{1-x}\right)^k = \left(\frac{x}{1-x}\right)^k\sum_{n=1}^{\infty}x^{nk}$$

$$= \frac{x^{2k}}{(1-x)^k(1-x^k)}.$$

3.75. The series equals $-\frac{x^k}{(1-x)^k}\ln(1-x^k)$. We have

$$\sum_{n=1}^{\infty}\frac{1}{n}\left(\frac{1}{1-x}-1-x-x^2-\cdots-x^n\right)^k = \sum_{n=1}^{\infty}\frac{1}{n}\left(\frac{x^{n+1}}{1-x}\right)^k = \frac{x^k}{(1-x)^k}\sum_{n=1}^{\infty}\frac{x^{kn}}{n}$$

$$= -\frac{x^k}{(1-x)^k}\ln(1-x^k).$$

3.76. Since $1+t+\cdots+t^{n-1} = \frac{1-t^n}{1-t}$, we obtain by integration that

$$\sum_{k=1}^{n}\frac{x^k}{k} = -\int_0^x\frac{t^n}{1-t}dt - \ln(1-x).$$

Thus,

$$\sum_{n=1}^{\infty}\frac{1}{n}\left(\sum_{k=1}^{n}\frac{x^k}{k}-\ln\frac{1}{1-x}\right) = -\sum_{n=1}^{\infty}\frac{1}{n}\left(\int_0^x\frac{t^n}{1-t}dt\right) = -\int_0^x\frac{1}{1-t}\left(\sum_{n=1}^{\infty}\frac{t^n}{n}\right)dt$$

$$= \int_0^x\frac{\ln(1-t)}{1-t}dt = -\frac{\ln^2(1-x)}{2}.$$

Remark. The interchange of the order of summation and the order of integration is permitted since $\sum_{n=1}^{\infty}t^n/n = -\ln(1-t)$ converges on $-1 \le t < 1$, and hence, the series converges uniformly on $[0,x] \subset (-1,1)$. When $x = -1$, the equality follows from Abel's summation formula.

3.77. We have

$$\sum_{n=1}^{\infty}\frac{1}{n}\left(\sum_{k=0}^{n-1}x^{pk}-\frac{1}{1-x^p}\right) = \sum_{n=1}^{\infty}\frac{1}{n}\left(\frac{1-x^{pn}}{1-x^p}-\frac{1}{1-x^p}\right)$$

$$= -\frac{1}{1-x^p}\sum_{n=1}^{\infty}\frac{x^{pn}}{n}$$

$$= \frac{\ln(1-x^p)}{1-x^p}.$$

The second part of the problem follows from the first part by replacing x^p by $-x^p$.

3.78. The series equals

$$\frac{x}{2(1+x)} - \frac{1}{4}\ln\frac{1+x}{1-x}.$$

Let S be the sum of the series and let S_{2n} be its $2n$th partial sum. We have

$$S_{2n} = \sum_{k=1}^{2n}(-1)^{k-1}k\left(\ln(1-x)+x+\frac{x^2}{2}+\cdots+\frac{x^k}{k}\right)$$

$$= \sum_{k=1}^{n}(2k-1)\left(\ln(1-x)+x+\frac{x^2}{2}+\cdots+\frac{x^{2k-1}}{2k-1}\right)$$

$$-\sum_{k=1}^{n}2k\left(\ln(1-x)+x+\frac{x^2}{2}+\cdots+\frac{x^{2k}}{2k}\right)$$

$$= -\sum_{k=1}^{n}x^{2k}-\sum_{k=1}^{n}\left(\ln(1-x)+x+\frac{x^2}{2}+\cdots+\frac{x^{2k-1}}{2k-1}\right).$$

It follows that

$$S = -\sum_{k=1}^{\infty}x^{2k}-\sum_{k=1}^{\infty}\left(\ln(1-x)+x+\frac{x^2}{2}+\cdots+\frac{x^{2k-1}}{2k-1}\right)$$

$$= -\frac{x^2}{1-x^2}+\sum_{k=1}^{\infty}\sum_{m=0}^{\infty}\frac{x^{2k+m}}{2k+m}$$

$$= -\frac{x^2}{1-x^2}+\sum_{k=1}^{\infty}\sum_{m=0}^{\infty}\int_0^x t^{2k+m-1}dt$$

$$= -\frac{x^2}{1-x^2}+\int_0^x\left(\sum_{k=1}^{\infty}t^{2k-1}\right)\left(\sum_{m=0}^{\infty}t^m\right)dt$$

$$= -\frac{x^2}{1-x^2}+\int_0^x\frac{t}{(1-t)(1-t^2)}dt$$

$$= \frac{x}{2(1+x)}-\frac{1}{4}\ln\frac{1+x}{1-x}.$$

3.79. The series equals

$$-\ln^2(1-x)+\frac{1+x}{1-x}\ln(1-x^2)-\frac{2x}{1-x}\ln(1-x).$$

We have

$$\ln\frac{1}{1-x} - x - \frac{x^2}{2} - \cdots - \frac{x^n}{n} = \sum_{k=1}^{\infty}\frac{x^{k+n}}{k+n} = \sum_{k=1}^{\infty}\int_0^x t^{k+n-1}dt = \int_0^x \frac{t^n}{1-t}dt.$$

Thus,

$$\sum_{n=1}^{\infty}\left(\ln\frac{1}{1-x} - x - \frac{x^2}{2} - \cdots - \frac{x^n}{n}\right)^2 = \sum_{n=1}^{\infty}\int_0^x \frac{t^n}{1-t}dt\int_0^x \frac{u^n}{1-u}du$$

$$= \int_0^x\int_0^x \frac{1}{(1-t)(1-u)}\sum_{n=1}^{\infty}(tu)^n dtdu$$

$$= \int_0^x\int_0^x \frac{tu}{(1-tu)(1-t)(1-u)}dtdu$$

$$= \int_0^x \frac{t}{1-t}\left(\int_0^x \frac{u}{(1-tu)(1-u)}du\right)dt.$$

On the other hand, a calculation shows that

$$\int_0^x \frac{u}{(1-tu)(1-u)}du = \frac{1}{1-t}\left(-\ln(1-x) + \frac{1}{t}\ln(1-tx)\right).$$

It follows that

$$\int_0^x \frac{t}{1-t}\left(\int_0^x \frac{u}{(1-tu)(1-u)}du\right)dt = \int_0^x \frac{t}{1-t}\left(\frac{-\ln(1-x)}{1-t} + \frac{\ln(1-xt)}{t(1-t)}\right)dt$$

$$= -\ln(1-x)\int_0^x \frac{t}{(1-t)^2}dt + \int_0^x \frac{\ln(1-xt)}{(1-t)^2}dt$$

$$= -\ln(1-x)\left(\ln(1-x) + \frac{x}{1-x}\right) + \left(\left.\frac{\ln(1-xt)}{1-t}\right|_{t=0}^{t=x} + x\int_0^x \frac{dt}{(1-tx)(1-t)}\right)$$

$$= -\ln(1-x)\left(\ln(1-x) + \frac{x}{1-x}\right) + \frac{\ln(1-x^2)}{1-x} + x\int_0^x \frac{1}{x-1}\left(\frac{x}{1-tx} - \frac{1}{1-t}\right)dt$$

$$= -\ln(1-x)\left(\ln(1-x) + \frac{x}{1-x}\right) + \frac{(1+x)\ln(1-x^2)}{1-x} - \frac{x}{1-x}\ln(1-x)$$

$$= -\ln^2(1-x) + \frac{1+x}{1-x}\ln(1-x^2) - \frac{2x}{1-x}\ln(1-x).$$

3.80. The series equals

$$\frac{x^2}{(1-x)^2}\left(\frac{\ln(1-x^2)}{x} - \ln(1-x)\right).$$

We have

$$\sum_{n=1}^{\infty}\left(\ln\frac{1}{1-x}-x-\frac{x^2}{2}-\cdots-\frac{x^n}{n}\right)\left(\frac{1}{1-x}-1-x-x^2-\cdots-x^n\right)$$

$$=\sum_{n=1}^{\infty}\left(\sum_{k=1}^{\infty}\frac{x^{n+k}}{n+k}\right)\left(\sum_{p=1}^{\infty}x^{p+n}\right)$$

$$=\sum_{n=1}^{\infty}\left(\sum_{k=1}^{\infty}\int_0^x t^{n+k-1}dt\right)\left(\sum_{p=1}^{\infty}x^{n+p}\right)$$

$$=\int_0^x\left(\sum_{n=1}^{\infty}(tx)^n\sum_{k=1}^{\infty}t^{k-1}\sum_{p=1}^{\infty}x^p\right)dt$$

$$=\int_0^x\frac{tx}{1-tx}\cdot\frac{1}{1-t}\cdot\frac{x}{1-x}dt$$

$$=\frac{x^2}{1-x}\int_0^x\frac{t}{(1-tx)(1-t)}dt$$

$$=\frac{x^2}{1-x}\int_0^x\frac{1}{x-1}\left(\frac{1}{1-tx}-\frac{1}{1-t}\right)dt$$

$$=\frac{x^2}{(1-x)^2}\left(\frac{\ln(1-x^2)}{x}-\ln(1-x)\right).$$

Remark. One can also prove that if $a\in[-1,1]$ and $a\neq0$, then

$$\sum_{n=1}^{\infty}a^n\left(\ln\frac{1}{1-x}-x-\frac{x^2}{2}-\cdots-\frac{x^n}{n}\right)\left(\frac{1}{1-x}-1-x-x^2-\cdots-x^n\right)$$

$$=\frac{ax^2}{(1-x)(ax-1)}\left(\ln(1-x)-\frac{\ln(1-ax^2)}{ax}\right).$$

3.81. (a) Let

$$T_k(x)=\sum_{n=1}^{\infty}\frac{e^{inx}}{n(n+1)(n+2)\cdots(n+k)}.$$

First we note that

$$\frac{e^{inx}}{n(n+1)(n+2)\cdots(n+k)}$$

$$=\frac{1}{k}\left(\frac{e^{inx}}{n(n+1)(n+2)\cdots(n+k-1)}-\frac{e^{inx}}{(n+1)(n+2)\cdots(n+k)}\right),$$

and it follows that

$$
\begin{aligned}
T_k(x) &= \frac{1}{k}\left(T_{k-1}(x) - \sum_{n=1}^{\infty} \frac{e^{inx}}{(n+1)(n+2)\cdots(n+k)} \right) \\
&\overset{n+1=m}{=} \frac{1}{k}\left(T_{k-1}(x) - \sum_{m=2}^{\infty} \frac{e^{i(m-1)x}}{m(m+1)(m+2)\cdots(m+k-1)} \right) \\
&= \frac{1}{k}\left(T_{k-1}(x) - e^{-ix}\left(T_{k-1}(x) - \frac{e^{ix}}{k!} \right) \right) \\
&= \frac{1-e^{-ix}}{k} T_{k-1}(x) + \frac{1}{k\cdot k!}.
\end{aligned}
$$

Let $a = 1 - e^{-ix} = 2\sin\frac{x}{2}\cdot e^{\frac{i(\pi-x)}{2}}$ and let $x_k = k!T_k(x)/a^k$. The recurrence relation implies that $x_k = x_{k-1} + 1/ka^k$, from which it follows that $x_k = x_0 + \sum_{j=1}^{k} 1/ja^j$. On the other hand, $x_0 = T_0(x) = \sum_{n=1}^{\infty} e^{ixn}/n$. Thus,

$$
T_k(x) = \frac{a^k}{k!}\left(\sum_{n=1}^{\infty} \frac{e^{ixn}}{n} + \sum_{j=1}^{k} \frac{1}{ja^j} \right).
$$

It follows that

$$
\begin{aligned}
&\text{It follows that} \\
T_k(x) &= \frac{(2\sin\frac{x}{2})^k}{k!} e^{\frac{i(\pi-x)}{2}k}\left(\sum_{n=1}^{\infty} \frac{e^{ixn}}{n} + \sum_{j=1}^{k} \frac{e^{-\frac{ij(\pi-x)}{2}}}{j(2\sin\frac{x}{2})^j} \right).
\end{aligned}
$$

Thus, for calculating $S_k(x)$ we need to calculate the real part of $T_k(x)$. Hence,

$$
S_k(x) = \Re\left(T_k(x)\right) = \Re\left(\frac{(2\sin\frac{x}{2})^k}{k!} e^{\frac{i(\pi-x)}{2}k}\left(\sum_{n=1}^{\infty} \frac{e^{ixn}}{n} + \sum_{j=1}^{k} \frac{e^{-\frac{ij(\pi-x)}{2}}}{j(2\sin\frac{x}{2})^j} \right) \right). \tag{3.25}
$$

We need the following two well-known Fourier series for sine and cosine [62, 1.441(1), 1.441(2), p. 44]:

$$
\sum_{n=1}^{\infty} \frac{\sin nx}{n} = \frac{\pi - x}{2}, \quad 0 < x < 2\pi,
$$

and

$$
\sum_{n=1}^{\infty} \frac{\cos nx}{n} = -\ln\left(2\sin\frac{x}{2}\right), \quad 0 < x < 2\pi.
$$

A calculation, based on (3.25) and the previous formulae, shows that

$$S_k(x) = \frac{(2\sin\frac{x}{2})^k}{k!}$$

$$\times \left(-\cos\frac{(\pi-x)k}{2}\ln\left(2\sin\frac{x}{2}\right) - \frac{\pi-x}{2}\sin\frac{(\pi-x)k}{2} + \sum_{j=1}^{k}\frac{\cos\frac{(\pi-x)(j-k)}{2}}{j\left(2\sin\frac{x}{2}\right)^j} \right).$$

(b) To solve the second part of the problem, we let $x = \pi$ and we get that

$$S_k(\pi) = \sum_{n=1}^{\infty}\frac{(-1)^n}{n(n+1)(n+2)\cdots(n+k)} = \frac{2^k}{k!}\left(-\ln 2 + \sum_{j=1}^{k}\frac{1}{j2^j}\right).$$

3.82. See [26, Proposition 4].

3.83. We have

$$\sum_{n=1}^{\infty}\sum_{m=1}^{\infty}\left(\frac{1}{1-x}-1-x-x^2-\cdots-x^{n+m}\right) = \sum_{n=1}^{\infty}\sum_{m=1}^{\infty}\sum_{p=0}^{\infty}x^{n+m+p+1}$$

$$= \sum_{n=1}^{\infty}x^n\sum_{m=1}^{\infty}x^m\sum_{p=0}^{\infty}x^{p+1}$$

$$= \left(\frac{x}{1-x}\right)^3.$$

On the other hand,

$$\sum_{n=1}^{\infty}\sum_{m=1}^{\infty}\left(\ln\frac{1}{1-x}-x-\frac{x^2}{2}-\cdots-\frac{x^{n+m}}{n+m}\right) = \sum_{n=1}^{\infty}\sum_{m=1}^{\infty}\left(\sum_{p=0}^{\infty}\frac{x^{n+m+p+1}}{n+m+p+1}\right)$$

$$= \sum_{n=1}^{\infty}\sum_{m=1}^{\infty}\left(\sum_{p=0}^{\infty}\int_0^x t^{n+m+p}dt\right)$$

$$= \int_0^x\left(\sum_{n=1}^{\infty}t^n\right)\left(\sum_{m=1}^{\infty}t^m\right)\left(\sum_{p=0}^{\infty}t^p\right)dt$$

$$= \int_0^x\frac{t^2}{(1-t)^3}dt$$

$$= \frac{3x^2-2x}{2(1-x)^2}-\ln(1-x).$$

3.84. Let $S(x)$ be the sum of the series. We prove that

$$S(x) = \begin{cases} x + \frac{x^3}{3!} + \cdots + \frac{x^p}{p!} - \sinh x & \text{if } p \text{ is odd,} \\ 1 + \frac{x^2}{2!} + \cdots + \frac{x^p}{p!} - \cosh x & \text{if } p \text{ is even.} \end{cases}$$

Let $S_n(x) = \sum_{k=1}^{n} (-1)^k \left(e^x - 1 - \frac{x}{1!} - \frac{x^2}{2!} - \cdots - \frac{x^{k+p}}{(k+p)!} \right)$. We apply Abel's summation formula (A.1), with

$$a_k = (-1)^k \quad \text{and} \quad b_k = e^x - 1 - \frac{x}{1!} - \frac{x^2}{2!} - \cdots - \frac{x^{k+p}}{(k+p)!},$$

and we get that

$$S_n(x) = \left((-1) + (-1)^2 + \cdots + (-1)^n \right) \left(e^x - 1 - \frac{x}{1!} - \frac{x^2}{2!} - \cdots - \frac{x^{n+1+p}}{(n+1+p)!} \right)$$

$$+ \sum_{k=1}^{n} \left((-1) + (-1)^2 + \cdots + (-1)^k \right) \frac{x^{k+p+1}}{(k+p+1)!}.$$

This implies that

$$S(x) = - \sum_{\substack{k=1 \\ k=\text{odd}}}^{\infty} \frac{x^{k+p+1}}{(k+p+1)!} = - \sum_{l=0}^{\infty} \frac{x^{2l+p+2}}{(2l+p+2)!}$$

$$= \begin{cases} x + \frac{x^3}{3!} + \cdots + \frac{x^p}{p!} - \sinh x & \text{if } p \text{ is odd,} \\ 1 + \frac{x^2}{2!} + \cdots + \frac{x^p}{p!} - \cosh x & \text{if } p \text{ is even.} \end{cases}$$

3.85. The series equals

$$\frac{e^x}{4} - \left(\frac{x}{2} + \frac{1}{4} \right) e^{-x}.$$

Let $S(x)$ be the sum of the series. We have

$$S'(x) = \sum_{n=1}^{\infty} (-1)^{n-1} n \left(e^x - 1 - x - \frac{x^2}{2!} - \cdots - \frac{x^{n-1}}{(n-1)!} \right)$$

$$= S(x) + \sum_{n=1}^{\infty} (-1)^{n-1} \frac{x^n}{(n-1)!}$$

$$= S(x) + xe^{-x}.$$

Thus, $S'(x) = S(x) + xe^{-x}$, and it follows that $S(x) = Ce^x - (x/2 + 1/4)e^{-x}$, where C is the constant of integration. Since $S(0) = 0$, we obtain that $C = 1/4$, and the problem is solved.

3.86. The series equals

$$\left(\frac{x^3}{3}+\frac{x^2}{2}\right)e^x.$$

Let $S(x)$ be the sum of the series. We have

$$S'(x) = \sum_{n=1}^{\infty} n^2\left(e^x - 1 - \frac{x}{1!} - \frac{x^2}{2!} - \cdots - \frac{x^{n-1}}{(n-1)!}\right)$$

$$= S(x) + \sum_{n=1}^{\infty} n^2 \cdot \frac{x^n}{n!}$$

$$= S(x) + \sum_{n=2}^{\infty} \frac{x^n}{(n-2)!} + \sum_{n=1}^{\infty} \frac{x^n}{(n-1)!}.$$

It follows that $S'(x) = S(x) + (x^2+x)e^x$, and hence $S(x) = (x^3/3+x^2/2)\,e^x + Ce^x$, where C is the constant of integration. Since $S(0) = 0$, we obtain that $C = 0$ and the problem is solved.

3.87. The sum equals

$$\begin{cases} e^{-x} - \left(1+\frac{x^2}{2!}+\cdots+\frac{x^p}{p!}\right) & \text{if } p \text{ is even,} \\ -\left(x+\frac{x^3}{3!}+\cdots+\frac{x^p}{p!}\right) & \text{if } p \text{ is odd.} \end{cases}$$

Let $S_n(x) = \sum_{k=1}^n (-1)^k \left(\frac{x^k}{k!} + \frac{x^{k+1}}{(k+1)!} + \cdots + \frac{x^{k+p}}{(k+p)!}\right)$. Using Abel's summation formula (A.1), with

$$a_k = (-1)^k \quad \text{and} \quad b_k = \frac{x^k}{k!} + \frac{x^{k+1}}{(k+1)!} + \cdots + \frac{x^{k+p}}{(k+p)!},$$

we get that

$$S_n(x) = \left((-1)+(-1)^2+\cdots+(-1)^n\right)\left(\frac{x^{n+1}}{(n+1)!} + \frac{x^{n+2}}{(n+2)!} + \cdots + \frac{x^{n+1+p}}{(n+1+p)!}\right)$$

$$+ \sum_{k=1}^n \left((-1)+(-1)^2+\cdots+(-1)^k\right)\left(\frac{x^k}{k!} - \frac{x^{k+p+1}}{(k+p+1)!}\right).$$

Thus,

$$\lim_{n\to\infty} S_n(x) = \sum_{\substack{k=1 \\ k=2l+1}}^{\infty} \left(\frac{x^{k+p+1}}{(k+p+1)!} - \frac{x^k}{k!}\right)$$

$$= \sum_{l=0}^{\infty} \left(\frac{x^{2l+p+2}}{(2l+p+2)!} - \frac{x^{2l+1}}{(2l+1)!} \right)$$

$$= \begin{cases} e^{-x} - \left(1 + \frac{x^2}{2!} + \cdots + \frac{x^p}{p!}\right) & \text{if } p \text{ is even,} \\ -\left(x + \frac{x^3}{3!} + \cdots + \frac{x^p}{p!}\right) & \text{if } p \text{ is odd.} \end{cases}$$

3.89. (a) Let

$$y(x) = \sum_{n=0}^{\infty} f^{(n)}(0) \left(e^x - 1 - \frac{x}{1!} - \frac{x^2}{2!} - \cdots - \frac{x^n}{n!} \right).$$

We have

$$y'(x) = f(0)e^x + \sum_{n=1}^{\infty} f^{(n)}(0) \left(e^x - 1 - \frac{x}{1!} - \frac{x^2}{2!} - \cdots - \frac{x^{n-1}}{(n-1)!} \right)$$

$$= f(0)e^x + \sum_{n=1}^{\infty} f^{(n)}(0) \left(e^x - 1 - \frac{x}{1!} - \frac{x^2}{2!} - \cdots - \frac{x^{n-1}}{(n-1)!} - \frac{x^n}{n!} + \frac{x^n}{n!} \right)$$

$$= (e^x - 1)f(0) + \sum_{n=1}^{\infty} f^{(n)}(0) \left(e^x - 1 - \frac{x}{1!} - \frac{x^2}{2!} - \cdots - \frac{x^{n-1}}{(n-1)!} - \frac{x^n}{n!} \right)$$

$$+ f(0) + \sum_{n=1}^{\infty} f^{(n)}(0) \cdot \frac{x^n}{n!}$$

$$= \sum_{n=0}^{\infty} f^{(n)}(0) \left(e^x - 1 - \frac{x}{1!} - \frac{x^2}{2!} - \cdots - \frac{x^{n-1}}{(n-1)!} - \frac{x^n}{n!} \right) + f(x)$$

$$= y(x) + f(x).$$

The solution of this linear differential equation, with initial condition $y(0) = 0$, is

$$y(x) = \int_0^x e^{x-t} f(t) dt,$$

and part (a) of the problem is solved.

For an alternative solution of this part of the problem, which is based on a recurrence formula, see [103].

(b) Let $f(x) = e^{\alpha x}$. Since $f^{(n)}(0) = \alpha^n$, it follows in view of part (a) of the problem that

$$\sum_{n=0}^{\infty} \alpha^n \left(e^x - 1 - \frac{x}{1!} - \frac{x^2}{2!} - \cdots - \frac{x^n}{n!} \right) = e^x \int_0^x e^{(\alpha-1)t} dt = \begin{cases} \frac{e^{\alpha x} - e^x}{\alpha - 1} & \text{if } \alpha \neq 1, \\ xe^x & \text{if } \alpha = 1. \end{cases}$$

Remark. When $\alpha = -1$ we get that

$$\sum_{n=0}^{\infty} (-1)^n \left(e^x - 1 - \frac{x}{1!} - \frac{x^2}{2!} - \cdots - \frac{x^n}{n!} \right) = \sinh x$$

and, when $\alpha = 1$ and $x = 1$ we have that

$$\sum_{n=0}^{\infty} \left(e - 1 - \frac{1}{1!} - \frac{1}{2!} - \cdots - \frac{1}{n!} \right) = e.$$

3.90. (a) Let

$$y(x) = \sum_{n=1}^{\infty} n f^{(n)}(0) \left(e^x - 1 - \frac{x}{1!} - \frac{x^2}{2!} - \cdots - \frac{x^n}{n!} \right).$$

We have

$$y'(x) = \sum_{n=1}^{\infty} n f^{(n)}(0) \left(e^x - 1 - \frac{x}{1!} - \frac{x^2}{2!} - \cdots - \frac{x^{n-1}}{(n-1)!} \right)$$

$$= \sum_{n=1}^{\infty} n f^{(n)}(0) \left(e^x - 1 - \frac{x}{1!} - \frac{x^2}{2!} - \cdots - \frac{x^{n-1}}{(n-1)!} - \frac{x^n}{n!} + \frac{x^n}{n!} \right)$$

$$= \sum_{n=1}^{\infty} n f^{(n)}(0) \left(e^x - 1 - \frac{x}{1!} - \frac{x^2}{2!} - \cdots - \frac{x^{n-1}}{(n-1)!} - \frac{x^n}{n!} \right)$$

$$+ \sum_{n=1}^{\infty} f^{(n)}(0) \cdot \frac{x^n}{(n-1)!}$$

$$= y(x) + x f'(x).$$

The solution of this linear differential equation, with initial condition $y(0) = 0$, is given by

$$y(x) = \int_0^x e^{x-t} t f'(t) dt.$$

(b) Let $f(x) = e^{\alpha x}$. Since $f^{(n)}(0) = \alpha^n$, it follows from part (a) of the problem that

$$\sum_{n=1}^{\infty} n \alpha^n \left(e^x - 1 - \frac{x}{1!} - \frac{x^2}{2!} - \cdots - \frac{x^n}{n!} \right) = \alpha e^x \int_0^x t e^{(\alpha-1)t} dt$$

$$= \begin{cases} \frac{\alpha}{(\alpha-1)^2} \left(e^{\alpha x} \alpha x - x e^{\alpha x} - e^{\alpha x} + e^x \right) & \text{if } \alpha \neq 1, \\ \frac{x^2 e^x}{2} & \text{if } \alpha = 1. \end{cases}$$

Remark. When $\alpha = -1$, we get that

$$\sum_{n=1}^{\infty} n(-1)^n \left(e^x - 1 - \frac{x}{1!} - \frac{x^2}{2!} - \cdots - \frac{x^n}{n!} \right) = \frac{xe^{-x}}{2} - \frac{\sinh x}{2}$$

and when $\alpha = 1$ and $x = 1$, we have

$$\sum_{n=1}^{\infty} n \left(e - 1 - \frac{1}{1!} - \frac{1}{2!} - \cdots - \frac{1}{n!} \right) = \frac{e}{2}.$$

3.91. A solution of this problem is suggested in [86].

3.92. The series equals

$$\frac{f(x) + f(-x)}{2} - f(0).$$

Let $S(x)$ be the value of the series and let $S_n(x)$ be its nth partial sum. An application of Abel's summation formula (A.1), with $a_k = (-1)^{k-1}$ and $b_k = f(x) - T_k(x)$, shows that

$$S_n(x) = \left((-1)^0 + (-1)^1 + \cdots + (-1)^{n-1} \right) (f(x) - T_{n+1}(x))$$

$$+ \sum_{k=1}^{n} \left((-1)^0 + (-1)^1 + \cdots + (-1)^{k-1} \right) \frac{x^{k+1} f^{(k+1)}(0)}{(k+1)!}.$$

Thus,

$$S(x) = \sum_{n=1}^{\infty} \left((-1)^0 + (-1)^1 + \cdots + (-1)^{n-1} \right) \frac{x^{n+1} f^{(n+1)}(0)}{(n+1)!}$$

$$= \sum_{p=1}^{\infty} \frac{x^{2p} f^{(2p)}(0)}{(2p)!} = \frac{f(x) + f(-x)}{2} - f(0).$$

A generalization of this problem is given in [90].

3.93. For a solution of this problem see [86].

3.9.6 Multiple Series

A proof tells us where to concentrate our doubts.

Kline Morris

3.94. The series equals $-\ln 2 - \pi^2/12 + 3\zeta(3)/4$ [77]. An alternative solution, which is based on an application of Problem **3.96**, can be found in [54].

3.95. The series equals $2\zeta(\alpha-3)+3\zeta(\alpha-2)-\zeta(\alpha-1)-\zeta(\alpha)$. We have

$$\sum_{n=1}^{\infty}\sum_{m=1}^{\infty}\frac{1}{(\lfloor\sqrt{n+m}\rfloor)^{\alpha}}=\sum_{k=1}^{\infty}\left(\sum_{k^2\leq n+m<(k+1)^2}\frac{1}{(\lfloor\sqrt{n+m}\rfloor)^{\alpha}}\right)$$

$$=\sum_{k=1}^{\infty}\frac{1}{k^{\alpha}}\left(\sum_{k^2\leq n+m<(k+1)^2}1\right).$$

On the other hand,

$$\sum_{k^2\leq n+m<(k+1)^2}1=\sum_{n+m=k^2}1+\sum_{n+m=k^2+1}1+\cdots+\sum_{n+m=(k+1)^2-1}1$$

$$=(k^2-1)+k^2+\cdots+\left((k+1)^2-2\right)$$

$$=2k^3+3k^2-k-1.$$

It follows that

$$\sum_{n=1}^{\infty}\sum_{m=1}^{\infty}\frac{1}{(\lfloor\sqrt{n+m}\rfloor)^{\alpha}}=\sum_{k=1}^{\infty}\frac{2k^3+3k^2-k-1}{k^{\alpha}}$$

$$=2\zeta(\alpha-3)+3\zeta(\alpha-2)-\zeta(\alpha-1)-\zeta(\alpha).$$

3.96. For positive integers v and n, we let $A_v=\sum_{m=1}^{\infty}a_{v+m}$ and $s_n=\sum_{v=1}^{n}A_v$. Likewise, for every positive integer n we let $\sigma_n=\sum_{k=1}^{n}a_k$ and $\tilde{\sigma}_n=\sum_{k=1}^{n}ka_k$. Since $A_v=\sigma-\sigma_v$, it follows that

$$s_n=n\sigma-\sum_{v=1}^{n}\sigma_v=n\sigma-\sum_{k=1}^{n}(n+1-k)a_k=n\sigma+\tilde{\sigma}_n-(n+1)\sigma_n=\frac{u_n}{v_n},\quad(3.26)$$

where $u_n=\sigma-\sigma_n+(\tilde{\sigma}_n-\sigma_n)/n$ and $v_n=1/n$. On the other hand, it is straightforward to show that

$$\frac{u_{n+1}-u_n}{v_{n+1}-v_n}=\tilde{\sigma}_{n+1}-\sigma_{n+1}.$$

Since

$$\lim_{n\to\infty}\frac{u_{n+1}-u_n}{v_{n+1}-v_n}=\tilde{\sigma}-\sigma,$$

an application of Stolz–Cesàro lemma (the $0/0$ case) implies that $s=\lim_{n\to\infty}s_n=\lim_{n\to\infty}u_n/v_n=\tilde{\sigma}-\sigma$.

3.97. Let $w_n=na_{2n}+\tilde{\sigma}_{2n-1}$. With the same notation as in the previous problem, we have, based on (3.26), that $\sigma+s_{2n}=(2n+1)(\sigma-\sigma_{2n})+\tilde{\sigma}_{2n}=x_n/y_n$, where $x_n=\sigma-\sigma_{2n}+\frac{1}{2n+1}\tilde{\sigma}_{2n}$ and $y_n=1/(2n+1)$. On the other hand,

$$\frac{x_{n+1}-x_n}{y_{n+1}-y_n}=\frac{2n+1}{2}a_{2n+2}+\tilde{\sigma}_{2n+1}=w_{n+1}-\frac{1}{2}a_{2n+2}.$$

Since $a_n \to 0$ as $n \to \infty$, we have, based on Stolz–Cesàro lemma (the $0/0$ case) and $\lim_{n\to\infty}(na_{2n} + \tilde{\sigma}_{2n-1}) = l$, that $\lim_{n\to\infty}(\sigma + s_{2n}) = \ell$, and it follows that $\lim_{n\to\infty} s_{2n} = \ell - \sigma$. On the other hand, since $s_{2n+1} = s_{2n} + A_{2n+1} = s_{2n} + \sigma - \sigma_{2n+1}$, we get that $\lim_{n\to\infty} s_{2n+1} = \lim_{n\to\infty} s_{2n} = \ell - \sigma$. Thus, $\lim_{n\to\infty} s_n = \ell - \sigma$.

Remark. Problems **3.96** and **3.97** are due to Tiberiu Trif, and they refer to the special case when a double iterated series equals a difference of two single series [54].

3.98. See [55].

3.99. The series equals $(\pi^2 \ln 2)/6 - (2\ln^3 2)/3 - \zeta(3)/4$. We have, based on (3.17), that

$$\sum_{n=1}^{\infty}\sum_{m=1}^{\infty}(-1)^{n+m}\frac{H_{n+m}}{n\cdot m} = -\sum_{n=1}^{\infty}\sum_{m=1}^{\infty}(-1)^{n+m}\frac{n+m}{n\cdot m}\int_0^1 x^{n+m-1}\ln(1-x)dx$$

$$= -\int_0^1 \frac{\ln(1-x)}{x}\sum_{n=1}^{\infty}\sum_{m=1}^{\infty}(-x)^{n+m}\left(\frac{1}{n}+\frac{1}{m}\right)dx$$

$$= -2\int_0^1 \frac{\ln(1-x)}{x}\sum_{n=1}^{\infty}\frac{(-x)^n}{n}\sum_{m=1}^{\infty}(-x)^m dx$$

$$= -2\int_0^1 \frac{\ln(1-x)\ln(1+x)}{1+x}dx.$$

It follows, in view of (3.18), that

$$\sum_{n=1}^{\infty}\sum_{m=1}^{\infty}(-1)^{n+m}\frac{H_{n+m}}{n\cdot m} = -2\int_0^1 \frac{\ln(1-x)\ln(1+x)}{1+x}dx$$

$$= 2\int_0^1 \ln(1-x)\sum_{n=1}^{\infty}H_n(-x)^n dx$$

$$= 2\sum_{n=1}^{\infty}(-1)^n \cdot H_n \int_0^1 x^n \ln(1-x)dx$$

$$\overset{(3.17)}{=} -2\sum_{n=1}^{\infty}(-1)^n \frac{H_n H_{n+1}}{n+1}$$

$$= 2\sum_{n=1}^{\infty}(-1)^{n+1}\left(H_{n+1}-\frac{1}{n+1}\right)\frac{H_{n+1}}{n+1}$$

$$= 2\left(\sum_{n=1}^{\infty}(-1)^{n+1}\frac{H_{n+1}^2}{n+1} - \sum_{n=1}^{\infty}(-1)^{n+1}\frac{H_{n+1}}{(n+1)^2}\right),$$

and the result follows based on Problems **3.56** and **3.57**.

3.100. The sum equals $\zeta(3)/4 - (\ln^3 2)/3 + (\ln^2 2)/2 + \ln 2 - 1$. Replacing x by $-x$ in (3.18), one has that

$$-\frac{\ln(1+x)}{1+x} = \sum_{n=1}^{\infty}(-1)^n \cdot H_n \cdot x^n, \quad -1 < x < 1. \tag{3.27}$$

Thus,

$$\sum_{n=1}^{\infty}\sum_{m=1}^{\infty}(-1)^{n+m}\frac{H_n H_m}{n+m} = \sum_{n=1}^{\infty}\sum_{m=1}^{\infty}(-1)^{n+m}H_n H_m \int_0^1 x^{n+m-1}dx$$

$$= \int_0^1 \frac{1}{x}\left(\sum_{n=1}^{\infty}(-1)^n H_n x^n\right)^2 dx \overset{(3.27)}{=} \int_0^1 \frac{\ln^2(1+x)}{x(1+x)^2}dx$$

$$= \int_0^1 \ln^2(1+x)\left(\frac{1}{x} - \frac{1}{1+x} - \frac{1}{(1+x)^2}\right)dx$$

$$= \int_0^1 \frac{\ln^2(1+x)}{x}dx - \int_0^1 \frac{\ln^2(1+x)}{1+x}dx - \int_0^1 \frac{\ln^2(1+x)}{(1+x)^2}dx$$

$$= \frac{\zeta(3)}{4} - \frac{\ln^3(1+x)}{3}\Big|_0^1 - \left(-\frac{\ln^2(1+x) + 2\ln(1+x) + 2}{1+x}\right)\Big|_0^1$$

$$= \frac{\zeta(3)}{4} - \frac{\ln^3 2}{3} + \frac{\ln^2 2}{2} + \ln 2 - 1,$$

since $\int_0^1 \ln^2(1+x)/x\,dx = \zeta(3)/4$.

3.101. We have

$$\sum_{n=1}^{\infty}\sum_{m=1}^{\infty}(-1)^{n+m}\frac{H_n \cdot H_m}{n+m+1} = \sum_{n=1}^{\infty}\sum_{m=1}^{\infty}(-1)^{n+m}H_n \cdot H_m \int_0^1 x^{n+m}dx$$

$$= \int_0^1 \left(\sum_{n=1}^{\infty}(-x)^n H_n\right)\cdot\left(\sum_{m=1}^{\infty}(-x)^m H_m\right)dx$$

$$\overset{(3.27)}{=} \int_0^1 \frac{\ln^2(1+x)}{(1+x)^2}dx$$

$$= \left(-\frac{\ln^2(1+x)}{1+x} - \frac{2\ln(1+x)}{1+x} - \frac{2}{1+x}\right)\Big|_0^1$$

$$= -\frac{\ln^2 2}{2} - \ln 2 + 1.$$

3.102. Exactly as in the solution of Problem **3.101**, we have that the series equals

$$(-1)^k \int_0^1 \frac{\ln^k(1+x)}{(1+x)^k}dx = (-1)^k \int_1^2 \frac{\ln^k y}{y^k}dy,$$

and the last integral can be calculated by parts.

3.103.–3.105. See [52].

3.106.[3] Let $y_m = (-1)+(-1)^2+\cdots+(-1)^m$. We apply Abel's summation formula (A.2), with

$$a_m = (-1)^m \quad \text{and} \quad b_m = 1+\frac{1}{2}+\cdots+\frac{1}{n+m} - \ln(n+m) - \gamma,$$

and we get that

$$\sum_{m=1}^{\infty}(-1)^m \left(1+\frac{1}{2}+\cdots+\frac{1}{n+m} - \ln(n+m) - \gamma\right)$$

$$= \lim_{m\to\infty} y_m \left(1+\frac{1}{2}+\cdots+\frac{1}{n+m+1} - \ln(n+m+1) - \gamma\right)$$

$$+ \sum_{m=1}^{\infty} y_m \left(\ln(n+m+1) - \ln(n+m) - \frac{1}{n+m+1}\right)$$

$$\overset{m:=2m-1}{=} - \sum_{m=1}^{\infty} \left(\ln(n+2m) - \ln(n+2m-1) - \frac{1}{n+2m}\right).$$

It follows that

$$S = \sum_{n=1}^{\infty}\sum_{m=1}^{\infty}(-1)^{n+m} \left(1+\frac{1}{2}+\cdots+\frac{1}{n+m} - \ln(n+m) - \gamma\right)$$

$$= - \sum_{n=1}^{\infty}(-1)^n \sum_{m=1}^{\infty} \left(\ln(n+2m) - \ln(n+2m-1) - \frac{1}{n+2m}\right).$$

We apply Abel's summation formula (A.2) one more time, with

$$a_n = (-1)^n \quad \text{and} \quad b_n = \sum_{m=1}^{\infty} \left(\ln(n+2m) - \ln(n+2m-1) - \frac{1}{n+2m}\right),$$

---

[3]The solution of this problem is based on a joint work with H. Qin.

and we have that

$$-S = \lim_{n \to \infty} y_n \sum_{m=1}^{\infty} \left(\ln(n+2m+1) - \ln(n+2m) - \frac{1}{n+1+2m} \right)$$

$$+ \sum_{n=1}^{\infty} y_n \sum_{m=1}^{\infty} \left(\ln \frac{(n+2m)^2}{(n+2m-1)(n+1+2m)} - \frac{1}{n+2m} + \frac{1}{n+2m+1} \right)$$

$$\overset{n:=2n-1}{=} - \sum_{n=1}^{\infty} \sum_{m=1}^{\infty} a_{n,m},$$

where $a_{n,m}$ denotes the sequence

$$2\ln(2n+2m-1) - \ln(2n+2m-2) - \ln(2n+2m) - \frac{1}{2n+2m-1} + \frac{1}{2n+2m}.$$

We used that (prove it!)

$$\lim_{n \to \infty} y_n \sum_{m=1}^{\infty} \left(\ln(n+2m+1) - \ln(n+2m) - \frac{1}{n+1+2m} \right) = 0.$$

Thus,

$$S = \sum_{n=1}^{\infty} \sum_{m=1}^{\infty} a_{n,m}$$

$$= \sum_{k=2}^{\infty} \left(\sum_{n+m=k} \left(2\ln(2k-1) - \ln(2k-2) - \ln(2k) - \frac{1}{2k-1} + \frac{1}{2k} \right) \right)$$

$$= \sum_{k=2}^{\infty} (k-1) \left(2\ln(2k-1) - \ln(2k-2) - \ln(2k) - \frac{1}{2k-1} + \frac{1}{2k} \right).$$

Let S_n be the nth partial sum of the preceding series. First, we note that

$$(k-1) \left(2\ln(2k-1) - \ln(2k-2) - \ln(2k) - \frac{1}{2k-1} + \frac{1}{2k} \right)$$

$$= (2k-1)\ln(2k-1) - \ln(2k-1) - 2(k-1)\ln 2 - (k-1)\ln(k-1)$$

$$- k\ln k + \ln k + \frac{1}{2(2k-1)} - \frac{1}{2k},$$

and it follows that

$$S_n = \sum_{k=2}^{n} (2k-1)\ln(2k-1) - \ln\frac{(2n)!}{2^n n!} - n(n-1)\ln 2 - \sum_{k=2}^{n} (k-1)\ln(k-1).$$

$$-\sum_{k=2}^{n} k\ln k + \ln n! + \frac{1}{2}\sum_{k=2}^{n}\frac{1}{2k-1} - \frac{1}{2}\sum_{k=2}^{n}\frac{1}{k}$$

$$= B_{2n} - 4B_n - (2n^2 - n)\ln 2 + \ln\frac{(n!)^2}{(2n)!} + n\ln n + \frac{H_{2n}}{2} - \frac{3H_n}{4},$$

where $B_n = \sum_{k=1}^{n} k\ln k$. On the other hand,

$$H_n = \gamma + \ln n + O\left(\frac{1}{n}\right)$$

and, by Stirling's formula,

$$\ln\frac{(n!)^2}{(2n)!} = \frac{1}{2}\ln n - 2n\ln 2 + \frac{1}{2}\ln\pi + O\left(\frac{1}{n}\right).$$

Thus,

$$S_n = B_{2n} - 4B_n - \left(2n^2 + n - \frac{1}{2}\right)\ln 2 + \left(n + \frac{1}{4}\right)\ln n - \frac{\gamma}{4} + \frac{\ln\pi}{2} + O\left(\frac{1}{n}\right).$$

Let

$$A_n = \frac{1^1 2^2 \cdots n^n}{n^{n^2/2 + n/2 + 1/12} e^{-n^2/4}},$$

and we note that $B_n = \ln A_n + (n^2/2 + n/2 + 1/12)\ln n - n^2/4$. Thus,

$$S_n = \ln A_{2n} - 4\ln A_n + \frac{7}{12}\ln 2 - \frac{\gamma}{4} + \frac{\ln\pi}{2} + O\left(\frac{1}{n}\right),$$

and the problem is solved.

Remark. The problem is motivated by an alternating series due to Hardy [20, Problem 45, p. 277] which states that

$$\sum_{n=1}^{\infty}(-1)^{n-1}\left(1 + \frac{1}{2} + \cdots + \frac{1}{n} - \ln n - \gamma\right) = \frac{1}{2}(\ln\pi - \gamma).$$

3.107. The series equals $\frac{1}{8}(\ln 32 - \pi)$. Let

$$T = \sum_{n=1}^{\infty}(-1)^n \sum_{m=1}^{\infty}(-1)^m\left(\ln 2 - \frac{1}{n+m+1} - \frac{1}{n+m+2} - \cdots - \frac{1}{2(n+m)}\right)$$

and let S_m be the mth partial sum of the inner series. We have

$$S_m = \sum_{k=1}^{m}(-1)^k\left(\ln 2 - \frac{1}{n+k+1} - \frac{1}{n+k+2} - \cdots - \frac{1}{2(n+k)}\right).$$

An application of Abel's summation formula (A.1), with

$$a_k = (-1)^k \quad \text{and} \quad b_k = \ln 2 - \frac{1}{n+k+1} - \frac{1}{n+k+2} - \cdots - \frac{1}{2(n+k)},$$

shows that

$$S_m = \sum_{k=1}^{m} (-1)^k \left(\ln 2 - \frac{1}{n+m+2} - \frac{1}{n+m+3} - \cdots - \frac{1}{2(n+m+1)} \right)$$

$$+ \sum_{k=1}^{m} \left((-1) + (-1)^2 + \cdots + (-1)^k \right) \left(\frac{1}{2n+2k+1} - \frac{1}{2n+2k+2} \right).$$

Thus,

$$\lim_{m \to \infty} S_m = \sum_{k=1}^{\infty} \left((-1) + (-1)^2 + \cdots + (-1)^k \right) \left(\frac{1}{2n+2k+1} - \frac{1}{2n+2k+2} \right)$$

$$\overset{k=2p+1}{=\!=} \sum_{p=0}^{\infty} \left(\frac{1}{2n+4p+4} - \frac{1}{2n+4p+3} \right)$$

$$= \sum_{p=0}^{\infty} \int_0^1 \left(x^{2n+4p+3} - x^{2n+4p+2} \right) dx$$

$$= - \int_0^1 \frac{x^{2n+2}}{1+x+x^2+x^3} dx.$$

This implies that

$$\sum_{m=1}^{\infty} (-1)^m \left(\ln 2 - \frac{1}{n+m+1} - \frac{1}{n+m+2} - \cdots - \frac{1}{2(n+m)} \right) = \int_0^1 \frac{-x^{2n+2} dx}{1+x+x^2+x^3},$$

and it follows that

$$T = \sum_{n=1}^{\infty} (-1)^{n+1} \int_0^1 \frac{x^{2n+2} dx}{1+x+x^2+x^3} = \int_0^1 \frac{x^4}{1+x+x^2+x^3} \sum_{n=1}^{\infty} (-x^2)^{n-1} dx$$

$$= \int_0^1 \frac{x^4 dx}{(1+x+x^2+x^3)(1+x^2)} = \int_0^1 \frac{x^4 dx}{(1+x)(1+x^2)^2} = \frac{1}{8} (\ln 32 - \pi).$$

3.108. The series behaves the same like the integral

$$I = \int_0^{\infty} \cdots \int_0^{\infty} \frac{dx_1 \cdots dx_n}{\left(1 + x_1^{p_1} + \cdots + x_n^{p_n} \right)^p}.$$

Changing variables, $x_i = y_i^{1/p_i}$, $i = 1, \ldots, n$, we get that $I = J/(p_1 \cdots p_n)$, where

$$J = \int_0^\infty \cdots \int_0^\infty \frac{y_1^{\frac{1}{p_1}-1} \cdots y_n^{\frac{1}{p_n}-1}}{(1+y_1+y_2+\cdots+y_n)^p} dy_1 \cdots dy_n.$$

For calculating J, we use the following equality:

$$\frac{1}{(1+y_1+y_2+\cdots+y_n)^p} = \frac{1}{\Gamma(p)} \int_0^\infty e^{-u(1+y_1+y_2+\cdots+y_n)} u^{p-1} du.$$

Thus,

$$J = \int_0^\infty \cdots \int_0^\infty y_1^{\frac{1}{p_1}-1} \cdots y_n^{\frac{1}{p_n}-1} \left(\frac{1}{\Gamma(p)} \int_0^\infty e^{-u(1+y_1+y_2+\cdots+y_n)} u^{p-1} du \right) dy_1 \cdots dy_n$$

$$= \frac{1}{\Gamma(p)} \int_0^\infty u^{p-1} e^{-u} \left(\int_0^\infty e^{-uy_1} y_1^{\frac{1}{p_1}-1} dy_1 \right) \cdots \left(\int_0^\infty e^{-uy_n} y_n^{\frac{1}{p_n}-1} dy_n \right) du.$$

It follows that

$$J = \frac{1}{\Gamma(p)} \int_0^\infty u^{p-1} e^{-u} \left(\frac{\Gamma(1/p_1)}{u^{1/p_1}} \right) \cdots \left(\frac{\Gamma(1/p_n)}{u^{1/p_n}} \right) du$$

$$= \frac{\Gamma(1/p_1) \cdots \Gamma(1/p_n)}{\Gamma(p)} \int_0^\infty e^{-u} u^{p-1-(1/p_1+1/p_2+\cdots+1/p_n)} du,$$

and this integral converges precisely when $p - (1/p_1 + \cdots + 1/p_n) > 0$, since

$$\int_0^\infty e^{-u} u^{p-1-(1/p_1+1/p_2+\cdots+1/p_n)} du = \Gamma\left(p - \frac{1}{p_1} - \frac{1}{p_2} - \cdots - \frac{1}{p_n} \right).$$

Remark. The problem was motivated by problem [9, Problem 36K, p. 315], which states that if $p > 1$, then the double series $\sum_{n=1}^\infty \sum_{m=1}^\infty 1/(m^2 + n^2)^p$ converges. Various versions of this problem appeared in the literature; see, for example, [133, p. 52] and [123, p. 142]. Two closely related problems [138, Problems 8 and 9, p. 189] are about proving that

$$\sum_{n=1}^N \sum_{m=1}^N \frac{1}{n^2 + m^2} \sim \ln N \quad \text{as} \quad N \to \infty$$

and

$$\sum_{n^2+m^2>N^2} \frac{1}{(n^2 + m^2)^{3/2}} \sim \frac{1}{N} \quad \text{as} \quad N \to \infty.$$

More generally, one can prove [46] that if $l_i \geq 0$ are real numbers and p, p_i are positive numbers, $i = 1, 2, \ldots, n$, then

$$\sum_{k_1,\ldots,k_n=1}^{\infty} \frac{k_1^{l_1} k_2^{l_2} \cdots k_n^{l_n}}{(k_1^{p_1} + \cdots + k_n^{p_n})^p} < \infty \quad \Leftrightarrow \quad \frac{1+l_1}{p_1} + \cdots + \frac{1+l_n}{p_n} < p.$$

3.109. We consider only the case when $a = -1$. The sum equals

$$(-1)^k \left(\ln 2 - \frac{1}{2^{k-1}(k-1)} - \frac{1}{2^{k-2}(k-2)} - \cdots - \frac{1}{2} \right).$$

We have

$$\sum_{n_1,\ldots,n_k=1}^{\infty} \frac{(-1)^{n_1+n_2+\cdots+n_k}}{n_1 + n_2 + \cdots + n_k} = -\sum_{n_1,\ldots,n_k=1}^{\infty} \int_0^1 (-x)^{n_1+n_2+\cdots+n_k-1} dx$$

$$= -\int_0^1 \left(\sum_{n_1,\ldots,n_k=1}^{\infty} (-x)^{n_1+n_2+\cdots+n_k-1} \right) dx$$

$$= -\int_0^1 \left(\sum_{n_1=1}^{\infty} (-x)^{n_1} \right) \cdots \left(\sum_{n_k=1}^{\infty} (-x)^{n_k-1} \right) dx$$

$$= (-1)^k \int_0^1 \frac{x^{k-1}}{(1+x)^k} dx.$$

Let

$$J = \int_0^1 \frac{x^{k-1}}{(1+x)^k} dx.$$

Using the substitution $x/(1+x) = z$, we get that

$$J = -\int_0^{\frac{1}{2}} \frac{z^{k-1}}{z-1} dz = -\int_0^{\frac{1}{2}} \left(z^{k-2} + z^{k-3} + \cdots + z + 1 + \frac{1}{z-1} \right) dz$$

$$= -\left(\frac{z^{k-1}}{k-1} + \frac{z^{k-2}}{k-2} + \cdots + z \right) \Big|_0^{\frac{1}{2}} + \ln 2$$

$$= \ln 2 - \left(\frac{1}{2^{k-1}(k-1)} + \frac{1}{2^{k-2}(k-2)} + \cdots + \frac{1}{2} \right).$$

The case when $a \in (-1, 1)$ is solved similarly.

For an alternative solution see [83].

3.110. Use that $1/(n_1 + \cdots + n_k) = \int_0^1 x^{n_1 + \cdots + n_k - 1} dx$.

3.111. We prove that

$$S(p) = \begin{cases} k!\zeta(k+1) & \text{if } p = 0, \\ k!\sum_{l=0}^{p-1}(-1)^l\binom{p-1}{l}\frac{1}{(l+1)^{k+1}} & \text{if } p \geq 1. \end{cases}$$

We have

$$S(p) = \sum_{m_1,\ldots,m_k=1}^{\infty} \frac{1}{m_1 m_2 \cdots m_k} \int_0^1 x^{m_1 + \cdots + m_k + p - 1} dx$$

$$= \int_0^1 \left(\sum_{m_1=1}^{\infty} \frac{x^{m_1}}{m_1}\right) \cdots \left(\sum_{m_k=1}^{\infty} \frac{x^{m_k}}{m_k}\right) x^{p-1} dx$$

$$= \int_0^1 (-\ln(1-x))^k x^{p-1} dx$$

$$= (-1)^k \int_0^1 \ln^k(1-x) x^{p-1} dx$$

$$= (-1)^k \int_0^1 \ln^k x (1-x)^{p-1} dx.$$

We distinguish here two cases.

Case $p = 0$. It is elementary to prove, using integration by parts, that

$$\int_0^1 x^l \ln^k x \, dx = (-1)^k \frac{k!}{(l+1)^{k+1}}.$$

It follows that

$$S(0) = (-1)^k \int_0^1 \frac{\ln^k x}{1-x} dx = (-1)^k \sum_{l=0}^{\infty} \int_0^1 x^l \ln^k x \, dx = \sum_{l=0}^{\infty} \frac{k!}{(l+1)^{k+1}} = k!\zeta(k+1).$$

Case $p \geq 1$. We have, based on the Binomial Theorem, that

$$S(p) = (-1)^k \int_0^1 \ln^k x (1-x)^{p-1} dx$$

$$= (-1)^k \int_0^1 \ln^k x \left(\sum_{l=0}^{p-1} \binom{p-1}{l}(-x)^l\right) dx$$

$$= (-1)^k \sum_{l=0}^{p-1} \binom{p-1}{l} (-1)^l \int_0^1 x^l \ln^k x \, dx$$

$$= k! \sum_{l=0}^{p-1} (-1)^l \binom{p-1}{l} \frac{1}{(l+1)^{k+1}}.$$

Remark. We mention that the case when $k = p = 2$ is Problem 3.99 on page 27 in [14] and if $p = 0$, one obtains the elegant form of the Riemann zeta function value $\zeta(k+1)$ as a multiple iterated series [98, Entry 25.6.5, p. 605].

3.112. We have

$$\sum_{n,m=1}^{\infty} \frac{1}{(n+m)!} = \sum_{k=2}^{\infty} \left(\sum_{n+m=k} \frac{1}{(n+m)!} \right) = \sum_{k=2}^{\infty} \frac{k-1}{k!} = \sum_{k=2}^{\infty} \left(\frac{1}{(k-1)!} - \frac{1}{k!} \right) = 1.$$

An alternative solution, which is based on a Beta function technique, is outlined in the solutions of the next problems.

3.113. The series equals $5e/6$. We use a Beta function technique by noting that, for integers $n, m \geq 1$, one has

$$\int_0^1 x^n (1-x)^{m-1} dx = \frac{n!(m-1)!}{(n+m)!}.$$

Thus,

$$\sum_{n=1}^{\infty} \sum_{m=1}^{\infty} \frac{n^2}{(n+m)!} = \sum_{n=1}^{\infty} \sum_{m=1}^{\infty} \frac{n^2}{n!(m-1)!} \int_0^1 x^n (1-x)^{m-1} dx$$

$$= \int_0^1 \left(\sum_{n=1}^{\infty} \frac{n x^n}{(n-1)!} \right) \left(\sum_{m=1}^{\infty} \frac{(1-x)^{m-1}}{(m-1)!} \right) dx$$

$$= \int_0^1 e^x (x^2 + x) e^{1-x} dx$$

$$= \frac{5e}{6}.$$

3.114. We have

$$\sum_{n=1}^{\infty} \sum_{m=1}^{\infty} \frac{n \cdot m}{(n+m)!} = \sum_{n=1}^{\infty} \sum_{m=1}^{\infty} \frac{n}{(n-1)!} \cdot \frac{1}{(m-1)!} \int_0^1 x^m (1-x)^{n-1} dx$$

$$= \int_0^1 \left(\sum_{n=1}^{\infty} \frac{n}{(n-1)!} (1-x)^{n-1} \right) \cdot \left(\sum_{m=1}^{\infty} \frac{x^m}{(m-1)!} \right) dx$$

$$= \int_0^1 (2-x)e^{1-x} \cdot xe^x dx$$

$$= e \int_0^1 (2-x)x dx$$

$$= \frac{2e}{3}.$$

3.115. We have

$$\sum_{n=1}^{\infty} \sum_{m=1}^{\infty} \frac{a^n b^m}{(n+m)!} = \sum_{n=1}^{\infty} \sum_{m=1}^{\infty} \frac{a^n}{(n-1)!} \cdot \frac{b^m}{m!} \int_0^1 x^m (1-x)^{n-1} dx$$

$$= a \int_0^1 \left(\sum_{n=1}^{\infty} \frac{(a(1-x))^{n-1}}{(n-1)!} \right) \sum_{m=1}^{\infty} \frac{(bx)^m}{m!} dx$$

$$= a \int_0^1 e^{a(1-x)} (e^{bx} - 1) dx$$

$$= \begin{cases} \frac{a(e^b - e^a)}{b-a} + 1 - e^a & \text{if } a \neq b, \\ (a-1)e^a + 1 & \text{if } a = b. \end{cases}$$

The second series can be calculated either directly or by differentiating the first series with respect to a and b.

Remark. Similarly, one can prove that if x is a real number, then

$$\sum_{n=1}^{\infty} \sum_{m=1}^{\infty} \frac{n \cdot m}{(n+m)!} x^{n+m} = x^2 e^x \left(\frac{x}{6} + \frac{1}{2} \right). \qquad (3.28)$$

3.116. Differentiate the first series from Problem **3.115.**

3.117. The series equals $5e/24$. We have, based on (3.28), that

$$\sum_{n,m,p=1}^{\infty} \frac{nm}{(n+m+p)!} = \sum_{n,m,p=1}^{\infty} \frac{nm}{(n+m)!(p-1)!} \cdot \frac{(n+m)!(p-1)!}{(n+m+p)!}$$

$$= \sum_{n,m,p=1}^{\infty} \frac{nm}{(n+m)!(p-1)!} \int_0^1 x^{n+m} (1-x)^{p-1} dx$$

$$= \int_0^1 \left(\sum_{n=1}^{\infty} \sum_{m=1}^{\infty} \frac{nmx^{n+m}}{(n+m)!} \right) \left(\sum_{p=1}^{\infty} \frac{(1-x)^{p-1}}{(p-1)!} \right) dx$$

$$= \int_0^1 x^2 e^x \left(\frac{x}{6} + \frac{1}{2} \right) e^{1-x} dx$$

$$= e \int_0^1 \left(\frac{x^3}{6} + \frac{x^2}{2} \right) dx$$

$$= \frac{5e}{24}.$$

3.118. We have, based on (3.28), that

$$\sum_{n,m,p=1}^{\infty} \frac{nmp}{(n+m+p)!} = \sum_{n,m,p=1}^{\infty} \frac{nmp}{(n+m)!(p-1)!} \cdot \frac{(n+m)!(p-1)!}{(n+m+p)!}$$

$$= \sum_{n,m,p=1}^{\infty} \frac{nmp}{(n+m)!(p-1)!} \int_0^1 x^{n+m}(1-x)^{p-1} dx$$

$$= \int_0^1 \left(\sum_{n=1}^{\infty} \sum_{m=1}^{\infty} \frac{nmx^{n+m}}{(n+m)!} \right) \left(\sum_{p=1}^{\infty} \frac{p(1-x)^{p-1}}{(p-1)!} \right) dx$$

$$= \int_0^1 x^2 e^x \left(\frac{x}{6} + \frac{1}{2} \right) e^{1-x}(2-x) dx$$

$$= e \int_0^1 \left(\frac{x^3}{6} + \frac{x^2}{2} \right)(2-x) dx$$

$$= \frac{31e}{120}.$$

3.119. We have

$$\sum_{n_1,\dots,n_k=1}^{\infty} \frac{1}{(n_1 + n_2 + \cdots + n_k)!} = \sum_{j=k}^{\infty} \left(\sum_{n_1+n_2+\cdots+n_k=j} \frac{1}{(n_1 + n_2 + \cdots + n_k)!} \right)$$

$$= \sum_{j=k}^{\infty} \binom{j-1}{k-1} \frac{1}{j!} = \frac{1}{(k-1)!} \sum_{j=k}^{\infty} \frac{1}{j(j-k)!}. \quad (3.29)$$

To calculate this sum we let

$$S(x) = \sum_{j=k}^{\infty} \frac{x^j}{j(j-k)!}.$$

Taking derivatives we get that

$$S'(x) = \sum_{j=k}^{\infty} \frac{x^{j-1}}{(j-k)!} = x^{k-1} \sum_{j=k}^{\infty} \frac{x^{j-k}}{(j-k)!} = x^{k-1} e^x,$$

and it follows that $S(x) = \int_0^x t^{k-1} e^t dt + C$. Since $S(0) = 0$ we get that

$$S(x) = \int_0^x t^{k-1} e^t dt. \qquad (3.30)$$

It follows from (3.29) and (3.30) that

$$\sum_{n_1,\ldots,n_k=1}^{\infty} \frac{1}{(n_1 + n_2 + \cdots + n_k)!} = \frac{1}{(k-1)!} \int_0^1 t^{k-1} e^t dt. \qquad (3.31)$$

Let $I_{k-1} = \int_0^1 t^{k-1} e^t dt$. A calculation shows that $I_{k-1} = e - (k-1)I_{k-2}$. This implies that $x_{k-1} = (-1)^{k-1} e/(k-1)! + x_{k-2}$, where $x_k = (-1)^k I_k/k!$. Thus, $x_{k-1} = \sum_{j=1}^{k-1} (-1)^j e/j! + x_0$. On the other hand, since $x_0 = e - 1$, we get that

$$I_{k-1} = (-1)^{k-1}(k-1)! \left(e \sum_{j=0}^{k-1} \frac{(-1)^j}{j!} - 1 \right). \qquad (3.32)$$

Combining (3.31) and (3.32), we get that the result follows and the problem is solved.

3.120. We have, due to symmetry reasons, that for $i, j = 1, \ldots, k$,

$$\sum_{n_1,\ldots,n_k=1}^{\infty} \frac{n_i}{(n_1 + n_2 + \cdots + n_k)!} = \sum_{n_1,\ldots,n_k=1}^{\infty} \frac{n_j}{(n_1 + n_2 + \cdots + n_k)!}.$$

Hence,

$$\sum_{n_1,\ldots,n_k=1}^{\infty} \frac{n_i}{(n_1 + n_2 + \cdots + n_k)!} = \frac{1}{k} \sum_{n_1,\ldots,n_k=1}^{\infty} \frac{1}{(n_1 + n_2 + \cdots + n_k - 1)!}.$$

On the other hand,

$$\sum_{n_1,\ldots,n_k=1}^{\infty} \frac{1}{(n_1 + n_2 + \cdots + n_k - 1)!} = \sum_{j=k}^{\infty} \left(\sum_{n_1 + \cdots + n_k = j} \frac{1}{(n_1 + n_2 + \cdots + n_k - 1)!} \right)$$

$$= \sum_{j=k}^{\infty} \binom{j-1}{k-1} \frac{1}{(j-1)!}$$

$$= \frac{1}{(k-1)!} \sum_{j=k}^{\infty} \frac{1}{(j-k)!}$$

$$= \frac{e}{(k-1)!}.$$

3.121. See the solution of Problem **3.122**.

3.122. and **3.124.** These problems are special cases of the following two lemmas.

Lemma 3.2. *Let $k \geq 1$ be a natural number and let $(a_n)_{n\in\mathbb{N}}$ be a sequence of real numbers such that $\sum_{n=1}^{\infty} a_n n^{k-1}$ converges absolutely. Then,*

$$\sum_{n_1,\ldots,n_k=1}^{\infty} a_{n_1+n_2+\cdots+n_k} = \frac{1}{(k-1)!} \sum_{i=1}^{k} s(k,i) \left(\sum_{p=k}^{\infty} a_p p^{i-1} \right),$$

where $s(k,i)$ are the Stirling numbers of the first kind.

Proof. We have

$$\sum_{n_1,\ldots,n_k=1}^{\infty} a_{n_1+n_2+\cdots+n_k} = \sum_{p=k}^{\infty} \left(\sum_{n_1+n_2+\cdots+n_k=p} a_{n_1+\cdots+n_k} \right)$$

$$= \sum_{p=k}^{\infty} a_p \left(\sum_{n_1+n_2+\cdots+n_k=p} 1 \right)$$

$$= \sum_{p=k}^{\infty} a_p \binom{p-1}{k-1}$$

$$= \frac{1}{(k-1)!} \sum_{p=k}^{\infty} \frac{a_p}{p} \left(p(p-1)(p-2)\cdots(p-k+1) \right)$$

$$= \frac{1}{(k-1)!} \sum_{p=k}^{\infty} \frac{a_p}{p} \left(\sum_{i=0}^{k} s(k,i) p^i \right)$$

$$= \frac{1}{(k-1)!} \sum_{i=0}^{k} s(k,i) \left(\sum_{p=k}^{\infty} a_p p^{i-1} \right)$$

$$= \frac{1}{(k-1)!} \sum_{i=1}^{k} s(k,i) \left(\sum_{p=k}^{\infty} a_p p^{i-1} \right),$$

since $s(k,0) = 0$ for $k \in \mathbb{N}$, and the lemma is proved.

Lemma 3.3. *Let $1 \leq i \leq k$ be fixed natural numbers and let $(a_n)_{n\in\mathbb{N}}$ be a sequence of positive numbers such that $\sum_{n=1}^{\infty} a_n n^k$ converges. Then,*

$$\sum_{n_1,\ldots,n_k=1}^{\infty} n_i \cdot a_{n_1+n_2+\cdots+n_k} = \frac{1}{k!} \sum_{i=1}^{k} s(k,i) \left(\sum_{p=k}^{\infty} a_p p^i \right),$$

where $s(k,i)$ are the Stirling numbers of the first kind.

Proof. We have, based on symmetry reasons, that for $i, j = 1, \ldots, k$, one has that

$$\sum_{n_1,\ldots,n_k=1}^{\infty} n_i \cdot a_{n_1+n_2+\cdots+n_k} = \sum_{n_1,\ldots,n_k=1}^{\infty} n_j \cdot a_{n_1+n_2+\cdots+n_k}$$

and hence

$$\sum_{n_1,\ldots,n_k=1}^{\infty} n_i \cdot a_{n_1+n_2+\cdots+n_k} = \frac{1}{k} \sum_{n_1,\ldots,n_k=1}^{\infty} (n_1 + n_2 + \cdots + n_k) \cdot a_{n_1+n_2+\cdots+n_k},$$

and the result follows from the previous lemma applied to the sequence $(na_n)_{n\in\mathbb{N}}$.

3.123. (a) We have, based on the integral equality

$$\frac{1}{a^m} = \frac{1}{\Gamma(m)} \int_0^{\infty} e^{-at} t^{m-1} dt$$

and the power series

$$\frac{1}{(1-x)^{\lambda}} = \sum_{n=0}^{\infty} \frac{\Gamma(n+\lambda)}{n!\Gamma(\lambda)} x^n, \quad -1 < x < 1, \quad \lambda > 0,$$

that

$$\sum_{n_1,\ldots,n_k=1}^{\infty} \frac{n_1 n_2 \cdots n_k}{(n_1 + n_2 + \cdots + n_k)^m} = \sum_{n_1,\ldots,n_k=1}^{\infty} \frac{n_1 n_2 \cdots n_k}{\Gamma(m)} \int_0^{\infty} e^{-(n_1+\cdots+n_k)t} t^{m-1} dt$$

$$= \frac{1}{\Gamma(m)} \int_0^{\infty} t^{m-1} \left(\sum_{n_1=1}^{\infty} n_1 e^{-n_1 t} \right) \cdots \left(\sum_{n_k=1}^{\infty} n_k e^{-n_k t} \right) dt$$

$$= \frac{1}{\Gamma(m)} \int_0^{\infty} t^{m-1} \left(\frac{e^{-t}}{(1-e^{-t})^2} \right)^k dt$$

$$= \frac{1}{\Gamma(m)} \int_0^{\infty} t^{m-1} e^{-kt} \frac{1}{(1-e^{-t})^{2k}} dt$$

$$= \frac{1}{\Gamma(m)} \int_0^{\infty} t^{m-1} e^{-kt} \sum_{n=0}^{\infty} \frac{\Gamma(n+2k)}{\Gamma(2k)} \cdot \frac{e^{-nt}}{n!} dt$$

$$= \frac{1}{\Gamma(m)\Gamma(2k)} \sum_{n=0}^{\infty} \frac{\Gamma(n+2k)}{n!} \int_0^{\infty} t^{m-1} e^{-(k+n)t} dt$$

$$= \frac{1}{\Gamma(2k)} \sum_{n=0}^{\infty} \frac{\Gamma(n+2k)}{n!(n+k)^m}.$$

(b) When $k = 2$ and $m > 4$, one has that

$$\sum_{n_1=1}^{\infty}\sum_{n_2=1}^{\infty}\frac{n_1 n_2}{(n_1+n_2)^m} = \frac{1}{\Gamma(4)}\sum_{n=0}^{\infty}\frac{\Gamma(n+4)}{n!(n+2)^m} = \frac{1}{6}(\zeta(m-3)-\zeta(m-1)).$$

3.125. We have

$$\sum_{n=1}^{\infty}\sum_{m=1}^{\infty}\frac{\zeta(n+m)-1}{n+m} = \sum_{k=2}^{\infty}\left(\sum_{n+m=k}\frac{\zeta(k)-1}{k}\right)$$

$$= \sum_{k=2}^{\infty}(k-1)\frac{\zeta(k)-1}{k}$$

$$= \sum_{k=2}^{\infty}(\zeta(k)-1) - \sum_{k=2}^{\infty}\frac{\zeta(k)-1}{k}$$

$$= \gamma,$$

since (see [122, Formula 3, p. 142] and [122, Formula 135, p. 173])

$$\sum_{k=2}^{\infty}(\zeta(k)-1) = 1 \quad \text{and} \quad \sum_{k=2}^{\infty}\frac{\zeta(k)-1}{k} = 1 - \gamma.$$

3.126. The series equals $\gamma/2$. We have, based on symmetry reasons, that

$$S = \sum_{n=1}^{\infty}\sum_{m=1}^{\infty}\frac{m(\zeta(n+m)-1)}{(n+m)^2} = \sum_{m=1}^{\infty}\sum_{n=1}^{\infty}\frac{n(\zeta(n+m)-1)}{(n+m)^2}.$$

It follows, based on Problem **3.125**, that

$$S = \frac{1}{2}(S+S) = \frac{1}{2}\sum_{n=1}^{\infty}\sum_{m=1}^{\infty}\frac{\zeta(n+m)-1}{n+m} = \frac{\gamma}{2}.$$

3.127. The series equals $3/8 - \ln 2/2$. Let S be the value of the series. We have, based on symmetry reasons, that

$$S = \sum_{n=1}^{\infty}\sum_{m=1}^{\infty}\frac{m(\zeta(2n+2m)-1)}{(n+m)^2} = \sum_{m=1}^{\infty}\sum_{n=1}^{\infty}\frac{n(\zeta(2n+2m)-1)}{(n+m)^2}.$$

Thus,

$$S = \frac{1}{2}(S+S)$$

$$= \frac{1}{2}\sum_{n=1}^{\infty}\sum_{m=1}^{\infty}\frac{\zeta(2n+2m)-1}{n+m}$$

$$= \frac{1}{2} \sum_{k=2}^{\infty} \left(\sum_{n+m=k} \frac{\zeta(2k)-1}{k} \right)$$

$$= \frac{1}{2} \sum_{k=2}^{\infty} (k-1) \cdot \frac{\zeta(2k)-1}{k}$$

$$= \frac{1}{2} \left(\sum_{k=2}^{\infty} (\zeta(2k)-1) - \sum_{k=2}^{\infty} \frac{\zeta(2k)-1}{k} \right)$$

$$= \frac{3}{8} - \frac{\ln 2}{2},$$

since (see [122, Formula 193, p. 178] and [122, Formula 152, p. 174])

$$\sum_{k=1}^{\infty} (\zeta(2k)-1) = \frac{3}{4} \quad \text{and} \quad \sum_{k=1}^{\infty} \frac{\zeta(2k)-1}{k} = \ln 2.$$

3.128. We have

$$S_k = \sum_{n_1,\dots,n_k=1}^{\infty} \frac{1}{n_1 + \cdots + n_k} \sum_{p=2}^{\infty} \frac{1}{p^{n_1 + \cdots + n_k}}$$

$$= \sum_{p=2}^{\infty} \left(\sum_{n_1,\dots,n_k=1}^{\infty} \frac{1}{p^{n_1 + \cdots + n_k}} \cdot \frac{1}{n_1 + \cdots + n_k} \right)$$

$$= \sum_{p=2}^{\infty} \left(\sum_{n_1,\dots,n_k=1}^{\infty} \int_0^{\frac{1}{p}} x^{n_1 + \cdots + n_k - 1} dx \right)$$

$$= \sum_{p=2}^{\infty} \int_0^{\frac{1}{p}} x^{k-1} \left(\sum_{n_1=1}^{\infty} x^{n_1-1} \right) \cdots \left(\sum_{n_k=1}^{\infty} x^{n_k-1} \right) dx$$

$$= \sum_{p=2}^{\infty} \int_0^{\frac{1}{p}} \frac{x^{k-1}}{(1-x)^k} dx$$

$$\overset{\frac{x}{1-x}=y}{=} \sum_{p=2}^{\infty} \int_0^{\frac{1}{p-1}} \frac{y^{k-1}}{1+y} dy.$$

Let

$$I_k = \int_0^{\frac{1}{p-1}} \frac{y^{k-1}}{1+y} dy.$$

For calculating I_k we distinguish between the cases when k is an even or an odd integer.

When $k = 2$ one has that

$$I_2 = \frac{1}{p-1} - \ln \frac{p}{p-1}. \tag{3.33}$$

Let $k = 2l$ with $l \geq 2$ be an integer. Since

$$\frac{y^{2l-1}}{1+y} = y^{2l-2} - y^{2l-3} + \cdots + (-y) + 1 - \frac{1}{1+y},$$

we obtain that

$$I_{2l} = \frac{1}{(2l-1)(p-1)^{2l-1}} - \frac{1}{(2l-2)(p-1)^{2l-2}} + \cdots - \frac{1}{2(p-1)^2}$$

$$+ \frac{1}{p-1} - \ln \frac{p}{p-1}. \tag{3.34}$$

On the other hand, if $k = 2l + 1$ with $l \geq 1$, we have that

$$\frac{y^{2l}}{1+y} = y^{2l-1} - y^{2l-2} + y^{2l-3} - \cdots - y^2 + y - 1 + \frac{1}{1+y},$$

and hence

$$I_{2l+1} = \frac{1}{2l(p-1)^{2l}} - \frac{1}{(2l-1)(p-1)^{2l-1}} + \cdots + \frac{1}{2(p-1)^2} - \frac{1}{p-1} + \ln \frac{p}{p-1}. \tag{3.35}$$

It follows, based on (3.33)–(3.35), that

$$S_2 = \sum_{p=2}^{\infty} \left(\frac{1}{p-1} - \ln \frac{p}{p-1} \right) = \gamma,$$

$$S_{2l} = \frac{\zeta(2l-1)}{2l-1} - \frac{\zeta(2l-2)}{2l-2} + \cdots - \frac{\zeta(2)}{2} + \gamma,$$

and

$$S_{2l+1} = \frac{\zeta(2l)}{2l} - \frac{\zeta(2l-1)}{2l-1} + \cdots - \frac{\zeta(3)}{3} + \frac{\zeta(2)}{2} - \gamma.$$

3.129. and 3.130. Exactly as in the solution of Problem **3.128**, we obtain that

$$A_k = (-1)^k \sum_{p=2}^{\infty} \int_0^{\frac{1}{p}} \frac{x^{k-1}}{(1+x)^k} dx \overset{\frac{x}{1+x}=y}{=} (-1)^{k+1} \sum_{p=2}^{\infty} \int_0^{\frac{1}{p+1}} \frac{y^{k-1}}{y-1} dy. \tag{3.36}$$

Let

$$J_k = \int_0^{\frac{1}{p+1}} \frac{y^{k-1}}{y-1} \, dy.$$

A calculation shows that

$$J_2 = \frac{1}{p+1} + \ln \frac{p}{p+1} \tag{3.37}$$

and

$$J_k = \frac{1}{k-1} \cdot \frac{1}{(p+1)^{k-1}} + \frac{1}{k-2} \cdot \frac{1}{(p+1)^{k-2}} + \cdots + \frac{1}{2(p+1)^2} + \frac{1}{p+1} + \ln \frac{p}{p+1}. \tag{3.38}$$

Combining (3.36)–(3.38), we obtain that the desired results follow and the problems are solved.

3.131. We have

$$\sum_{n_1,\ldots,n_k=1}^{\infty} (\zeta(n_1 + \cdots + n_k) - 1) = \sum_{n_1,\ldots,n_k=1}^{\infty} \left(\sum_{p=2}^{\infty} \frac{1}{p^{n_1+\cdots+n_k}} \right)$$

$$= \sum_{p=2}^{\infty} \left(\sum_{n_1=1}^{\infty} \frac{1}{p^{n_1}} \right) \cdots \left(\sum_{n_k=1}^{\infty} \frac{1}{p^{n_k}} \right)$$

$$= \sum_{p=2}^{\infty} \frac{1}{(p-1)^k}$$

$$= \zeta(k).$$

On the other hand,

$$\sum_{n_1,\ldots,n_k=1}^{\infty} (-1)^{n_1+n_2+\cdots+n_k} \left(\zeta(n_1 + \cdots + n_k) - 1 \right)$$

$$= \sum_{n_1,\ldots,n_k=1}^{\infty} (-1)^{n_1+\cdots+n_k} \sum_{p=2}^{\infty} \frac{1}{p^{n_1+\cdots+n_k}}$$

$$= \sum_{p=2}^{\infty} \left(\sum_{n_1=1}^{\infty} \frac{(-1)^{n_1}}{p^{n_1}} \right) \cdots \left(\sum_{n_k=1}^{\infty} \frac{(-1)^{n_k}}{p^{n_k}} \right)$$

$$= \sum_{p=2}^{\infty} \left(\frac{-1}{p+1} \right)^k$$

$$= (-1)^k \left(\zeta(k) - 1 - \frac{1}{2^k} \right).$$

3.132. (a) We have, based on Problem **3.8**, that

$$\sum_{n=2}^{\infty}(n-\zeta(2)-\zeta(3)-\cdots-\zeta(n)) = \sum_{n=2}^{\infty}\sum_{k=1}^{\infty}\frac{1}{k(k+1)^n}$$

$$= \sum_{k=1}^{\infty}\frac{1}{k}\left(\sum_{n=2}^{\infty}\frac{1}{(k+1)^n}\right)$$

$$= \sum_{k=1}^{\infty}\frac{1}{k^2(k+1)}$$

$$= \zeta(2)-1.$$

(b) Let T_k be the value of the multiple series. We have

$$T_k = \sum_{n_1,\ldots,n_k=1}^{\infty}\left(\sum_{p=1}^{\infty}\frac{1}{p(p+1)^{n_1+n_2+\cdots+n_k}}\right)$$

$$= \sum_{p=1}^{\infty}\frac{1}{p}\left(\sum_{n_1=1}^{\infty}\frac{1}{(p+1)^{n_1}}\right)\cdots\left(\sum_{n_k=1}^{\infty}\frac{1}{(p+1)^{n_k}}\right)$$

$$= \sum_{k=1}^{\infty}\frac{1}{p^{k+1}}$$

$$= \zeta(k+1).$$

3.133. We have, based on Problem **3.8**, that

$$T_j = \sum_{n_1,\ldots,n_k=1}^{\infty}n_1 n_2\cdots n_j\left(\sum_{p=1}^{\infty}\frac{1}{p(p+1)^{n_1+n_2+\cdots+n_k}}\right)$$

$$= \sum_{p=1}^{\infty}\frac{1}{p}\left(\sum_{m=1}^{\infty}\frac{m}{(p+1)^m}\right)^j\left(\sum_{s=1}^{\infty}\frac{1}{(p+1)^s}\right)^{k-j}$$

$$= \sum_{p=1}^{\infty}\frac{(p+1)^j}{p^{k+j+1}}$$

$$= \sum_{p=1}^{\infty}\frac{1}{p^{k+j+1}}\left(\sum_{m=0}^{j}\binom{j}{m}p^m\right)$$

$$= \sum_{m=0}^{j}\binom{j}{m}\sum_{p=1}^{\infty}\frac{1}{p^{k+j+1-m}}$$

$$= \sum_{m=0}^{j}\binom{j}{m}\zeta(k+j+1-m).$$

Remark. When $j = 0$ we have

$$\sum_{n_1,\ldots,n_k=1}^{\infty} (n_1 + n_2 + \cdots + n_k - \zeta(2) - \zeta(3) - \cdots - \zeta(n_1 + n_2 + \cdots + n_k)) = \zeta(k+1).$$

When $j = 1$ we have that

$$\sum_{n_1,\ldots,n_k=1}^{\infty} n_1 (n_1 + n_2 + \cdots + n_k - \zeta(2) - \zeta(3) - \cdots - \zeta(n_1 + n_2 + \cdots + n_k))$$

equals

$$\zeta(k+2) + \zeta(k+1),$$

and when $j = k$ one has that

$$\sum_{n_1,\ldots,n_k=1}^{\infty} n_1 \cdots n_k (n_1 + n_2 + \cdots + n_k - \zeta(2) - \zeta(3) - \cdots - \zeta(n_1 + n_2 + \cdots + n_k))$$

equals

$$\sum_{m=0}^{k} \binom{k}{m} \zeta(2k+1-m).$$

Appendix A
Elements of Classical Analysis

A.1 Exotic Constants

He who understands Archimedes and Apollonius will admire less the achievements of the foremost men of later times.

Gottfried Wilhelm Leibniz (1646–1716)

The Euler–Mascheroni constant, γ, is the special constant defined by

$$\gamma = \lim_{n \to \infty} \left(1 + \frac{1}{2} + \cdots + \frac{1}{n} - \ln n \right) = 0.57721\,56649\,01532\,86060\ldots.$$

In other words γ *measures* the amount by which the partial sum of the harmonic series differs from the logarithmic term $\ln n$. The connection between $1 + 1/2 + 1/3 + \cdots + 1/n$ and $\ln n$ was first established in 1735 by Euler [41], who used the notation C for it and stated that was worthy of *serious consideration*. It is worth mentioning that it was the Italian geometer Lorenzo Mascheroni (1750–1800) who actually introduced the symbol γ for the constant (although there is controversy about this claim) and also computed, though with error, the first 32 digits [36]. Unsurprisingly, establishing the existence of γ has attracted many proofs [65, pp. 69–73], [133, p. 235] to mention a few. However, a short and elegant proof is based on showing that the sequence $(1 + 1/2 + 1/3 + \cdots + 1/n - \ln n)_{n \geq 1}$ is strictly decreasing and bounded below by 0.

The Euler–Mascheroni constant, γ, considered to be the third important mathematical constant next to π and e, has appeared in a variety of mathematical formulae involving series, products, and integrals [43, pp. 28–32], [65, p. 109], [122, pp. 4–6], [123, pp. 13–22]. For an interesting survey paper on γ, as well as other reference material, the reader is referred to [34]. We mention that it is still an *open problem* to determine whether γ is a rational or an irrational number.

O. Furdui, *Limits, Series, and Fractional Part Integrals: Problems in Mathematical Analysis*, Problem Books in Mathematics, DOI 10.1007/978-1-4614-6762-5,
© Springer Science+Business Media New York 2013

The nth harmonic number, H_n, is defined for $n \geq 1$ by

$$H_n = 1 + \frac{1}{2} + \cdots + \frac{1}{n}.$$

In other texts, it is also defined as the partial sum of the *harmonic series*

$$\sum_{k=1}^{\infty} \frac{1}{k} = \lim_{n \to \infty} \left(\sum_{k=1}^{n} \frac{1}{k} \right) = \infty.$$

The nth harmonic number verifies the recurrence formula $H_n = H_{n-1} + 1/n$ for $n \geq 2$ with $H_1 = 1$ which can be used to establish the *generating function*

$$-\frac{\ln(1-x)}{1-x} = \sum_{n=1}^{\infty} H_n x^n, \quad -1 \leq x < 1.$$

Significantly, one of the most important properties on the nth harmonic number refers to its divergence. The first proof of the divergence of the harmonic series is attributed, surprisingly, to Nicholas Oresme (1323–1382), one of the greatest philosophers of the Middle Ages, and similar proofs were discovered in later centuries by the Bernoulli brothers. Even Euler proved this result by evaluating an improper integral in two different ways [65, p. 23]. Another curious property is that H_n is *nonintegral*. This means that even though H_n increases without bound, it avoids all integers, apart from $n = 1$. The nth harmonic number has been involved in a variety of mathematical results: from summation formulae involving the binomial coefficients to a nice formula that expresses the nth harmonic number in terms of the *unsigned Stirling numbers* of the first kind [13]. A detailed history on the nth harmonic number as well as its connection to γ can be found in Havil's book [65]. *The Glaisher–Kinkelin* constant, A, is defined by the limit

$$A = \lim_{n \to \infty} n^{-n^2/2 - n/2 - 1/12} e^{n^2/4} \prod_{k=1}^{n} k^k = 1.28242\,71291\,00622\,63687\ldots.$$

Equivalently,

$$\ln A = \lim_{n \to \infty} \left(\sum_{k=1}^{n} k \ln k - \left(\frac{n^2}{2} + \frac{n}{2} + \frac{1}{12} \right) \ln n + \frac{n^2}{4} \right).$$

Another remarkable formula, involving factorials, in which A appears is the following:

$$\lim_{n \to \infty} \frac{1! \cdot 2! \cdots (n-1)!}{e^{-\frac{3}{4}n^2} (2\pi)^{\frac{1}{2}n} n^{\frac{1}{2}n^2 - \frac{1}{12}}} = \frac{e^{\frac{1}{12}}}{A}.$$

These formulae are due to Kinkelin [72], Jeffery [71], and Glaisher [59–61].

The constant A, which plays the same role as $\sqrt{2\pi}$ plays in Stirling's formula, has the following closed-form expression [43, p. 135]:

$$A = \exp\left(\frac{1}{12} - \zeta'(-1)\right) = \exp\left(-\frac{\zeta'(2)}{2\pi^2} + \frac{\ln(2\pi) + \gamma}{12}\right),$$

where ζ' denotes the derivative of the Riemann zeta function and γ is the Euler–Mascheroni constant.

Many beautiful formulae including A exist in the literature, from infinite products and definite integrals [43, pp. 135–136] to the evaluation of infinite series involving Riemann zeta function [122]. Other formulae involving infinite series and products, believed to be new in the literature, in which A appears, are recorded as problems in this book.

The Stieltjes constants, γ_n, are the special constants defined by

$$\gamma_n = \lim_{m\to\infty}\left(\sum_{k=1}^{m}\frac{(\ln k)^n}{k} - \frac{(\ln m)^{n+1}}{n+1}\right),$$

and, in particular, $\gamma_0 = \gamma$, the Euler–Mascheroni constant.

They occur in the *Laurent expansion* of the Riemann zeta function in a neighborhood of its simple pole at $z = 1$ (see [98, Entry 25.2.4, p. 602]):

$$\zeta(s) = \frac{1}{s-1} + \sum_{n=0}^{\infty}\frac{(-1)^n}{n!}\gamma_n(s-1)^n, \quad \Re(s) > 0.$$

Interesting properties related to the *sign* of the Stieltjes constants as well as an open problem involving their magnitudes are recorded in [43, pp. 166–167].

The logarithmic constants, δ_n, are defined by

$$\delta_n = \lim_{m\to\infty}\left(\sum_{k=1}^{m}\ln^n k - \int_{1}^{m}\ln^n x\,dx - \frac{1}{2}\ln^n m\right) = (-1)^n(\zeta^{(n)}(0) + n!).$$

In particular $\delta_0 = 1/2$, $\delta_1 = 1/2\ln(2\pi) - 1$.

Sitaramachandrarao [118] proved that they appear in the Laurent expansion of the Riemann zeta function at the origin rather than at unity

$$\zeta(z) = \frac{1}{z-1} + \sum_{n=0}^{\infty}\frac{(-1)^n}{n!}\delta_n z^n.$$

We called these constants logarithmic for simplicity and because of the logarithmic terms involved in their definition.

The constants, which are also discussed in [43, p. 168], have been used in Problem **3.98** in order to evaluate a double logarithmic series.

A.2   Special Functions

> *Euler's integral appears everywhere and is inextricably bound*
> *to a host of special functions. Its frequency and simplicity make*
> *it fundamental.*
>
>   Philip J. Davis

Euler's Gamma function, which extends factorials to the set of complex numbers, is
a function of a complex variable defined by

$$\Gamma(z) = \int_0^\infty x^{z-1} e^{-x} dx, \quad \Re(z) > 0.$$

When $\Re(z) \leq 0$, the Gamma function is defined by analytic continuation. It is a
meromorphic function with no zeros and with *simple poles of residue* $(-1)^n/n!$ at
$z = -n$.

Historically, the Gamma function was first defined by Euler,[1] as the limit of a
product

$$\Gamma(z) = \lim_{n \to \infty} \frac{1 \cdot 2 \cdots (n-1)}{z(z+1)(z+2) \cdots (z+n-1)} n^z,$$

from which the infinite integral $\int_0^\infty x^{z-1} e^{-x} dx$ can be derived.

Weierstrass defined the Gamma function as an infinite product[2]

$$\frac{1}{\Gamma(z)} = z e^{\gamma z} \prod_{n=1}^\infty \left\{ \left(1 + \frac{z}{n}\right) e^{-\frac{z}{n}} \right\}.$$

The interesting and exotic Gamma function has attracted the interest of famous
mathematicians such us Stirling, Euler, Weierstrass, Legendre, and Schlömilch.
Even Gauss worked on it: his paper on Gamma function was a part of a larger work
on hypergeometric series published in 1813. There are lots of interesting properties
that are satisfied by the Gamma function; for example, it satisfies the *difference
equation* $\Gamma(z+1) = z \cdot \Gamma(z)$ and *Euler's reflection formula*

---

[1] Euler did not introduce a notation for the Gamma function. The notation Γ was introduced by
Legendre, in 1814, while Gauss used Π.

[2] Weierstrass defined the Gamma function by this product in 1856. In fact, it was the German
mathematician Schlömilch who discovered the formula $\Gamma(m) = e^{-\gamma m}/m \prod_{k=1}^\infty (1 + m/k)^{-1} e^{m/k}$
even earlier, in 1843 (see [110, p. 457]).

$$\Gamma(z) \cdot \Gamma(1-z) = \frac{\pi}{\sin \pi z}, \quad z \neq 0, \pm 1, \pm 2, \dots$$

to mention a few.

The Gamma function is one of the basic special functions that plays a significant part in the development of the theory of infinite products and the calculation of certain infinite series and integrals. More information, as well as historical accounts on the Gamma function can be found in [110, pp. 444–475] and [133, pp. 235–264]. *The Beta function* is defined by

$$B(a,b) = \int_0^1 x^{a-1}(1-x)^{b-1} dx, \quad \Re(a) > 0, \quad \Re(b) > 0.$$

In his work on Gamma function, Euler discovered the important theorem that $B(a,b) \cdot \Gamma(a+b) = \Gamma(a) \cdot \Gamma(b)$, which connects the Beta with the Gamma function.

Other integral representations of the Beta function as well as related properties and formulae can be found in [110] and [123].

The Riemann zeta function is defined by

$$\zeta(z) = \sum_{n=1}^{\infty} \frac{1}{n^z} = 1 + \frac{1}{2^z} + \cdots + \frac{1}{n^z} + \cdots, \quad \Re(z) > 1.$$

Elsewhere $\zeta(z)$ is defined by analytic continuation. It is a meromorphic function whose only singularity in $\mathbb{C}$ is a simple pole as $z = 1$, with residue 1.

Although the function was known to Euler, its most remarkable properties were not discovered before Riemann[3] who discussed it in his memoir on prime numbers. The Riemann zeta function is perhaps the most important special function in mathematics that arises in lots of formulae involving integrals and series ([98, pp. 602–606], [122]) and is intimately related to deep results surrounding the Prime Number Theorem [39]. While many of the properties of this function have been investigated, there remain important fundamental conjectures, most notably being the *Riemann hypothesis* which states that all *nontrivial zeros* of ζ lie on the vertical line $\Re(z) = 1/2$.

The Psi (or Digamma) function, also known as the logarithmic derivative of the Gamma function, is defined by

$$\psi(z) = \frac{\Gamma'(z)}{\Gamma(z)}, \quad \Re(z) > 0.$$

Additional properties as well as special values of the digamma function can be found in [122, pp. 13–22].

---

[3] Berliner Monatsberichte, 671–680 (1859); Ges. Werke, 136–144 (1876).

The Polylogarithm function $\mathrm{Li}_n(z)$ *is defined, for* $|z| \leq 1$ *and* $n \neq 1, 2$, *by*

$$\mathrm{Li}_n(z) = \sum_{k=1}^{\infty} \frac{z^k}{k^n} = \int_0^z \frac{\mathrm{Li}_{n-1}(t)}{t} dt.$$

When $n = 1$, we define $\mathrm{Li}_1(z) = -\ln(1-z)$ and when $n = 2$, we have that $\mathrm{Li}_2(z)$, also known as the Dilogarithm function, is defined by

$$\mathrm{Li}_2(z) = \sum_{n=1}^{\infty} \frac{z^n}{n^2} = -\int_0^z \frac{\ln(1-t)}{t} dt.$$

Clearly $\mathrm{Li}_n(1) = \zeta(n)$ for $n \geq 2$. Two special values of the Dilogarithm function are $\mathrm{Li}_2(1) = \zeta(2)$ and $\mathrm{Li}_2(1/2) = \pi^2/12 - \ln^2 2/2$. The last equality follows as a consequence of *Landen's formula* (see [123, Formula 10, p. 177]):

$$\mathrm{Li}_2(z) + \mathrm{Li}_2(1-z) = \frac{\pi^2}{6} - \ln z \ln(1-z).$$

A.3 Lemmas and Theorems

Men pass away, but their deeds abide.

Augustin-Louis Cauchy (1789–1857)

Lemma A.1 (Abel's Summation Formula). *Let* $(a_n)_{n \geq 1}$ *and* $(b_n)_{n \geq 1}$ *be two sequences of real numbers and let* $A_n = \sum_{k=1}^{n} a_k$. *Then,*

$$\sum_{k=1}^{n} a_k b_k = A_n b_{n+1} + \sum_{k=1}^{n} A_k (b_k - b_{k+1}). \qquad (A.1)$$

The proof of this lemma, which is elementary, is given in [15, Theorem 2.20, p. 55]. When needed, we will be using the following version of the lemma

$$\sum_{k=1}^{\infty} a_k b_k = \lim_{n \to \infty} (A_n b_{n+1}) + \sum_{k=1}^{\infty} A_k (b_k - b_{k+1}). \qquad (A.2)$$

Lemma A.2 (The Wallis Product Formula). *The following limit holds*

$$\lim_{n \to \infty} \frac{2}{1} \cdot \frac{4}{3} \cdots \frac{2n}{2n-1} \cdot \frac{1}{\sqrt{2n+1}} = \sqrt{\frac{\pi}{2}}.$$

Theorem A.1 (Stirling's Approximation Formula). *Stirling's formula, for esti-mating large factorials, states that*

$$n! \sim \sqrt{2\pi n} \left(\frac{n}{e}\right)^n \quad (n \to \infty)$$

or, more generally, that

$$\Gamma(x+1) \sim \sqrt{2\pi x} \left(\frac{x}{e}\right)^x \quad (x \to \infty; \; x \in \mathbb{R}).$$

One of the most well-known improvements is

$$n! = n^n e^{-n} \sqrt{2\pi n} \left(1 + \frac{1}{12n} + \frac{1}{288n^2} - \frac{139}{51840n^3} - \frac{571}{2488320n^4} + \cdots\right).$$

For proofs and other details on Stirling's formula the reader is referred to [22,33], [65, p. 87], [68, 91, 95].

We will be using throughout this book the following version of Stirling's formula:

$$\ln n! = \frac{1}{2} \ln(2\pi) + \frac{2n+1}{2} \ln n - n + O\left(\frac{1}{n}\right). \tag{A.3}$$

Theorem A.2 (Cauchy–d'Alembert Criteria). *Let* $(x_n)_{n \in \mathbb{N}}$ *be a sequence of pos-itive real numbers such that the limit* $\lim_{n \to \infty} x_{n+1}/x_n = l$ *exists. Then, the limit* $\lim_{n \to \infty} \sqrt[n]{x_n}$ *also exists and is equal to l.*

Theorem A.3 (Dirichlet's Theorem). *Let* $\zeta : (1, \infty) \to \mathbb{R}$, *denote the Riemann zeta function defined by* $\zeta(x) = \sum_{n=1}^{\infty} 1/n^x$. *Then*

$$\lim_{x \to 0} \frac{x\zeta(1+x) - 1}{x} = \gamma.$$

See [62, Entry 9.536, p. 1028] and [133, Corollary, p. 271].

Theorem A.4 (Bernoulli's Integral Inequality). *Let* $f : [a,b] \to (0, \infty)$ *be an integrable function. Then*

$$\left(\int_a^b f(x)dx\right)^\alpha \leq (b-a)^{\alpha-1} \int_a^b f^\alpha(x)dx \quad \text{if} \quad \alpha \in (-\infty, 0) \cup (1, \infty),$$

$$\left(\int_a^b f(x)dx\right)^\alpha \geq (b-a)^{\alpha-1} \int_a^b f^\alpha(x)dx \quad \text{if} \quad \alpha \in (0,1).$$

This theorem can be proved by using Jensen's inequality[4] [21, Theorem 1, p. 370], [112, p. 62] for convex functions. It states that if $f : [a,b] \to [\alpha, \beta]$ is an integrable function and $\varphi : [\alpha, \beta] \to \mathbb{R}$ is a convex continuous function, then

$$\varphi \left(\frac{1}{b-a} \int_a^b f(x)\mathrm{d}x \right) \le \frac{1}{b-a} \int_a^b \varphi(f(x))\mathrm{d}x.$$

Now the result follows by taking $\varphi(x) = x^{\alpha}$.

Theorem A.5 (Lagrange's Mean Value Theorem). *Let $f : [a,b] \to \mathbb{R}$ be a function continuous on $[a,b]$ and differentiable on (a,b). Then there exists $c \in (a,b)$ such that*

$$f(b) - f(a) = (b-a)f'(c).$$

For a proof of this theorem, which is also known as *Lagrange's finite-increment theorem*, see [136, Theorem 1, p. 216].

Geometrically, this theorem states that there exists a suitable point $(c, f(c))$ on the graph of f such that the tangent to the curve $y = f(x)$ is parallel to the straight line through the points $(a, f(a))$ and $(b, f(b))$.

Theorem A.6 (First Mean Value Theorem for the Integral). *Let f be continuous on $[a,b]$ and let g be a nonnegative integrable function on $[a,b]$. Then there is $c \in [a,b]$ such that*

$$\int_a^b f(x)g(x)\mathrm{d}x = f(c) \int_a^b g(x)\mathrm{d}x.$$

A proof of this theorem is given in [136, Theorem 5, p. 352]. To avoid confusion, we mention that the first mean value theorem for the integral is often considered to be the result which states that if f is a continuous function on $[a,b]$, then there exists $c \in [a,b]$ such that

$$\int_a^b f(x)\mathrm{d}x = f(c)(b-a).$$

However, we followed Zorich [136] and reserved the name for this somewhat more general theorem.

---

[4]Johan Jensen (1859–1925) was a largely self-taught Danish mathematician. Jensen made significant contributions to the theory of complex analytic functions; however, he is best known for the discovery of a famous inequality involving convex functions. Motivated by an idea of Cauchy that was used to prove the AM–GM inequality, Jensen noticed that Cauchy's method could be generalized to convex functions [110, p. 85]. He proved that if ϕ is a convex function on $[a,b]$, then for any n numbers $x_1, x_2, \ldots, x_n$ in $[a,b]$,

$$\phi \left(\frac{x_1 + x_2 + \cdots + x_n}{n} \right) \le \frac{\phi(x_1) + \phi(x_2) + \cdots + \phi(x_n)}{n},$$

an inequality which is known as *Jensen's inequality*.

Theorem A.7 (The Riemann–Lebesgue Lemma). *If* $f : [a,b] \to \mathbb{R}$ *is an integrable function, then*

$$\lim_{n \to \infty} \int_a^b f(x) \sin nx dx = \lim_{n \to \infty} \int_a^b f(x) \cos nx dx = 0.$$

This classical result, due to Riemann, is recorded in many books that cover the standard topics on integration theory. A proof of this lemma, in a slightly modified form, can be found in [112, Sect. 5.14, p. 103]. It also appears as a problem in various other texts (see [104, Problem 105, p. 60], [111, Problem 16, p. 94], [124, Problem 22, p. 94]). Various generalizations of the Riemann–Lebesgue lemma are discussed in [108, pp. 400, 401], and a nice historical comment on this result and how Riemann derived his lemma is given in [110, Sect. 22.5, p. 436].

Theorem A.8 (Monotone Convergence Theorem). *Let* $(f_n)_{n \in \mathbb{N}}$ *be an increasing sequence of nonnegative measurable functions, and let* $f = \lim_{n \to \infty} f_n$ *a.e. Then*

$$\int f = \lim \int f_n.$$

For a proof see [111, Theorem 10, p. 87].

Theorem A.9 (Bounded Convergence Theorem). *Let* (f_n) *be a sequence of measurable functions defined on a set E of finite measure, and suppose that there is a real number M such that* $|f_n(x)| \le M$ *for all n and all x. If* $\lim_{n \to \infty} f_n(x) = f(x)$ *for each x in E, then*

$$\lim \int_E f_n = \int_E f.$$

See [111, Proposition 6, p. 84]. The interested reader is also referred to a recent paper by Nadish de Silva [117] which gives an original method, which does not involve the measure theory techniques, for proving the celebrated Arzela's Bounded Convergence Theorem.

Theorem A.10 (Fatou's Lemma). *If* (f_n) *is a sequence of nonnegative measurable functions and* $f_n(x) \to f(x)$ *almost everywhere on a set E, then*

$$\int_E f \le \underline{\lim} \int_E f_n.$$

See [111, Theorem 9, p. 86].

Theorem A.11 (Lebesgue Convergence Theorem). *Let g be an integrable function over E and let* (f_n) *be a sequence of measurable functions such that* $|f_n| \le g$ *on E and for almost all x in E we have* $f(x) = \lim_{n \to \infty} f_n(x)$. *Then*

$$\int_E f = \lim \int_E f_n.$$

See [111, Theorem 16, p. 91].

Appendix B
Stolz–Cesàro Lemma

The most practical solution is a good theory.

Albert Einstein (1879–1955)

One of the most powerful tools of analysis for evaluating limits of sequences is Stolz–Cesàro lemma, the discrete version of l'Hôpital's rule. This lemma is attributed to the mathematicians Otto Stolz (1842–1905) and Ernesto Cesàro (1859–1906), and it proves to be an efficient method for calculating limits of indeterminate form of types ∞/∞ and $0/0$. In one of its simplest form, this lemma refers to the existence of the limits $\lim_{n\to\infty} a_n/b_n$ and $\lim_{n\to\infty}(a_{n+1}-a_n)/(b_{n+1}-b_n)$ where $(a_n)_{n\in\mathbb{N}}$ and $(b_n)_{n\in\mathbb{N}}$ are sequences of real numbers that verify certain conditions.

The lemma was first published in [126] and, since then, has been recorded in various texts that have topics on sequences and infinite series. It appears, as problem 70 on page 11, in the famous problem book of Pólya and Szegö [104] and as problem 6 on page 56 in [57]. The lemma has lots of applications in analysis, where it is used for calculating limits involving the Cesàro mean and is connected to the study of the speed of convergence and the order of growth to infinity of sequences of real numbers. Another recent beautiful application of this lemma is related to the calculation of the coefficients of the polynomial determined by the sum of the integer powers $S_k(n) = \sum_{i=1}^{n} i^k$ (see [81]). New forms of this lemma and examples are discussed in [92], and a stronger version of Stolz–Cesàro lemma and some of its most important applications are given in [94].

Theorem B.1 (Stolz–Cesàro Lemma, the ∞/∞ case). *Let $(a_n)_{n\in\mathbb{N}}$ and $(b_n)_{n\in\mathbb{N}}$ be two sequences of real numbers such that*

(a) $0 < b_1 < b_2 < \cdots < b_n < \cdots$ and $\lim_{n\to\infty} b_n = \infty$.
(b) $\lim_{n\to\infty} \frac{a_{n+1}-a_n}{b_{n+1}-b_n} = l \in \overline{\mathbb{R}}$.

Then $\lim_{n\to\infty} a_n/b_n$ exists, and moreover, $\lim_{n\to\infty} a_n/b_n = l$.

O. Furdui, *Limits, Series, and Fractional Part Integrals: Problems in Mathematical Analysis*, Problem Books in Mathematics, DOI 10.1007/978-1-4614-6762-5, © Springer Science+Business Media New York 2013

Proof. First we consider the case when l is finite. Let $\varepsilon > 0$ and let N be a natural number such that for all $n > N$ one has

$$\left| \frac{a_{n+1} - a_n}{b_{n+1} - b_n} - l \right| < \frac{\varepsilon}{2}. \tag{B.1}$$

This implies, since $b_{n+1} - b_n > 0$, that $(l - \varepsilon/2)(b_{n+1} - b_n) < a_{n+1} - a_n < (l + \varepsilon/2)(b_{n+1} - b_n)$ for all $n > N$. Now we apply these inequalities to each of the parenthesis in the numerator of the right-hand side fraction

$$\frac{a_n - a_{N+1}}{b_n - b_{N+1}} = \frac{(a_n - a_{n-1}) + (a_{n-1} - a_{n-2}) + \cdots + (a_{N+2} - a_{N+1})}{b_n - b_{N+1}}$$

and we get that for all $n > N + 1$ one has

$$\left| \frac{a_n - a_{N+1}}{b_n - b_{N+1}} - l \right| < \frac{\varepsilon}{2}.$$

On the other hand, a calculation shows that

$$\frac{a_n}{b_n} - l = \frac{a_{N+1} - l b_{N+1}}{b_n} + \left(1 - \frac{b_{N+1}}{b_n} \right) \left(\frac{a_n - a_{N+1}}{b_n - b_{N+1}} - l \right),$$

and it follows that

$$\left| \frac{a_n}{b_n} - l \right| \leq \left| \frac{a_{N+1} - l b_{N+1}}{b_n} \right| + \left| \frac{a_n - a_{N+1}}{b_n - b_{N+1}} - l \right|. \tag{B.2}$$

Since $\lim_{n \to \infty} b_n = \infty$, there is a natural number N' such that

$$\left| \frac{a_{N+1} - l b_{N+1}}{b_n} \right| < \frac{\varepsilon}{2} \quad \text{for all} \quad n > N'. \tag{B.3}$$

Let $N'' = \max \{ N + 1, N' \}$. We have, based on (B.1)–(B.3), that

$$\left| \frac{a_n}{b_n} - l \right| < \varepsilon \quad \text{for all} \quad n > N'',$$

and this implies that $\lim_{n \to \infty} a_n / b_n = l$.

Now we consider the case when $l = \infty$. Since

$$\lim_{n \to \infty} \frac{a_{n+1} - a_n}{b_{n+1} - b_n} = \infty,$$

we get that there is a natural number N such that $a_{n+1} - a_n > b_{n+1} - b_n$ for all $n > N$. This implies that the sequence $(a_n)_{n \geq N+1}$ increases and $\lim_{n \to \infty} a_n = \infty$. Now we apply the previous case to the sequence b_n/a_n and we get, since the limit of the quotient of the difference of two consecutive terms is finite, that

$$\lim_{n\to\infty} \frac{b_n}{a_n} = \lim_{n\to\infty} \frac{b_{n+1}-b_n}{a_{n+1}-a_n} = 0.$$

This implies that $\lim_{n\to\infty} a_n/b_n = \infty$ and the lemma is proved.

The reciprocal of Theorem B.1 does not hold. To see this, we let $a_n = 3n-(-1)^n$ and $b_n = 3n+(-1)^n$. Then, a calculation shows that $\lim_{n\to\infty} a_n/b_n = 1$ and

$$\frac{a_{n+1}-a_n}{b_{n+1}-b_n} = \frac{3+2(-1)^n}{3+2(-1)^{n+1}},$$

which implies that $\lim_{n\to\infty}(a_{n+1}-a_n)/(b_{n+1}-b_n)$ does not exist.

Theorem B.2 (Stolz–Cesàro Lemma, the $0/0$ case). *Let $(a_n)_{n\in\mathbb{N}}$ and $(b_n)_{n\in\mathbb{N}}$ be two sequences of real numbers such that*

(a) $\lim_{n\to\infty} a_n = \lim_{n\to\infty} b_n = 0$.
(b) (b_n) *is strictly decreasing.*
(c) $\lim_{n\to\infty} \frac{a_{n+1}-a_n}{b_{n+1}-b_n} = l \in \overline{\mathbb{R}}$.

Then $\lim_{n\to\infty} a_n/b_n$ exists, and moreover, $\lim_{n\to\infty} a_n/b_n = l$.

Proof. First, we consider the case when l is finite. For $\varepsilon > 0$, there is a natural number n_ε such that

$$\left| \frac{a_{n+1}-a_n}{b_{n+1}-b_n} - l \right| < \frac{\varepsilon}{2} \quad \text{for all} \quad n > n_\varepsilon.$$

This implies, since $b_n - b_{n+1} > 0$, that $(l - \varepsilon/2)(b_n - b_{n+1}) < a_n - a_{n+1} < (l+\varepsilon/2)(b_n - b_{n+1})$ for all $n > n_\varepsilon$. Let p be a positive integer and we apply the preceding inequalities to each of the parenthesis in the numerator of the right-hand side fraction:

$$\frac{a_n - a_{n+p}}{b_n - b_{n+p}} = \frac{(a_n - a_{n+1}) + (a_{n+1} - a_{n+2}) + \cdots + (a_{n+p-1} - a_{n+p})}{b_n - b_{n+p}}$$

and we get that for $n > n_\varepsilon$, we have

$$l - \frac{\varepsilon}{2} < \frac{a_n - a_{n+p}}{b_n - b_{n+p}} < l + \frac{\varepsilon}{2}.$$

Letting p tend to infinity, we obtain that $|a_n/b_n - l| \le \varepsilon/2 < \varepsilon$ for all $n > n_\varepsilon$.

Now we consider the case when $l = \infty$. Since $\lim_{n\to\infty}(a_{n+1}-a_n)/(b_{n+1}-b_n) = \infty$ we have that for $\varepsilon > 0$ there is a positive integer N_ε such that, for all $n > N_\varepsilon$, we have $(a_{n+1} - a_n)/(b_{n+1} - b_n) > 2\varepsilon$. This in turn implies that $a_n - a_{n+1} > 2\varepsilon(b_n - b_{n+1})$

for all $n > N_\varepsilon$. Let m and n be positive integers with $n, m > N_\varepsilon$ and $m > n$. We have $a_n - a_m = \sum_{k=n}^{m-1}(a_k - a_{k+1}) > 2\varepsilon \sum_{k=n}^{m-1}(b_k - b_{k+1}) = 2\varepsilon(b_n - b_m)$. Dividing this inequality by $b_n > 0$ we get that

$$\frac{a_n}{b_n} > 2\varepsilon\left(1 - \frac{b_m}{b_n}\right) + \frac{a_m}{b_n}.$$

Passing to the limit, as $m \to \infty$, in the preceding inequality, we get that $a_n/b_n \geq 2\varepsilon > \varepsilon$ for all $n > N_\varepsilon$.

The proof of the case $l = -\infty$ is left, as an exercise, to the reader.

The reciprocal of Theorem B.2 does not hold. To see this, we let $a_n = 1/(3n - (-1)^n)$ and $b_n = 1/(3n + (-1)^n)$. Then, a calculation shows that $\lim_{n \to \infty} a_n/b_n = 1$ and

$$\frac{a_{n+1} - a_n}{b_{n+1} - b_n} = \frac{(3n + 3 + (-1)^{n+1})(3n + (-1)^n)}{(3n + 3 - (-1)^{n+1})(3n - (-1)^n)} \cdot \frac{-3 - 2\cdot(-1)^n}{-3 + 2\cdot(-1)^n},$$

which implies that $\lim_{n \to \infty}(a_{n+1} - a_n)/(b_{n+1} - b_n)$ does not exist.

However, under certain conditions on the sequence $(b_n)_{n \in \mathbb{N}}$ we prove that the reciprocal of Stolz–Cesàro Lemma is valid.

Theorem B.3 (Reciprocal of Stolz–Cesàro Lemma). *Let $(a_n)_{n \in \mathbb{N}}$ and $(b_n)_{n \in \mathbb{N}}$ be two sequences of real numbers such that*

(a) $0 < b_1 < b_2 < \cdots < b_n < \cdots$ *and* $\lim_{n \to \infty} b_n = \infty$.
(b) $\lim_{n \to \infty} a_n/b_n = l \in \mathbb{R}$.
(c) $\lim_{n \to \infty} b_n/b_{n+1} = L \in \mathbb{R} \setminus \{1\}$.

Then the limit

$$\lim_{n \to \infty} \frac{a_{n+1} - a_n}{b_{n+1} - b_n},$$

exists and is equal to l.

Proof. We have

$$\frac{a_{n+1}}{b_{n+1}} = \frac{a_{n+1} - a_n}{b_{n+1} - b_n}\left(1 - \frac{b_n}{b_{n+1}}\right) + \frac{a_n}{b_n} \cdot \frac{b_n}{b_{n+1}}.$$

Passing to the limit, as $n \to \infty$, in the preceding equality we get that

$$l = (1 - L) \cdot \lim_{n \to \infty} \frac{a_{n+1} - a_n}{b_{n+1} - b_n} + l \cdot L,$$

and the result follows.

Theorem B.3 shows that if $\lim_{n \to \infty} b_n/b_{n+1}$ exists and is not equal to 1, the reciprocal of Stolz–Cesàro Lemma is valid. We stop our line of investigation here and invite the reader to study further the additional conditions required such that the reciprocal of Stolz–Cesàro Lemma remains valid in the trouble case $\lim_{n \to \infty} b_n/b_{n+1} = 1$.

References

1. Abel, U., Furdui, O., Gavrea, I., Ivan, M.: A note on a problem arising from risk theory. Czech. Math. J. **60**(4), 1049–1053 (2010)
2. Abel, U., Ivan, M., Lupaş, A.: The asymptotic behavior of a sequence considered by I. J. Schoenberg. J. Anal. **8**, 179–193 (2000)
3. Aldenhoven, N., Wanrooy, D.: Problem 2008/1-A. Nieuw Arch. Wiskd. **9**(3), 305 (2008)
4. Anderson, J.: Iterated exponentials. Am. Math. Mon. **111**(8), 668–679 (2004)
5. Andreoli, M.: Problem 819, Problems and Solutions. Coll. Math. J. **37**(1), 60 (2006)
6. Armstrong Problem Solvers, Perfetti, P., Seaman, W.: Evaluating an integral limit. Coll. Math. J. **43**(1), 96–97 (2012)
7. Artin, E.: The Gamma Function. Holt, Rinehart and Winston, New York (1964)
8. Basu, A., Apostol, T.M.: A new method for investigating Euler sums. Ramanujan J. **4**, 397–419 (2000)
9. Bartle, R.G.: The Elements of Real Analysis, 2nd edn. Wiley, New York (1976)
10. Bataille, M.: Problem 1784, Problems and Solutions. Math. Mag. **81**(5), 379 (2008)
11. Bataille, M., Cibes, M., G.R.A.20 Problem Solving Group, Seaman, W.: Problem 844, Problems and Solutions. Coll. Math. J. **39**(1), 71–72 (2008)
12. Benito, M., Ciaurri, O.: Problem 3301. Crux with Mayhem **35**(1), 47–49 (2009)
13. Benjamin, A.T., Preston, G.O.: A Stirling encounter with harmonic numbers. Math. Mag. **75**(2), 95–103 (2002)
14. Biler, P., Witkowski, A.: Problems in Mathematical Analysis. Marcel Dekker, New York (1990)
15. Bonar, D.D., Khoury, M.J.: Real Infinite Series, Classroom Resource Materials. The Mathematical Association of America, Washington, DC (2006)
16. Boros, G., Moll, V.H.: Irresistible Integrals, Symbolics, Analysis and Experiments in the Evaluation of Integrals. Cambridge University Press, Cambridge (2004)
17. Borwein D., Borwein J.: On an intriguing integral and some series related to $\zeta(4)$. Proc. Am. Math. Soc. **123**(4), 1191–1198 (1995)
18. Boyadzhiev, K.N.: Problem H-691, Advanced Problems and Solutions. Fibonacci Q. **50**(1), 90–92 (2012)
19. Boyadzhiev, K.N.: On a series of Furdui and Qin and some related integrals, p. 7, August 2012. Available at http://arxiv.org/pdf/1203.4618.pdf
20. Bromwich, T.J.I'A.: An Introduction to the Theory of Infinite Series, 3rd edn. AMS Chelsea Publishing, Providence (1991)
21. Bullen, P.S.: Handbook of Means and Their Inequalities. Kluwer, Dordrecht (2003)
22. Burckel, R.M.: Stirling's formula via Riemann sums. Coll. Math. J. **37**(4), 300–307 (2006)
23. Chen, H.: Problem 854, Problems and Solutions. Coll. Math. J. **39**(3), 243–245 (2008)

O. Furdui, *Limits, Series, and Fractional Part Integrals: Problems in Mathematical Analysis*, Problem Books in Mathematics, DOI 10.1007/978-1-4614-6762-5, © Springer Science+Business Media New York 2013

24. Chen, H.: Evaluation of some variant Euler sums. J. Integer Seq. **9**, Article 06.2.3 (2006)
25. Choi, J., Srivastava, H.M.: Explicit evaluation of Euler and related sums. Ramanujan J. **10**, 51–70 (2005)
26. Coffey, M.W.: Integral and series representation of the digamma and polygamma functions, p. 27, August 2010. Available at http://arxiv.org/pdf/1008.0040.pdf
27. Coffey, M.W.: Expressions for two generalized Furdui series. Analysis, Munchen **31**(1), 61–66 (2011)
28. Cohen, E.: Problem 809, Problems and Solutions. Coll. Math. J. **37**(4), 314–316 (2006)
29. Comtet, L.: Advanced Combinatorics, Revised and Enlarged Edition. D. Reidel Publishing Company, Dordrecht (1974)
30. Curtis, C.: TOTTEN–02, TOTTEN SOLUTIONS. Crux with Mayhem **36**(5), 320–321 (2010)
31. Daquila, R., Seaman, W., Tamper, B.: Problem 905, Problems and Solutions. Coll. Math. J. **41**(3), 246–247 (2010)
32. Davis, B.: Problem 899, Problems and Solutions. Coll. Math. J. **41**(2), 167–168 (2010)
33. Deeba, E.Y., Rodrigues, D.M.: Stirling's series and Bernoulli numbers. Am. Math. Mon. **98**(5), 423–426 (1991)
34. Dence, T.P., Dence, J.B.: A survey of Euler's constant. Math. Mag. **82**(4), 255–265 (2009)
35. Dobiński, G.: Summirung der Reihe $\sum n^m/n!$ für $m = 1, 2, 3, 4, 5, \ldots$. Arch. für Mat. und Physik **61**, 333–336 (1877)
36. Dunham, W.: Euler: The Master of Us All, pp. 35–36. The Mathematical Association of America, Washington, DC (1999)
37. Duemmel, J., Lockhart, J.: Problem 921, Problems and Solutions. Coll. Math. J. **42**(2), 152–155 (2011)
38. Dwight, H.B.: Table of Integrals and Other Mathematical Data. The Macmillan Company, New York (1934)
39. Edwards, H.M.: Riemann's Zeta Function. Dover, Mineola (2001)
40. Efthimiou, C.: Remark on problem 854, Problems and Solutions. Coll. Math. J. **40**(2), 140–141 (2009)
41. Euler, L.: De progressionibus harmonicus observationes. Commentarii Academiae Scientarum Imperialis Petropolitanae **7**, 150–161 (1734/1735)
42. Finbarr, H.: $\lim_{m\to\infty} \sum_{k=0}^m (k/m)^m = e/(e-1)$. Math. Mag. **83**(1), 51–54 (2010)
43. Finch, S.R.: Mathematical Constants. Encyclopedia of Mathematics and Its Applications 94. Cambridge University Press, New York (2003)
44. Freitas, P., Integrals of Polylogarithmic functions, recurrence relations, and associated Euler sums. Math. Comput. **74**(251), 1425–1440 (2005)
45. Furdui, O.: A class of fractional part integrals and zeta function values. Integral Transforms Spec. Funct. doi:10.1080/10652469.2012.708869 (to appear)
46. Furdui, O.: A convergence criteria for multiple harmonic series. Creat. Math. Inform. **18**(1), 22–25 (2009)
47. Furdui, O.: Series involving products of two harmonic numbers. Math. Mag. **84**(5), 371–377 (2011)
48. Furdui, O.: From Lalescu's sequence to a gamma function limit. Aust. Math. Soc. Gaz. **35**(5), 339–344 (2008)
49. Furdui, O.: Problem 63, Newsletter Eur. Math. Soc. Newsl. **77**, 63–64 (2010)
50. Furdui, O.: Problem 164, Solutions. Missouri J. Math. Sci. **19**(2), 4–6 (2007)
51. Furdui, O.: Problem 5073, Solutions. Sch. Sci. Math. **109**(8), 10–11 (2009)
52. Furdui, O.: Closed form evaluation of a multiple harmonic series. Automat. Comput. Appl. Math. **20**(1), 19–24 (2011)
53. Furdui, O.: TOTTEN–03, TOTTEN SOLUTIONS. Crux with Mayhem **36**(5), 321–324 (2010)
54. Furdui, O., Trif, T.: On the summation of certain iterated series. J. Integer Seq. **14**, Article 11.6.1 (2011)
55. Furdui, O., Qin, H.: Problema 149, Problemas y Soluciones. Gac. R. Soc. Mat. Esp. **14**(1), 103–106 (2011)

56. Garnet, J.B.: Bounded Analytic Functions. Graduate Text in Mathematics, vol. 236. Springer, New York (2007)
57. Garnir, H.G.: Fonctions de variables réelles, Tome I. Librairie Universitaire, Louvain and Gauthier-Villars, Paris (1963)
58. Geupel, O.: Problem 11456, Problems and Solutions. Am. Math. Mon. **118**(2), 185 (2011)
59. Glaisher, J.W.L.: On a numerical continued product. Messenger Math. **6**, 71–76 (1877)
60. Glaisher, J.W.L.: On the product $1^1 2^2 \cdots n^n$. Messenger Math. **7**, 43–47 (1878)
61. Glaisher, J.W.L.: On certain numerical products. Messenger Math. **23**, 143–147 (1893)
62. Gradshteyn, I.S., Ryzhik, I.M.: Table of Integrals, Series, and Products, 6th edn. Academic, San Diego (2000)
63. G.R.A.20 Problem Solving Group, Rome, Italy.: Problem 938, Problems and Solutions. Coll. Math. J. **42**(5), 410–411 (2011)
64. Harms, P.M.: Problem 5097, Solutions. Sch. Sci. Math. **110**(4), 15–16 (2010)
65. Havil, J.: GAMMA, Exploring Euler's Constant. Princeton University Press, Princeton (2003)
66. Herman, E.: Problem 920, Problems and Solutions. Coll. Math. J. **42**(1), 69–70 (2011)
67. Howard, R.: Problem 11447, Problems and Solutions. Am. Math. Mon. **118**(2), 183–184 (2011)
68. Impens, C.: Stirling's series made easy. Am. Math. Mon. **110**, 730–735 (2003)
69. Ivan, D.M.: Problem 11592, Problems and Solutions. Am. Math. Mon. **118**, 654 (2011)
70. Janous, W.: Around Apery's constant. JIPAM, J. Inequal. Pure Appl. Math. **7**(1), Article 35 (2006)
71. Jeffery, H.M.: On the expansion of powers of the trigonometrical ratios in terms of series of ascending powers of the variable. Q. J. Pure Appl. Math. **5**, 91–108 (1862)
72. Kinkelin, J.: Über eine mit der Gammafunction verwandte Transcendente und deren Anwendung auf die Integralrechnung. J. Reine Angew. Math. **57**, 122–158 (1860)
73. Klamkin, M.S.: A summation problem, Advanced Problem 4431. Am. Math. Mon. **58**, 195 (1951); Am. Math. Mon. **59**, 471–472 (1952)
74. Klamkin, M.S.: Another summation. Am. Math. Mon. **62**, 129–130 (1955)
75. Boyd, J.N., Hurd, C., Ross, K., Vowe, M.: Problem 475, Problems and Solutions. Coll. Math. J. **24**(2), 189–190 (1993)
76. Knopp, K.: Theorie und Anwendung der unendlichen Reihen, Grundlehren der Mathematischen Wissenschaften 2. Springer, Berlin (1922)
77. Kouba, O.: Problem 1849, Problems and Solutions. Math. Mag. **84**(3), 234–235 (2011)
78. Kouba, O.: The sum of certain series related to harmonic numbers, p. 14, March 2012. Available at http://arxiv.org/pdf/1010.1842.pdf
79. Kouba, O.: A Harmonic Series, SIAM Problems and Solutions, Classical Analysis, Sequences and series, Problem 06-007 (2007). http://www.siam.org/journals/categories/06-007.php
80. Krasopoulos, P.T.: Problem 166, Solutions. Missouri J. Math. Sci. **20**(3) (2008)
81. Kung, S.H.: Sums of integer powers via the Stolz–Cesàro theorem. Coll. Math. J. **40**(1), 42–44 (2009)
82. Lalescu, T.: Problema 579. Gazeta Matematică **6**(6), 148 (1900) (in Romanian)
83. Lau, K.W.: Problem 5013, Solutions. Sch. Sci. Math. **108**(6) (2008)
84. Lewin, L.: Polylogarithms and Associated Functions. Elsevier (North-Holand), New York (1981)
85. Lossers, O.P.: Problem 11338, Problems and Solutions. Am. Math. Mon. **116**(8), 750 (2009)
86. Mabry, R.: Problem 893, Problems and Solutions. Coll. Math. J. **41**(1), 67–69 (2010)
87. Manyama, S.: Problem 1, Solutions, Problem Corner of RGMIA (2010). Available electronically at http://www.rgmia.org/pc.php
88. Medarde, D.L.: Problema 150, Problemas Resueltos. Revista Escolar de la OIM, Núm. **31**, Marzo–Abril (2008)
89. Miller, A., Refolio, F.P.: Problem 895, Problems and Solutions. Coll. Math. J. **41**(1), 70–71 (2010)
90. Mortici, C.: Problem 3375, Solutions. Crux with Mayhem **35**(6), 415–416 (2009)

91. Mortici, C.: Product approximations via asymptotic integration. Am. Math. Mon. **117**, 434–441 (2010)
92. Mortici, C.: New forms of Stolz-Cesaró lemma. Int. J. Math. Educ. Sci. Technol. **42**(5), 692–696 (2011)
93. Mortini, R.: Problem 11456, Problems and Solutions. Am. Math. Mon. **116**, 747 (2009)
94. Nagy, G.: The Stolz-Cesaro theorem, Preprint, p. 4. Manuscript available electronically at http://www.math.ksu.edu/~nagy/snippets/stolz-cesaro.pdf
95. Namias, V.: A simple derivation of Stirling's asymptotic series. Am. Math. Mon. **93**(1), 25–29 (1986)
96. Northwestern University Math Prob Solving Group: Problem 1797. Math. Mag. **82**(3), 229 (2009)
97. Ogreid, M.O., Osland, P.: More series related to the Euler series. J. Comput. Appl. Math. **136**, 389–403 (2001)
98. Olver, F.W.J., Lozier, D.W., Boisvert, R.F., Clark, C.W.: In: Olver, F.W.J., Lozier, D.W., Boisvert, R.F., Clark, C.W. (eds.) NIST Handbook of Mathematical Functions. NIST National Institute of Standards and Technology, US Department of Commerce and Cambridge University Press, Cambridge (2010)
99. Perfetti, P.: Problem 5, Problems and Solutions. MATHPROBLEMS **1**(1), 2 (2010)
100. Perfetti, P.: Problem F07–4, Solutions. The Harvard Coll. Math. Rev. **2**(1), 99–100 (2008)
101. Perfetti, P.: Problema 155, Problemas Resueltos. Revista Escolar de la OIM, Núm. **32**, Mayo–Junio (2008)
102. Pinsky, M.A.: Introduction to Fourier Analysis and Wavelets. Graduate Studies in Mathematics, vol. 102. American Mathematical Society, Providence (2002)
103. Plaza, Á.: Problem 2011-2, Electronic Journal of Differential Equations, Problem section (2011). Available electronically at http://math.uc.edu/ode/odesols/s2011-2.pdf
104. Pólya, G., Szegö, G.: Aufgaben und Lehrsätze aus der Analysis, vol. 1. Verlag von Julius Springer, Berlin (1925)
105. de la Vallée Poussin, Ch.: Sur les valeurs moyennes de certaines fonctions arithmétiques. Annales de la Societe Scientifique de Bruxelles **22**, 84–90 (1898)
106. Qin, H.: Problem 150, Solutions. Missouri J. Math. Sci. **17**(3), Fall (2005)
107. Rassias, M.Th.: On the representation of the number of integral points of an elliptic curve modulo a prime number, p. 17, October 2012. Available at http://arxiv.org/pdf/1210.1439.pdf
108. Rădulescu, T.L., Rădulescu, D.V., Andreescu, T.: Problems in Real Analysis: Advanced Calculus on the Real Axis. Springer, Dordrecht (2009)
109. Remmert, R.: Classical Topics in Complex Function Theory. Graduate Text in Mathematics, vol. 172. Springer, New York (1998)
110. Roy, R.: Sources in the Development of Mathematics. Infinite Series and Products from the Fifteenth to the Twenty-first Century. Cambridge University Press, New York (2011)
111. Royden, H.L.: Real Analysis, 3rd edn. Prentice Hall, Englewood Cliffs (1987)
112. Rudin, W.: Real and Complex Analysis, 3rd edn. WCB/McGraw-Hill, New York (1987)
113. Schmeichel, E.: Problem 1764, Problems and Solutions. Math. Mag. **81**(1), 67 (2008)
114. Schoenberg, I.J.: Problem 640. Nieuw Arch. Wiskd. III Ser. **30**, 116 (1982)
115. Seaman, W., Vowe, M.: Problem 906, Problems and Solutions. Coll. Math. J. **41**(4), 330–331 (2010)
116. Seiffert, H.-J.: Problem H-653, Advanced Problems and Solutions. Fibonacci Q. **46/47**(2), 191–192 (2009)
117. de Silva, N.: A concise, elementary proof of Arzela's bounded convergence theorem. Am. Math. Mon. **117**(10), 918–920 (2010)
118. Sitaramachandrarao, R.: Maclaurin coefficients of the Riemann zeta function. Abstr. Am. Math. Soc. **7**, 280 (1986)
119. Sîntămărian, A.: A generalization of Euler's constant. Numer. Algorithms **46**(2), 141–151 (2007)
120. Sondow, J.: Double integrals for Euler's constant and $\ln 4/\pi$ and an analogue of Hadjicostas's formula. Am. Math. Mon. **112**, 61–65 (2005)

121. Spivey, Z.M.: The Euler-Maclaurin formula and sums of powers. Math. Mag. **79**, 61–65 (2006)

122. Srivastava, H.M., Choi, J.: Series Associated with the Zeta and Related Functions. Kluwer, Dordrecht (2001)

123. Srivastava, H.M., Choi, J.: Zeta and q-Zeta Functions and Associated Series and Integrals. Elsevier Insights, Amsterdam (2012)

124. Stein, E.M., Shakarchi, R.: Real Analysis, Measure Theory, Integration, & Hilbert Spaces, III. Princeton University Press, Princeton (2005)

125. Stieltjes, T.J.: Collected Papers. Springer, New York (1993)

126. Stolz, O.: Über die grenzwerte der quotienten. Math. Ann. **15**, 556–559 (1879)

127. Stong, R.: Problem 11371, Problems and Solutions. Am. Math. Mon. **117**(5), 461–462 (2010)

128. Summer, J., Kadic-Galeb, A.: Problem 873, Problems and Solutions. Coll. Math. J. **40**(2), 136–137 (2009)

129. Tejedor, V.J.: Problema 139, Problemas y Soluciones. Gac. R. Soc. Mat. Esp. **13**(4), 715–716 (2010)

130. Tyler, D.B.: Problem 11494, Problems and Solutions. Am. Math. Mon. **118**(9), 850–851 (2011)

131. Vara, G.C.J.: Problema 94, Problemas y Soluciones. Gac. R. Soc. Mat. Esp. **11**(4), 698–701 (2008)

132. Valderrama, J.M.M.: Problema 113, Problemas y Soluciones. Gac. R. Soc. Mat. Esp. **12**(3), 524–527 (2009)

133. Whittaker, E.T., Watson, G.N.: A Course of Modern Analysis, 4th edn. Cambridge University Press, London (1927)

134. Young, R.M.: Euler's constant. Math. Gaz. **472**, 187–190 (1991)

135. Yu, Y.: A multiple fractional integral, SIAM Problems and Solutions, Classical Analysis, Integrals, Problem 07-002 (2007). http://www.siam.org/journals/problems/downloadfiles/07-002s.pdf

136. Zorich, V.A.: Mathematical Analysis I. Universitext. Springer, Berlin (2004)

137. Zuazo, D.J.: Problema 159, Problemas y Soluciones. Gac. R. Soc. Mat. Esp. **14**(3), 515 (2011)

138. Zhu, K.: Analysis on Fock Spaces. Graduate Text in Mathematics, vol. 263. Springer, New York (2012)

Index

O. Furdui, *Limits, Series, and Fractional Part Integrals: Problems in Mathematical Analysis*, Problem Books in Mathematics, DOI 10.1007/978-1-4614-6762-5,
© Springer Science+Business Media New York 2013

Printed in the United States
By Bookmasters